Y0-BXV-575

ENTROPY

PHILIPS TECHNICAL LIBRARY

ENTROPY

The significance of the concept of entropy
and its applications
in science and technology

by

J. D. FAST

Chief Metallurgist Philips Research Laboratories
and Professor at the Technical University Eindhoven

SECOND EDITION

Revised and Enlarged

MACMILLAN

Original Dutch edition © D. B. Centen's Uitgeversmaatschappij, Hilversum 1960
English edition © N. V. Philips' Gloeilampenfabrieken, Eindhoven 1962, 1968, 1970

All rights reserved. No part of this publication may be reproduced or transmitted, in any form or by any means, without permission.

CAMROSE LUTHERAN COLLEGE
LIBRARY

First published 1962
Second edition, revised and enlarged, 1968
Reprint, revised and enlarged, 1970

Published by
MACMILLAN AND CO
London and Basingstoke
Associated companies in New York, Toronto Melbourne, Johannesburg, Dublin and Madras

PHILIPS

Trademarks of N. V. Philips' Gloeilampenfabrieken, Eindhoven, The Netherlands

Printed in Great Britain by The Whitefriars Press Ltd, London and Tonbridge

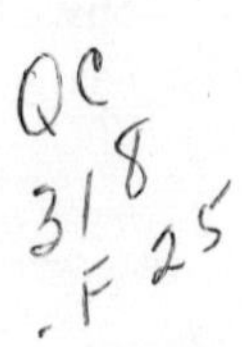

PREFACE TO THE FIRST ENGLISH EDITION

The entropy concept, which plays a role as important as that of the energy concept in every field of physics, chemistry and technology, generally presents many difficulties. This was the reason which induced me in 1947 to write a book about entropy in a form clear enough to be read without undue difficulty by the chemist and physicist. This work met with such success that a second edition became inevitable. This edition, in contrast to the first, appears not only in Dutch, but also in English, French, German, Spanish and Polish.

In the first edition I limited myself to discussing the basic principles of thermodynamics, statistical mechanics and quantum mechanics only so far as was necessary to explain the entropy concept and to calculate the entropies of monatomic and diatomic gases by the two methods provided by thermodynamics and statistical mechanics. The success of the first edition encouraged me to increase considerably, in the present edition, the number of examples of the applications of the entropy concept. Some of these examples lie in the domain of chemistry, others in that of the physics and physical chemistry of solids. The subjects discussed include paramagnetic and ferromagnetic substances, interstitial and substitutional solid solutions, internal friction, order-disorder transformations and precipitation in alloys, the structure of glass, the electronic theory of metals and semi-conductors, the structure of martensite, the elasticity of rubber, the specific heats of solids and the existence of vacancies in solids. Furthermore, attention is paid to solutions of polymers, the calculation of chemical equilibrium, the efficiency of heat engines, the attainment of extremely low temperatures and the radiation of heat and light. In contrast to the Dutch and German editions, the English edition contains an additional section on fuel cells and heat pumps (Section 3.14).

It is obviously impossible, and was certainly never my intention, to give a complete survey of all questions relating to the concept of entropy. For this reason the reader will search in vain for the thermodynamics of irreversible processes and the accompanying "production of entropy". Nor will he find anything about the use (and frequent abuse) of the entropy concept in information theory. This limitation and the considerable reduction of the chapters dedicated to gases have made it possible to limit this book to a reasonable number of pages.

It is probable that some readers will lack the time to study this work thoroughly. For their benefit, I have arranged the subject matter in such a way that the first part of the book (Chapters 1 to 3) forms a complete and coherent unit. They have only to look through this section to realize that the elementary concepts of thermodynamics and statistical mechanics explained in the first two chapters are sufficient to tackle the numerous problems dealt with in Chapter 3.

The principles of quantum mechanics and quantum statistics, which make possible a detailed discussion of the basis of the entropy concept, are not explained until Chapter 4. This chapter is scarcely more complicated or difficult than the preceding ones and the reader will be forced to admit that quantum statistics are logically more satisfactory and, for that reason, that they are "simpler" in many ways than classical statistics. The last two chapters have been devoted to rarefied monatomic and diatomic gases. It is a happy circumstance that the thermodynamic properties of these gases, as well as the chemical equilibria between these gases can be calculated exactly from spectroscopic data, while the calculations carried out for solids in the first chapters are only approximate, since the theory of the solid state is not yet sufficiently developed to permit of accurate calculations.

In accordance with the nature of this book, the treatment is not extremely rigorous. Without sacrificing more precision than necessary, I have endeavoured to reach all of those students and researchers to whom thermodynamics and statistical mechanics seem a little frightening, although a certain knowledge of these subjects is indispensible to them. With this aim in view, I have intentionally avoided mentioning several difficulties of a fundamental nature in discussing statistical mechanics. I am convinced that, for the majority of readers, this voluntary omission offers more advantages than it causes inconvenience and I dare to hope that the theoreticians will not take offence. Moreover, these principles have been set out in detail in several classic works (for example R. C. Tolman, The Principles of Statistical Mechanics, Oxford, 1938).

Several friends and colleagues have been so kind as to read part or all of the manuscript before going to press. I would like to take this opportunity to express my appreciation. Their criticisms, and in particular those of Professors J. L. Meijering, D. Polder and J. H. van Santen, and Drs. W. de Groot, H. J. G. Meyer and J. Smit, have enabled me to introduce several improvements in this book. I also wish to thank Mrs. Mulder-Woolcock, B.Sc., for translating the Dutch text into English and Mr. C. Hargreaves,

M.Sc., for checking the text as regards technical terms, as well as Mr. M. B. Verrijp for drawing the diagrams and correcting the proofs.

Eindhoven, March 1962 J. D. FAST

PREFACE TO THE SECOND EDITION

The greater part of the original printing has been retained. In addition to the correction of several errors and the introduction of a few minor improvements, the section dealing with *fuel cells* has been extended. Some new material has been added in an Appendix. The new topics include a discussion of "negative absolute temperatures" and *lasers*, and a treatment of the thermodynamics of *superconductors*.

Eindhoven, March 1968 J. D. FAST

NOTE TO 1970 REPRINT

This reprint contains an appendix explaining my omission from the book of a discussion of the entropy concept in information theory and communications engineering. I believe that the term entropy is misused in this context; information and entropy have different dimensions and the relationship between the two is only of a formal mathematical nature.

The reader is asked to add a negative sign to the right side of Equation (1.6.7) on page 16, in order to reconcile it with the convention concerning the sign of the electromotive force of a cell. The latter is called positive if the chemical reaction involved is so written that it proceeds spontaneously from left to right when the cell is short-circuited.

Eindhoven, April 1970 J. D. FAST

TABLE OF CONTENTS

GENERAL INTRODUCTION

The entropy concept springs from two roots. On the one hand, in classical thermodynamics, entropy is defined in an abstract manner as a thermodynamic variable of the system under consideration; on the other hand, in statistical mechanics, it is defined as a measure of the number of ways in which the elementary particles of the system may be arranged under the given circumstances.

It seems to me useful to give these two definitions (Equations (1) and (2) or (4)) in the general introduction, without requiring that the reader should "understand" them completely at the start.

ORIGIN OF THE ENTROPY CONCEPT

The concept of entropy originated in the development stage of classical thermodynamics, in the period when thermodynamics was still almost entirely concerned with the study of steam engines or, more generally, with the conditions under which heat can be converted into work.

One of the important conclusions established during the development of classical thermodynamics was the following: The quantity of heat $\int_1^2 dQ$, which is needed to bring a body or a system of bodies from a state 1 to a state 2, is not uniquely defined. It depends on the way in which the transition is accomplished or, in other words, on the path followed from 1 to 2. The question of how much heat is present in a body under particular conditions, is therefore meaningless. Even if some convention were agreed upon as to the "zero point" of the "heat content" of a body no unambiguous answer to the question is possible.

As will be seen in Chapter 1, however, there is another quantity, very closely related to the amount of heat introduced, which *is* uniquely defined by the state of a system. If a system is brought from a state 1 to a state 2, then $\int_1^2 dQ/T$, T being the absolute temperature, has a value which is independent of the path followed, provided that the path is reversible (see Section 1.5) and that the integration is so carried out that each elementary quantity of heat is divided by the temperature at which it is introduced. This integral, calculated from the zero point of absolute temperature, is called the *entropy* S of the body or system of bodies.

$$S = \int_0^T \frac{dQ_{rev}}{T} \tag{1}$$

As we shall see in Chapter 2, this definition may be applied only to systems which are in stable or metastable equilibrium, e.g. to chemically pure, ideally crystallized substances. Glasses and solid solutions, for example, must be attributed with a finite value of the entropy even at absolute zero: for such substances, S in Equation (1) must be replaced by $S - S_0$.

The entropy concept is the central concept in classical thermodynamics. According to the second law of thermodynamics, the entropy in an isolated system tends to a maximum, so that this variable is a criterion for the direction in which processes can take place.

As can be seen from the definition (1), the measurement of the entropy of a substance in classical thermodynamics depends on the measurement of quantities of heat, that is, on calorimetry.

THE ENTROPY CONCEPT IN STATISTICAL MECHANICS

The thermodynamics of the past century, in which the entropy concept originated, is concerned only with the macroscopic states of matter, i.e. with the experimentally observable properties and with the effects of heat and work in the interaction between systems and their surroundings. Thermodynamics, thus, does not enquire into the mechanism of the phenomena and is therefore unconcerned with what happens on an atomic scale.

The atomic picture, however, can help to give deeper meaning to the thermodynamic laws and concepts. The branch of science concerned with this aspect is *statistical mechanics*, the mechanics of such a large number of atoms or molecules that a specification of the state of each separate particle is impossible and one is obliged to resort to statistical methods.

The thermodynamic state of a system is known when a small number of macroscopic variables, such as pressure, temperature, volume, chemical composition etc, are known. It is clear that a description of this kind still leaves open many possibilities as regards the detailed state on an atomic scale. In the words of Gibbs [1]: "So when gases of different kinds are mixed,

---

[1] J. W. Gibbs, Trans. Connecticut Acad. Science **3**, 228 (1876) or "The Collected Works of J. W. Gibbs", New York, 1928, Vol. 1, p. 166.

if we ask what changes in external bodies are necessary to bring the system to its original state, we do not mean a state in which each particle shall occupy more or less exactly the same position as at some previous epoch, but only a state which shall be indistinguishable from the previous one in its sensible properties. It is to states of systems thus incompletely defined that the problems of thermodynamics relate".

One and the same state, in the thermodynamic sense, thus comprises very many states on the atomic scale, or, in other words: a thermodynamic state can be realized in many ways. One speaks of the possibilities of realization or *micro-states* of the thermodynamic state. If the number of these micro-states be denoted by g, then the statistical definition of entropy reads

$$S = k \ln g \tag{2}$$

where k is Boltzmann's constant, viz. the gas constant per molecule,

$$k = \frac{R}{N_0} \tag{3}$$

R being the gas constant per mole and N_0 Avogadro's number. In order to be able to apply Equation (2) directly, all the micro-states must have the same probability of occurring.

That the quantity S in Equation (2) may, indeed, be considered as the statistical interpretation of entropy, follows from the fact that this variable appears, on closer examination, to exhibit all the properties attributed to thermodynamic entropy.

That entropy tends to a maximum means, according to (2), a tendency towards the state with a maximum number of possibilities of realization, i.e. a tendency towards the most probable state. In fact, it will be seen in Chapter 2 that the number of micro-states is a direct measure of the probability of the thermodynamic state. Thus, if the entropy of a system has a value less than the attainable maximum, this means, according to (2), that the system is not in its most probable state and that, given sufficient molecular agitation, it is likely to pass into more probable states. The statistical meaning of this statement is as follows: Suppose that we have a very large number of independent specimens of the same system, which are identical in their macroscopic properties and thus have the same entropy (which is assumed to be less than the attainable maximum): the entropy of the overwhelming majority of these systems will spontaneously increase, while in only a negligibly small number of the systems will the entropy decrease.

According to this statistical interpretation of entropy the second law does not hold entirely rigidly. It is not absolutely certain, but only highly probable, that the entropy will increase in each process spontaneously taking place in an isolated system.

In statistical mechanics it is sometimes preferable to use a definition of entropy with more general validity than (2) [1]). It links up with the above case where a large number of specimens of the same system were considered. An "ensemble" of systems of this sort will be distributed over more different micro-states in proportion as the entropy is larger. The ensemble is defined when it is known what fraction of the systems is in each micro-state. If this fraction be denoted by p_i for the ith micro-state, the more general definition of the statistical entropy reads

$$S = -k \sum_i p_i \ln p_i \tag{4}$$

where k is again Boltzmann's constant. The terms of the sum are sometimes referred to as the paradox terms of Gibbs. If the ensemble is distributed over g micro-states, all of which are equally probable, then all the p_i's have the same value $1/g$. Equation (4) then becomes

$$S = -kg \frac{1}{g} \ln \frac{1}{g} = k \ln g$$

i.e. Equation (2).

As we have seen above, entropy is determined in a completely different manner in statistical mechanics than in thermodynamics. Where the determination of "calorimetric entropy" depended on the measurement of quantities of heat, the determination of "statistical entropy" depends on counting the number of micro-states.

It is one of the finest achievements of physics and chemistry that these two paths generally lead to the same result, while the divergences which appear can be explained by the theory in a completely satisfactory manner and even serve to endorse it. We shall demonstrate this in this book, for instance, when calculating the entropy of monatomic and diatomic gases.

With the help of examples from widely divergent fields of physics, chemistry and technology we shall show further that the concept of entropy has a significance no less fundamental than that of energy. On the relation between energy and entropy R. Emden [2]) writes: "As a student, I read with advantage a small book by F. Wald entitled The Mistress of the World and her Shadow. These meant energy and entropy. In the course of advancing knowledge the two seem to me to have exchanged places. In the huge manufactory of natural processes, the principle of entropy occupies the position

---

[1]) Cf. R. C. Tolman, "The Principles of Statistical Mechanics", Oxford, 1938.

[2]) R. Emden, Nature **141**, 908 (1938).

of manager, for it dictates the manner and method of the whole business, whilst the principle of energy merely does the book-keeping, balancing credits and debits".

CHAPTER 1

THE CONCEPT OF ENTROPY IN CLASSICAL THERMODYNAMICS

1.1. INTRODUCTION

All natural phenomena are governed by the two laws of thermodynamics. These laws not only exert their influence in every field of science, but also play a part in all industrial processes in which energy is transferred. In technology, therefore, they are of almost universal application. For example, they make it possible to calculate the maximum efficiency of steam engines and to predict the direction and maximum yield of chemical reactions. These and various other applications will be dealt with in this book, together with the more theoretical aspects of the subject.

The two laws can be formulated in many ways. Superficially, these various formulations often seem to bear little relation to each other, but essentially they are equivalent. Applied to an isolated system, i.e. a system without interaction with the outside world (see the following section), they can be stated as follows: First Law: The *energy* of an isolated system is constant. Second Law: The *entropy* of an isolated system tends to a maximum.

Before discussing the two laws and explaining the importance of the entropy concept, we shall first introduce several concepts which will be met with frequently in the following discussions.

1.2. SYSTEM, THERMODYNAMIC STATE, THERMODYNAMIC VARIABLE

The word *system* has already been used above. In thermodynamics it is used simply to denote that portion of the universe which one wishes to study. Although, in principle, one is free in the choice of the limits of a system, that choice sometimes determines whether or not a problem is soluble.

Once a particular portion of the universe has been chosen as the system, the rest of the universe is referred to as the *surroundings* of the system.

Example of choice: When studying a galvanic cell, the poles of which are connected by a resistance wire, it is useful to consider the cell as the system and the resistance wire as part of the surroundings.

A system and its surroundings can be so isolated from each other that they can perform no work on each other and that no interchange of heat or matter can occur. From experience we know that, sooner or later, no further appreciable change will take place in the state of such an isolated system. It is obvious that "state" does not here refer to the state on an atomic scale. Consider, for instance, a mixture of gases which is in equilibrium. It is clear that the arrangement of the molecules and the distribution of kinetic energy between the various molecules, will change continuously. A distinction must therefore be made between the "macro-state" of a system, which does not change in the case under consideration, and the "micro-state" which does change. As a branch of macroscopic (non-atomic) physics, classical thermodynamics cannot distinguish between the various micro-states in which a system can exist. States differ thermodynamically only when they are distinguishable by the senses or with the help of instruments which are not so sensitive that they react to fluctuation phenomena (see Chapter 2). States in the macroscopic sense described above, are called *thermodynamic states*. Each thermodynamic state comprises very many micro-states.

The thermodynamic states of a system can be completely described by means of a number of variables, which have a particular value for each of these states. They are referred to as thermodynamic variables or *thermodynamic functions*. If the system under consideration is a known quantity of gas (contained, for example, in a cylinder closed by a movable piston), then its thermodynamic state is given completely by two variables. As the determining variables may be chosen the pressure p and the volume V, but — as will appear later — the (internal) energy U and the entropy S may also be chosen.

As indicated above, the thermodynamic variables have only a macroscopic significance; they are *not* applicable to a "system" consisting of only a small number of molecules. In consequence, it is useless to consider very small changes in these variables. When, in thermodynamics, one speaks of the "differential" $\mathrm{d}V$ of the volume or of a very small or (less correctly) "infinitely small" or "infinitesimal" change in the volume, this always means a change which is very small with respect to the volume V itself, but very large with respect to V/N, where N is the total number of molecules composing the system. In fact, if $\mathrm{d}V$ is to have thermodynamic significance, then this "infinitesimal" volume must contain so many molecules that it can be considered as a very small macroscopic system. The "differential" $\mathrm{d}p$ of the pressure refers to a change in p which is very small with respect to the pressure itself, but very large with respect to the fluctuations which occur spontaneously all the time in the pressure p of each elementary volume $\mathrm{d}V$.

The difference in formulation when explaining the thermodynamic significance of dp as compared to that of dV, is based on the distinction which must be made between *extensive* and *intensive* thermodynamic variables. The meaning of these terms is given by the fact that the extensive thermodynamic quantities are halved when a system in equilibrium is divided into two equal parts, while the intensive quantities remain unchanged. Examples of extensive variables are the volume and — as will appear later — the entropy. Intensive variables are, for example, the pressure and the temperature.

1.3. THE CONCEPT OF TEMPERATURE

The concept of temperature has its origin in our physiological experience of "warm" and "cold". An important experimental fact made it possible to introduce temperature as a more scientific variable, viz. as thermodynamic variable of a system in the sense of the preceding section. This experimental fact is the following: If two systems are each individually in thermal equilibrium with a third system, then they will also be in thermal equilibrium with each other, i.e. their state does not change when they are brought into thermal contact with each other. The same value of a particular thermodynamic variable can thus be attached to all systems which are in thermal equilibrium with a particular reference system. This variable is called temperature.

To express the temperature of a body as a reproducible number, use can be made of all those properties of materials which change in a uniform and reproducible manner with the temperature, thus, not only of volume (liquid thermometer) or of pressure (gas thermometer), but also, for example, of the electrical resistance (resistance thermometer). In the first instance, the choice of a temperature scale is entirely arbitrary. The reader will undoubtedly be familiar with several thermometer scales that have been adopted in the course of time.

Physically speaking, the most satisfactory agreement for fixing a temperature scale on the basis of the properties of materials, was that in which the temperature is defined by the pressure p of a quantity of gas at constant volume. This gives the advantage that one is almost independent of the nature of the gas, provided the temperature is far above the condensation temperature and that the pressures are relatively low. In fact, if the quantity of gas chosen is 1 gram molecule (1 mole), it appears that to a first approximation *all* rarefied gases obey the relation

$$p = \frac{R}{v}(273.1 + t) \; . \tag{1.3.1}$$

Here t is the temperature in degrees Centigrade (°C), v is the volume of 1 mole of gas and R the so-called gas constant, which has the value

$$R = 8.314 \text{ joule/deg.mole} = 1.987 \text{ cal/deg.mole}$$

By introducing the "absolute temperature"

$$T = 273.1 + t \tag{1.3.2}$$

Equation (1.3.1) takes the simplified form

$$pv = RT \tag{1.3.3}$$

The absolute temperature is expressed in degrees Kelvin (°K).

No gas exists which obeys Equation (1.3.3) exactly. The hypothetical gas which behaves in that way is called a perfect or ideal gas. A perfect gas is also attributed with the property that its internal energy (see Section 1.6) is independent of the volume. In the atomic picture, this means that any two molecules in a perfect gas exert no forces on each other, except at the moment that they collide. As the pressure decreases, the behaviour of real gases approaches that of perfect gases.

The second law, still to be discussed, will make it possible to define absolute temperature in such a way that it is independent of the properties of any material whatsoever (see Section 1.14).

1.4. EXACT AND INEXACT DIFFERENTIALS

Let us consider a material, the thermodynamic state of which is completely specified when two of the three thermodynamic variables p, V and T are known. For a substance of this kind, there is a mathematical relationship between p, V and T, the so-called equation of state. For example, we may write

$$V = \mathrm{f}(p,T) \tag{1.4.1}$$

The differential of V (cf. Section 1.2) is given by

$$\mathrm{d}V = \mathrm{d}V_p + \mathrm{d}V_T = \left(\frac{\partial V}{\partial T}\right)_p \mathrm{d}T + \left(\frac{\partial V}{\partial p}\right)_T \mathrm{d}p \tag{1.4.2}$$

In contrast with the partial differentials $\mathrm{d}V_p$ and $\mathrm{d}V_T$, $\mathrm{d}V$ is called an *exact* differential.

In general, if z is a function of x and y

$$z = \mathrm{f}(x,y). \tag{1.4.3}$$

then the total change in z, the exact differential dz, is equal to the sum of the partial differentials:

$$\mathrm{d}z = \left(\frac{\partial z}{\partial y}\right)_x \mathrm{d}y + \left(\frac{\partial z}{\partial x}\right)_y \mathrm{d}x \tag{1.4.4}$$

For example, if $z = x^3y^4$, then from Equation (1.4.4):

$$\mathrm{d}z = 4x^3y^3\mathrm{d}y + 3x^2y^4\mathrm{d}x$$

The function f(x,y) may be differentiated first partially with respect to x and then partially with respect to y. It can be shown that the same result is obtained when these operations are carried out in the reverse order, i.e. that

$$\frac{\partial^2 z}{\partial x \partial y} = \frac{\partial^2 z}{\partial y \partial x} \tag{1.4.5}$$

In the above example of the function $z = x^3y^4$, we have

$$\frac{\partial z}{\partial x} = 3x^2y^4; \quad \frac{\partial\left(\frac{\partial z}{\partial x}\right)}{\partial y} = 12x^2y^3; \quad \frac{\partial z}{\partial y} = 4x^3y^3; \quad \frac{\partial\left(\frac{\partial z}{\partial y}\right)}{\partial x} = 12x^2y^3$$

The general form of the exact differential of a function of two independent variables according to the above, is

$$\mathrm{d}z = \left(\frac{\partial z}{\partial x}\right)_y \mathrm{d}x + \left(\frac{\partial z}{\partial y}\right)_x \mathrm{d}y = X\mathrm{d}x + Y\mathrm{d}y. \tag{1.4.6}$$

where X and Y are, in general, also functions of x and y.

Conversely, however, not all expressions of the form $X\mathrm{d}x + Y\mathrm{d}y$, in which X and Y are functions of x and y, are exact differentials. In other words, such expressions cannot *always* be regarded as differentials of some function $z = \mathrm{f}(x,y)$. One then speaks of *inexact* differentials. If $X\mathrm{d}x + Y\mathrm{d}y$ is an exact differential dz, then from (1.4.6.):

$$X = \left(\frac{\partial z}{\partial x}\right)_y \text{ and } Y = \left(\frac{\partial z}{\partial y}\right)_x.$$

Thus, from Equation (1.4.5), the condition that dz is an *exact differential* is

$$\frac{\partial X}{\partial y} = \frac{\partial Y}{\partial x}. \tag{1.4.7}$$

If $X\mathrm{d}x + Y\mathrm{d}y$ is not an exact differential, this equality does not apply and the *condition for an inexact differential* is thus

$$\frac{\partial X}{\partial y} \neq \frac{\partial Y}{\partial x} \tag{1.4.8}$$

By *cross-differentiation* we can thus tell whether the variable considered is an exact or inexact differential.

Example. The expression $3x^2y^4\mathrm{d}x + 4x^3y^3\mathrm{d}y$ is an exact differential because

$$\frac{\partial(3x^2y^4)}{\partial y} = \frac{\partial(4x^3y^3)}{\partial x} = 12x^2y^3$$

On the other hand, the expression $\mathrm{d}z = x^2y^4\mathrm{d}x + x^3y^3\mathrm{d}y$ is an inexact differential because

$$\frac{\partial(x^2y^4)}{\partial y} \neq \frac{\partial(x^3y^3)}{\partial x}$$

In this case there is no function z of which $\mathrm{d}z$ is the differential.

If $X\mathrm{d}x + Y\mathrm{d}y$ is an exact differential then, from the above, we may write

$$\int_1^2 (X\mathrm{d}x + Y\mathrm{d}y) = \int_1^2 \mathrm{d}z = z_2 - z_1.$$

since there is a function z of which $\mathrm{d}z$ is the differential. Thus, if an exact differential is integrated from the initial values x_1, y_1 to the final values x_2, y_2 of the variables, then the result is dependent only on the limits, and *not* on the path by which the integration is performed. As an example we will consider once more the exact differential $3x^2y^4\mathrm{d}x + 4x^3y^3\mathrm{d}y$. Obviously

$$\int_1^2 (3x^2y^4\mathrm{d}x + 4x^3y^3\mathrm{d}y) = x_2^3y_2^4 - x_1^3y_1^4$$

is independent of the path. If the variables have the same values at the beginning and end of the integration, i.e. if one integrates over a closed cycle, then from the above, the integral has the value zero:

$$\oint dz = 0 \qquad (dz = \text{exact differential})$$

If, on the other hand, $dz = Xdx + Ydy$ is an inexact differential then, as will be seen in the following sections, integration produces a result which is dependent on the path, while integration over a cycle produces, in general, a result which is not zero:

$$\oint dz \neq 0 \qquad (dz = \text{inexact differential})$$

1.5. REVERSIBLE AND IRREVERSIBLE CHANGES OF STATE

In thermodynamics it is of great importance always to make a clear distinction between reversible and irreversible changes of state.

The word reversible must be taken literally in the sense that all changes which occur during a reversible process in system *and* surroundings must be capable of being restored by reversing the direction of the process. On the other hand, with an irreversible process, the original state of the system cannot be restored without leaving changes in the surroundings.

A change of state of a system can only occur reversibly if the forces which system and surroundings exert on each other are continually balanced during the change and if heat exchange takes place without the presence of appreciable temperature differences. Strictly speaking, only processes which occur infinitely slowly can be completely reversible, since more rapid change requires finite forces and temperature differences. In nature, only irreversible processes occur. In the laboratory, it is often possible to approximate to the above conditions so closely that it is justifiable to speak of a reversible change.

Example 1. A gas may be expanded in a reversible manner to twice its volume with the help of a cylinder closed by a piston. The expansion is allowed to take place in a constant temperature enclosure and in such a way that the pressure of the piston only differs very slightly from the gas pressure. At each stage the process is reversible in the sense that an extremely small change in the piston pressure is sufficient to reverse the direction of the process. On the "way back" the system passes through the same states as on the "outward path".

However, we can also expand a gas to twice its volume in an irreversible manner, by suddenly connecting the container in which it is enclosed to an evacuated container of the same size. Only the initial and final conditions are known in this case. Now, too, the original state can be restored by performing work on the gas. While in the reversible process the work to be performed by the surroundings was the same as that received during the expansion of the system, the surroundings must now perform work without this compensation. To restore the state of the system, therefore, that of the surroundings must change.

Example 2. At normal pressure, ice is in equilibrium with water only at 0 °C. Ice at a pressure of 1 atm can thus be reversibly melted only at 0 °C, viz. by applying heat from a reservoir of which the temperature is only very slightly higher than 0 °C. A very small lowering of the temperature is then sufficient to reverse the direction of the process.

However, if ice is placed in water at, say, 10 °C, the melting process will be irreversible.

Example 3. In many cases it is possible to allow a chemical reaction, e.g. the reaction

$$CuSO_4\,aq + Zn \rightarrow ZnSO_4\,aq + Cu$$

to proceed reversibly in a galvanic cell. An opposing voltage must then be applied which is only minutely less than the electromotive force E of the cell, so that a very small increase in the opposing voltage is once more sufficient to reverse the direction of the reaction. On the other hand, the reaction will take place irreversibly if the cell is short-circuited or if the copper sulphate solution and the metallic zinc are simply put together in a beaker.

Even systems which are not in equilibrium can, in many cases, undergo changes which are reversible or nearly reversible. To understand this, consider a stoichiometric mixture of hydrogen and oxygen (2 volumes of H_2 to 1 volume O_2), a mixture which without any doubt is far removed from the equilibrium condition. At room temperature the reaction between the two gases does not occur at a noticeable rate, so that the mixture may be subjected to the reversible expansion described in example 1. Heat may also be removed from or added to the mixture in a reversible way. During these reversible processes, the system behaves as though the chemical reaction could not take place at all. Such a case is referred to as a (wholly) suppressed chemical reaction.

Work and heat may also be exchanged reversibly between a system and its surroundings if there exist in the system chemical equilibria which are rapidly established. In this case a reversible change of state consists of a continuous transition through a succession of equilibrium states. The equilibria referred to are known as non-suppressed chemical equilibria.

There are also cases in which a system is continually in equilibrium with respect to certain internal transitions while other transitions are almost completely suppressed. If the cylinder in example 1 should contain, for instance, a quantity of liquid CS_2 in equilibrium with its vapour, this equilibrium can continually adjust itself when the piston is slowly displaced or when heat is reversibly introduced. Complete establishment of the chemical equilibrium, however, requires a decomposition of the carbon disulphide into its constituent elements. Under normal conditions this dissociation reaction is almost completely suppressed. Also a system of this kind, in which some changes are suppressed and others are non-suppressed, can be subjected to reversible changes of state. If one also considers nuclear reactions, it is clear that for the systems with which the chemist is normally concerned, these reactions are suppressed.

The exchange of work and heat between a system and its surroundings is always *irreversible* if the internal chemical reactions of the system do *not* proceed either so slowly that they can be neglected, or so rapidly that chemical equilibrium is continually maintained.

1.6. THE FIRST LAW

The first law, mentioned in Section 1.1, and also known as the law of conservation of energy, originates in the experimental fact that heat and work can be transformed one into the other. If a system is not isolated from the rest of the world, we can introduce a quantity of heat $\mathrm{d}Q$ and perform work $\mathrm{d}W$ on it. According to the first law these quantities of energy must reappear wholly as an increase in the internal energy U of the system:

$$\mathrm{d}U = \mathrm{d}Q + \mathrm{d}W. \tag{1.6.1}$$

Strictly speaking, until we have left the domain of classical thermodynamics, the conception of internal energy only takes on a definite meaning through this mathematical formulation of the first law. Equation (1.6.1) is thus, at the same time, the defining equation for the internal energy. The definition implies that in classical thermodynamics only relative values of U are known, its absolute value depending on an unknown constant. In the atomic picture, of which classical thermodynamics is independent, the internal energy is the sum of the kinetic and potential energies of all the elementary particles of which the system is composed.

In accordance with an international sign convention $\mathrm{d}Q$ will be taken as positive when heat is *given to* the system under consideration and dW is taken as positive when work is done *on* the system. Heat given to the surroundings and work done on the surroundings are taken as negative. To avoid confusion, it should be noted that this agreement is not universally kept. In chemistry, for example, a heat of reaction is still often given a positive sign when the heat is given up by the system to the surroundings, i.e. if the reaction is exothermal. In accordance with the above-mentioned convention, exothermal reactions will always be allocated a negative heat of reaction in this book.

Work performed on a system may be of a mechanical, electrical, magnetic or chemical nature. If the system consists of a quantity of argon in a cylinder under a movable piston, then work may be performed on it by reducing the volume by means of the piston. If A is the area of the (weightless and frictionless) piston and if the external pressure is p, then the force which is exerted on the piston from outside is equal to pA. If the piston is displaced through a distance dx inwards, then the work done on the gas is

$$dW = -pA dx = -p dV. \tag{1.6.2}$$

The negative sign is in accordance with the convention for the sign of the amount of work, since work done on the system corresponds with a negative value of dV. If p in Equation (1.6.2) is taken to indicate the *internal* pressure, then it is only applicable to a reversible change of state. Only in that case are external and internal pressure equal.

Mechanical work can be performed on a liquid both by reducing the volume and by increasing the surface area. The work which must be performed to increase the free surface in a reversible and isothermal manner by unit area, is called the surface tension γ of the liquid. For a small increase dA of the surface, we have

$$dW = \gamma dA. \tag{1.6.3}$$

On a solid body in the form of a rod, mechanical work may be performed in a number of ways, e.g. by stretching it longitudinally under the influence of a gradually increasing tensile force K. If the length of the rod increases by dl at a particular value of K, then the work done will be

$$dW = K dl. \tag{1.6.4}$$

Per unit volume of the material of the rod, the work performed will be

$$dW = \sigma d\varepsilon \tag{1.6.4a}$$

where σ is the tensile stress, i.e. the tensile force K divided by the cross-section A of the rod, and $d\varepsilon$ the fractional increment of length dl/l.

Bodies which can be magnetized may have work performed on them via a magnetic field. If an external field H is required to produce a magnetic moment M_m in the body, then the work performed on the body when M_m increases by an amount dM_m is given by

$$dW = \mathrm{H} dM_m \tag{1.6.5}$$

Per unit volume of the body, the work performed is given by

$$dW = \mathrm{H} dI \tag{1.6.5a}$$

where I is the intensity of magnetization, i.e. the magnetic moment per unit volume.

A similar equation applies for electrical work performed on a dielectric. If M_e is the electric moment in an electric field E and if this moment increases by dM_e, then work

$$dW = E dM_e \tag{1.6.6}$$

is done on the body. Per unit volume the work done is given by

$$dW = E dP \tag{1.6.6a}$$

where P is the polarization, i.e. the electric moment per unit volume.

If the system considered is a galvanic cell (see Sections 1.2 and 1.5), "chemical work" may be performed on it, i.e. work which results in a chemical change. If F is the charge on a gram-ion of monovalent positive ions (ca. 96500 coulomb), then the charge on dn z valent positive gram-ions is $zFdn$. If this charge is moved through the cell under the influence of an applied opposing voltage, which is only very little greater than the electromotive force E of the cell, then the work performed on the cell is

$$dW = zFE dn \tag{1.6.7}$$

For the cell discussed in Example 3 in Section 1.5, $z = 2$. The performance of work dW on this cell results in this case in dn mole Cu going into solution and dn mole Zn being deposited.

All forms of work can be mutually transformed without appreciable losses with the help of pulleys, dynamos, etc. Each form of work can also always be converted wholly into heat, e.g. by friction. On the other hand, as we shall see in the following, heat can only be converted into work under very limited conditions.

Returning to the first law, let us suppose that, with accompanying exchanges of quantities of heat and work with the surroundings, a system is subjected to physical and chemical changes such that it finally returns to the original state. The algebraic sum of the quantities of heat and work must be zero:

$$\oint \mathrm{d}U = \oint \mathrm{d}Q + \oint \mathrm{d}W = 0 \tag{1.6.8}$$

If this were not so, we should have created or destroyed energy. Equation (1.6.8) is thus merely another formulation of the first law.

1.7. THE INTERNAL ENERGY IS A THERMODYNAMIC FUNCTION

If a system undergoes a finite change, in which it passes from an initial state 1 to a final state 2, it follows from (1.6.1) that

$$\int_1^2 \mathrm{d}U = \int_1^2 \mathrm{d}Q + \int_1^2 \mathrm{d}W \tag{1.7.1}$$

Here, the integral on the left is the only one which is dependent only on the initial and final states; the two other integrals are also dependent on the path followed. The truth of this assertion can be demonstrated as follows.

Consider two paths I and II, by which a system may be brought from state 1 to state 2. (Suppose, for example, that in state 1 we have 2 moles H_2 and 1 mole O_2 in an enclosure at temperature T_1 and in state 2 we have 2

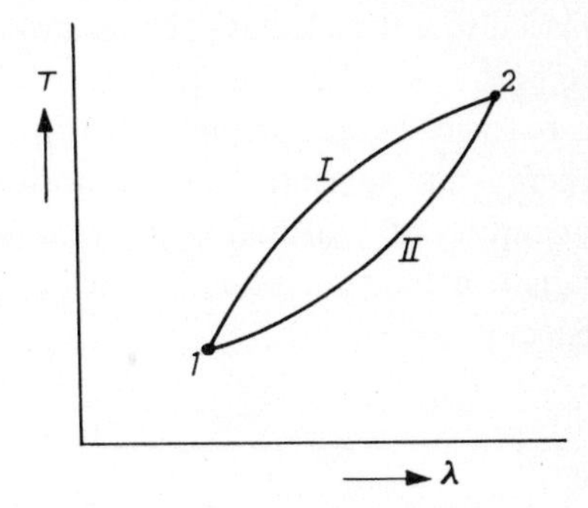

Fig. 1. A system can be transferred from a state 1 to a state 2 by various "paths" (e.g. I and II). In the case considered the state is characterized by the reaction co-ordinate λ and the temperature T.

moles H_2O at temperature T_2; see Fig. 1). The changes in internal energy along paths I and II will be called

$$ {}^{\mathrm{I}}\int_1^2 \mathrm{d}U \quad \text{and} \quad {}^{\mathrm{II}}\int_1^2 \mathrm{d}U $$

If we perform a cycle, from 1 to 2 along path I and back along path II, then from Equation (1.6.8):

$$ {}^{\mathrm{I}}\int_1^2 \mathrm{d}U + {}^{\mathrm{II}}\int_2^1 \mathrm{d}U = 0. $$

and thus

$$ {}^{\mathrm{I}}\int_1^2 \mathrm{d}U = {}^{\mathrm{II}}\int_1^2 \mathrm{d}U. $$

The internal energy of a system thus depends only on its state i.e. on its pressure, temperature, volume, chemical composition, structure, etc. The previous history of a system has no effect on the value of its internal energy. U is therefore a thermodynamic function in the sense of Section 1.2. W and Q, on the other hand, are *not* thermodynamic functions. Indeed, according to Equation (1.6.1), we can produce the same change $\mathrm{d}U$ in the internal energy of a system either by the addition of heat alone or by performing work alone. We shall now illustrate the assertion concerning W and Q by considering the expansion of a perfect gas.

A perfect gas is allowed to expand isothermally to n times its volume in two different ways, viz. (*a*) in an irreversible manner by connecting the container in which it is enclosed to an evacuated space of the required volume, (*b*) in a reversible manner by means of a cylinder, closed by a piston and placed in a thermostat. The increase in the internal energy will be the same in both cases, because in each of the changes of state the same final state is reached. In fact, we have seen from Section 1.3 that the change in internal energy due to isothermal expansion of a perfect gas is always zero.The quantities of heat and work introduced into the system are, however, different in the processes considered. In the irreversible change of state, we have

$$ \int \mathrm{d}W = -\int \mathrm{d}Q = 0, $$

on the other hand, for the reversible change

$$\int \mathrm{d}W = -\int p\mathrm{d}V,$$

and thus

$$\int \mathrm{d}Q = +\int p\mathrm{d}V.$$

Furthermore, from a p,V-diagram it is immediately obvious that even if the process is carried out reversibly, the work $W = -\int p\mathrm{d}V$ is dependent not only on the initial and final states, but also on the path followed. This is demonstrated in Fig. 2, in which the integral is shown as a shaded area for

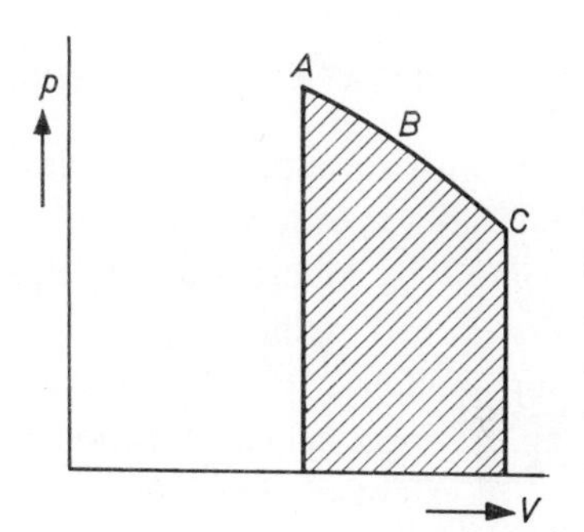

Fig. 2. When a gas is expanded reversibly by the "path" *ABC*, work is performed on the surroundings. The amount of work is given by the shaded area and is thus dependent on the path followed.

the integration path *ABC*. If the path is changed, the value of the integral is also changed. Only for a reversible *and* isothermal process is the work performed dependent exclusively on the initial and final states. In other words, only in the special case of a reversible and isothermal change of state is the work performed on the system equal to the change in a thermodynamic function. We shall see later that this thermodynamic function is given the name "free energy" (see Chapter 3).

It follows from the above that although we can speak of the internal energy of, for example, a given quantity of gas at a given temperature and pressure, we cannot speak of the work or the heat of this gas. For example, we may write

$$U = \mathrm{f}(p,T).$$

but similar expressions do not apply for W and Q. For the differential of U

(when p and T are chosen as independent variables) we have:

$$\mathrm{d}U = \left(\frac{\partial U}{\partial p}\right)_T \mathrm{d}p + \left(\frac{\partial U}{\partial T}\right)_p \mathrm{d}T \tag{1.7.2}$$

On the other hand, dW and dQ are not differentials, but only infinitesimal quantities of work and heat, i.e. inexact differentials in the sense of Section 1.4.

For $\int_1^2 \mathrm{d}U$ we can always write $U_2 - U_1$, the internal energy in the final state diminished by that in the initial state. For $\int_1^2 \mathrm{d}W$ and $\int_1^2 \mathrm{d}Q$, however, it is in general *not* possible to write $W_2 - W_1$ or $Q_2 - Q_1$ because no functions W and Q exist of which dW and dQ are the differentials.

1.8. APPLICATION OF THE FIRST LAW TO A HOMOGENEOUS PHASE

We wish to apply the first law to a constant quantity of a homogeneous mixture (or pure substance) in which possible chemical reactions are either completely suppressed or completely non-suppressed. It is supposed that the system can exchange quantities of heat and work (but not matter) with the surroundings. It is also supposed that these exchanges do not cause appreciable changes in magnetization, polarization, in elastic and plastic deformation or in extent of the surface of the system. In that case the thermodynamic state of the homogeneous phase is usually completely defined by two thermodynamic variables.

In the previous sentence, the word "usually" is essential, since thermodynamics gives no binding rules concerning the minimum number of variables necessary to characterize fully the state of a system. If the homogeneous phase is a quantity of water, for example, then in the region of maximum density pairs of states can be found such that the members of any pair have the same value of V and the same value of p, but different values of T. In the more usual case where the state of a homogeneous phase is completely specified by two variables, each thermodynamic variable is an unambiguous function of two others. If we choose V and T as independent variables, we can write, for instance

$$U = \mathrm{f}\,(V, T).$$

$$\mathrm{d}U = \left(\frac{\partial U}{\partial T}\right)_V \mathrm{d}T + \left(\frac{\partial U}{\partial V}\right)_T \mathrm{d}V. \tag{1.8.1}$$

When changes in magnetization, polarization, etc. do not have to be taken into account, the only possibility of performing work is that connected with changes in volume. For a very small change of state of our homogeneous phase we can, therefore, write the first law in the following simple form:

$$dQ = dU + p dV \tag{1.8.2}$$

and thus, substituting from (1.8.1):

$$dQ = \left(\frac{\partial U}{\partial T}\right)_V dT + \left\{\left(\frac{\partial U}{\partial V}\right)_T + p\right\} dV \tag{1.8.3}$$

This equation states that an amount of energy dQ given to a homogeneous phase in the form of heat is used in the following three ways: (*a*) for an increase in energy of the phase resulting from the rise in temperature produced, $(\partial U/\partial T)_V dT$; (*b*) for an increase in energy of the phase resulting from the change in volume, $(\partial U/\partial V)_T dV$; (*c*) for the work performed on the surroundings during that change in volume, $-dW = p dV$.

The ratio, dQ/dT, between the very small quantity of heat introduced and the very small rise in temperature which it produces is called the *heat capacity* C of the system. Its value depends on the conditions under which the temperature change takes place. If, for example, the volume is kept constant while the heat is introduced, it is known as the heat capacity at constant volume C_V. It is clear from (1.8.3) that this is given by

$$C_V = \left(\frac{dQ}{dT}\right)_V = \left(\frac{\partial U}{\partial T}\right)_V \tag{1.8.4}$$

On the other hand, the heat capacity at constant pressure C_p is seen from (1.8.2) and (1.7.2) to be

$$C_p = \left(\frac{dQ}{dT}\right)_p = \left(\frac{\partial U}{\partial T}\right)_p + p\left(\frac{\partial V}{\partial T}\right)_p \tag{1.8.5}$$

This may also be written

$$C_p = \left(\frac{\partial H}{\partial T}\right)_p \tag{1.8.6}$$

where

$$H = U + pV. \tag{1.8.7}$$

As a function purely of thermodynamic functions, H itself is a thermodynamic function. It is called *enthalpy*.

The heat capacity of 1 gram of a substance is known as the *specific heat* of that substance and the heat capacity of 1 gram-molecule is also called the specific heat or, where confusion with the former quantity is possible, the specific heat per mole or molar (specific) heat.

In this book molar thermodynamic variables will always be indicated by small letters, so that e.g. V is the volume of an arbitrary quantity, while v is the volume of 1 mole or the molar volume, U is the energy of an arbitrary quantity but u is the energy per mole or molar energy, etc.

By this convention the first law as formulated in (1.8.3) and (1.8.4) when applied to 1 mole of a homogeneous phase takes the following form:

$$\mathrm{d}Q = c_v \mathrm{d}T + \left\{ \left(\frac{\partial u}{\partial v}\right)_T + p \right\} \mathrm{d}v \tag{1.8.8}$$

1.9. APPLICATION OF THE FIRST LAW TO A PERFECT GAS

For a perfect gas $(\partial u/\partial v)_T = 0$, so that from (1.8.8) the first law for 1 mole of a perfect gas becomes

$$\mathrm{d}Q = c_v \mathrm{d}T + p\mathrm{d}v \tag{1.9.1}$$

For a constant value of the pressure $p\mathrm{d}v = R\mathrm{d}T$ and thus from (1.9.1)

$$\left(\frac{\mathrm{d}Q}{\mathrm{d}T}\right)_p = c_p = c_v + R \tag{1.9.2}$$

For an infinitely small *isothermal* change of state of a perfect gas ($\mathrm{d}T = 0$) we have, from equation (1.9.1)

$$\mathrm{d}Q = -\mathrm{d}W = p\mathrm{d}v$$

and thus, for a finite isothermal change of state, in which the volume increases from v_1 to v_2

$$Q = -W = \int_{v_1}^{v_2} p\mathrm{d}v. \tag{1.9.3}$$

Here p is the external pressure. If the isothermal change of state takes place reversibly, then p is always equal to the pressure of the gas *), so that, substituting from the equation of state, $pv = RT$, we can also write for (1.9.3):

$$Q = -W = RT \int_{v_1}^{v_2} \frac{dv}{v} = RT \ln \frac{v_2}{v_1}. \tag{1.9.4}$$

Next we shall consider a change of state which is not isothermal, but *adiabatic*, i.e. a process which takes place without the exchange of heat with the surroundings. To this end the gas is supposed to be enclosed in a cylinder the walls and piston of which are impermeable to heat. For a very small displacement of the piston, from (1.9.1):

$$c_v dT + p dv = 0.$$

If the adiabatic change of state occurs reversibly, then the external pressure will again be equal to the gas pressure, so that with the help of the equation $pv = RT$, we obtain

$$c_v dT + RT \frac{dv}{v} = 0,$$

i.e.

$$\frac{dT}{T} + \frac{R}{c_v} \frac{dv}{v} = 0.$$

Integrating, this equation becomes

$$\ln T + \frac{R}{c_v} \ln v = \text{constant}$$

Hence

$$Tv^{R/c_v} = \text{constant}$$

Combining with Equation (1.9.2) we can also write for this:

$$Tv^{(c_p/c_v)-1} = \text{constant} \tag{1.9.5}$$

or, since $pv = RT$:

$$pv^{c_p/c_v} = \text{constant}. \tag{1.9.6}$$

In a normal p,v diagram the isotherms are rectangular hyperbolae ($pv =$

*) If the process takes place irreversibly, it is impossible to speak of *one* particular gas pressure at any given moment.

constant), while the adiabatic lines, since $c_p/c_v > 1$, have a steeper slope (see Fig. 3).

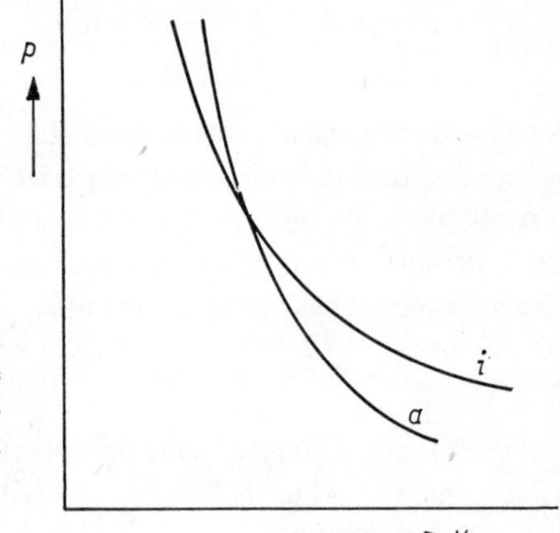

Fig. 3. Graphic representation of p as a function of v for an adiabatic-reversible change of state (a) and an isothermal-reversible change of state (i) of a perfect gas.

1.10. NON-MATHEMATICAL FORMULATION OF THE SECOND LAW

The first law, discussed above, is based in principle on experience; the experience that it is impossible to construct a *perpetuum mobile*, i.e. a machine which performs work without the help of a source of energy. This law appeared to be of such great general validity that it was placed as axiom at the head of thermodynamics. Applied to an *isolated* system it states that the energy U_2 after some process has taken place is equal to the energy U_1 in the initial state. The variable U gives no indication that a process has taken place. The first law thus contains no information about the *direction* in which a process will occur. To illustrate this, we shall consider two examples.

(1) If two bodies at different temperatures are brought into contact with one another, heat flows from the body at the higher temperature to that at the lower temperature. The reverse never happens, although this would not be in conflict with the first law.

(2) If we drop a stone from a certain height, the potential energy decreases and the kinetic energy increases. When the stone reaches the ground, all the potential energy has been converted into kinetic energy. On striking the ground, the kinetic energy disappears and is converted into an equivalent amount of heat. The first law also permits the reverse process: A stone lying on the ground takes up a quantity of heat from the surroundings and rises

to such a height that its potential energy is equal to the amount of heat taken up. This, however, never happens.

In general, experience shows that all changes which occur under constant external conditions have a very definite direction. The systems concerned tend, under these conditions, to a particular final state. When this is reached, they never return spontaneously to one of their previous states. As an example, let us consider a mixture of gases, enclosed in a glass bulb which is placed in a thermostat. If there was turbulence in the gas at first, this will disappear. If the temperature of various parts of the gas was different, it will become uniform. If the mixture was not homogeneous, it will become so. Possible chemical reactions will also proceed until an equilibrium state is reached, although this is sometimes an immeasurably slow process. All changes which occur spontaneously thus have a particular direction, i.e. they are irreversible (cf. Section 1.5). It is clear that it is of primary importance to discover the factors which determine this direction. From the examples just quoted it appears that for this purpose we need, in addition to the first law, another important principle, a *second law*.

In the words of Clausius, the second law states: *Heat can never, of itself, flow from a lower to a higher temperature*. The words "of itself" serve to express the fact that no feasible process can be devised in which heat is drawn from a reservoir *A* and given up to a reservoir *B* at a higher temperature, unless other simultaneous changes persist in the surroundings.

In the words of Thomson (Lord Kelvin) the second law may be stated thus: *It is impossible to extract heat from a reservoir and convert it wholly into work without causing other changes in the universe*. (This conversion without compensation would happen in the example of the stone which takes heat from its surroundings and rises).

A machine of the type prohibited by the Thomson statement would be as useful to mankind as a *perpetuum mobile*. It would enable us to convert technically useless heat, which is available in almost unlimited quantities, into work. Ships fitted with a machine of this kind would be able to sail the seas simply be extracting heat from the sea water and converting it into work. This imaginary machine has therefore been called a *perpetuum mobile* of the second kind. The second law might thus also be stated: *Perpetual motion of the second kind is impossible*.

The two formulations of the second law: (1) that heat cannot flow of itself from a lower to a higher temperature and (2) that heat at a given temperature cannot be converted wholly into work without compensation, are only superficially different. In fact, it is easy to show that if one of these so-called impossible processes were possible, the other would also be possible.

Suppose, for example, that in violation of (2), an amount of heat at temperature T_2 could be converted quantitatively into mechanical energy, then this mechanical energy could be converted by friction into heat at a higher temperature T_1. The final result would be that, in violation of (1), a quantity of heat had been transferred from T_2 to T_1 without any further change in the surroundings having taken place.

1.11. MATHEMATICAL FORMULATION OF THE SECOND LAW

The greatest benefit can be reaped from the second law when it is mathematically formulated. This formulation reads thus: *In every reversible cycle* $\oint \mathrm{d}Q/T = 0$, *while for an irreversible cycle* $\oint \mathrm{d}Q/T < 0$. *Cycles in which* $\oint \mathrm{d}Q/T > 0$ *are impossible.*

Like the first law, the second law has the character of an axiom (postulate), which has shown its worth in the truth of all the conclusions derived from it. In the following sections we shall indicate the historical path by which this form of the second law was reached. We shall first show that the mathematical statement contains the statements in the previous section.

(1) Suppose that we have a device which is able to extract a quantity of heat Q from a reservoir at the lower temperature T_2 and deliver it to another reservoir at the higher temperature T_1. Since after the delivery the imaginary device has returned to its original state (or in other words, it has completed a cycle), the quantity of heat Q has been transferred from a lower to a higher temperature without other changes persisting in the universe. In the cycle performed by the device, we have

$$\oint \frac{\mathrm{d}Q}{T} = \frac{Q}{T_2} - \frac{Q}{T_1} > 0 \text{ (since } T_1 > T_2\text{)}$$

Also according to the mathematical formulation of the second law, the transfer of heat from a lower to a higher temperature (without other changes persisting) is thus impossible.

(2) Suppose that we have a device which can execute a cycle in which it extracts heat from a reservoir at a temperature T and converts it wholly into work. In this cycle heat is only received by the device, none is given up, i.e.

$$\oint \frac{\mathrm{d}Q}{T} = \frac{Q}{T} > 0$$

Also according to the mathematical formulation of the second law, it is thus impossible to convert heat from one temperature wholly into work without causing other changes.

1.12. CARNOT CYCLES

Historically, the mathematical formulation of the second law was reached through the study of heat engines, i.e. engines in which work is obtained from heat. It has already been stated that it is impossible to convert heat at one particular temperature wholly into work without causing other changes. If changes are allowed complete conversion is possible. For example, a gas enclosed in a cylinder by a piston may be allowed to expand isothermally, thus performing work and (if the gas is perfect) extracting an equivalent amount of heat from a reservoir. However, this has not been accomplished without changing the state of the system, for the pressure of the gas has been reduced.

Only if infinitely long cylinders were available would it be possible to convert unlimited quantities of heat into work in the manner described. To operate a machine in practice, it is necessary to return the piston to its initial position. If the initial temperature is maintained while doing this, then the work obtained from the expansion will all be used up again. The only way to gain work is thus to carry out the compression at a lower temperature than the expansion. Indeed, consideration of the various kinds of heat engine shows that a necessary condition for their operation is always that heat (Q_1) is supplied to the working substance at a high temperature and a smaller quantity (Q_2) extracted from it at a lower temperature. Thus, for example, the operating principle of the steam engine is as follows: The steam passes through a cycle in which it takes up heat at a high temperature in the boiler and gives up heat at a low temperature in the condenser. The difference is available as work.

In 1824, Carnot attempted to calculate the maximum efficiency which could be attained by an ideal heat engine, from the consideration of cycles consisting of two isothermal changes of state, in which heat is exchanged with heat reservoirs, and two adiabatic changes of state, in which no heat is exchanged but the temperature changes. In the following examples, we imagine the cycles to be performed by a particular quantity of the substance under consideration (the working substance) and assume that the states through which it passes can be described by a p, V diagram.

As the first example of a Carnot cycle we shall consider an idealized steam engine. The cylinder contains water and steam at a temperature T_1 under a

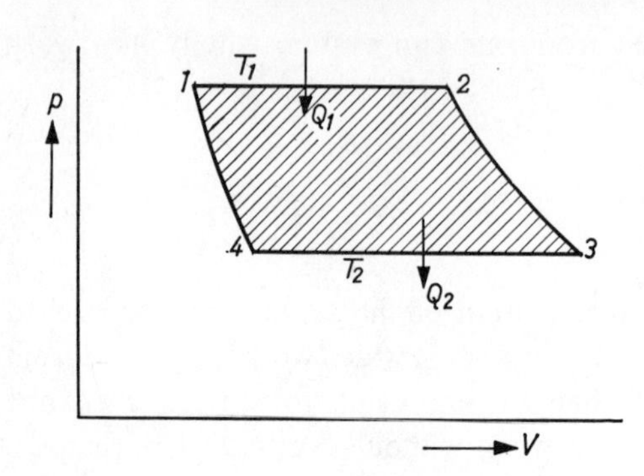

Fig. 4. Carnot cycle, performed on a two-phase system consisting of water and steam.

movable piston. The state of this heterogeneous system is represented by point 1 in Fig. 4. The volume is allowed to increase isothermally and reversibly, while the pressure remains equal to the vapour pressure p_1 at the temperature T_1 because a quantity of liquid evaporates. Thus in the p, V diagram we move along the horizontal line 1-2. In doing so, a quantity of heat Q_1 is extracted from a heat reservoir at constant temperature T_1. On reaching state 2, a wall impermeable to heat is slipped around the cylinder and the volume is allowed to increase further, adiabatically. The temperature will drop. The expansion is continued until state 3 is reached, at which point the temperature has dropped to T_2. The insulating wall is now removed and the cylinder is placed in contact with a reservoir at the temperature T_2 and the volume is reduced isothermally, so that a quantity of heat Q_2 is taken up by the reservoir. During this process the gas pressure remains constant and equal to the vapour pressure p_2 at the temperature T_2. The isothermal compression is stopped at a point 4 such that if the compression is continued adiabatically, point 1 will again be reached. As soon as point 1 is reached, the cycle is complete.

As a second example of a Carnot cycle, let us consider a process of this kind carried out with a perfect gas. As we know, in this case the isotherms are not horizontal lines, but rectangular hyperbolae (see Fig. 5). Along the isotherm 1-2 a quantity of heat Q_1' is absorbed by the ideal gas, along the isotherm 3-4 a quantity Q_2' is given up.

For the quantitative treatment of Carnot processes it is convenient to always consider the absolute values of the quantities of work and heat converted. These will be written $|W|$ and $|Q|$ respectively. The work obtained is given by $|W| = \int p \mathrm{d}V$ and thus, per cycle, by the shaded area in Figs. 4 and 5. This shaded area, according to the first law, is given by

$$|W| = |Q_1| - |Q_2|, \tag{1.12.1}$$

since the internal energy of the working substance returns to its initial value on completion of a cycle. The thermal efficiency is defined by

$$\eta = \frac{|W|}{|Q_1|} = \frac{|Q_1| - |Q_2|}{|Q_1|}. \tag{1.12.2}$$

If a Carnot cycle is carried out reversibly, the effect produced can be exactly cancelled by carrying out the process in the reverse direction. A quantity of work represented by the shaded area must then be supplied to the working substance. At the lower temperature T_2, a quantity of heat Q_2 is now absorbed by the substance, while at the higher temperature T_1 it gives up a quantity Q_1. This is equal to

$$|Q_1| = |W| + |Q_2|.$$

It is clear that we are here dealing with an idealized refrigerator (compare Section 3.14).

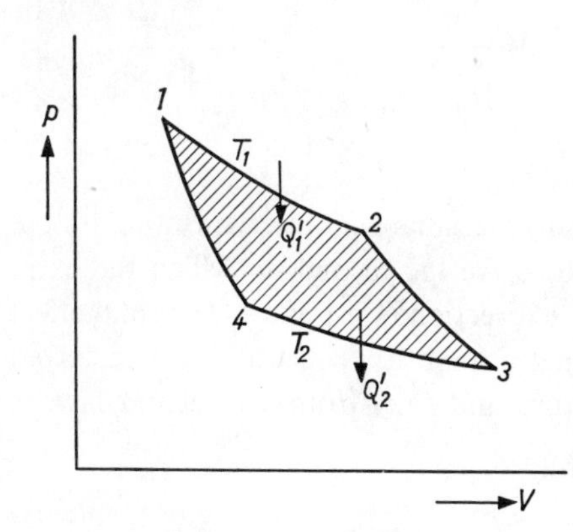

Fig. 5. Carnot cycle performed on a perfect gas.

1.13. CALCULATION OF THE EFFICIENCY OF CARNOT CYCLES

On the basis of the non-mathematical statement of the second law, we can easily show that the efficiency of a Carnot cycle, when carried out reversibly, is independent of the nature of the working substance. In order to do this, let us consider the two cycles in Figs. 4 and 5 and assume that they work between the same temperatures T_1 and T_2. Furthermore, care is taken that the two shaded areas are equal to one another, thus that

$$|W| = |Q_1| - |Q_2| = |Q_1'| - |Q_2'| \tag{1.13.1}$$

From Equation (1.12.2), the efficiencies of the two cycles are equal if $|Q_1| = |Q_1'|$. We begin by supposing that these two quantities are not equal, e.g.

$$|Q_1'| = |Q_1| + |k|$$

Then from Equation (1.13.1), we also have

$$|Q_2'| = |Q_2| + |k|.$$

In this case, thus, in the second cycle, in order to obtain the same amount of work as in the first one, an extra quantity of heat $|k|$ flows from the "upper" reservoir to the "lower" reservoir (see Fig. 6).

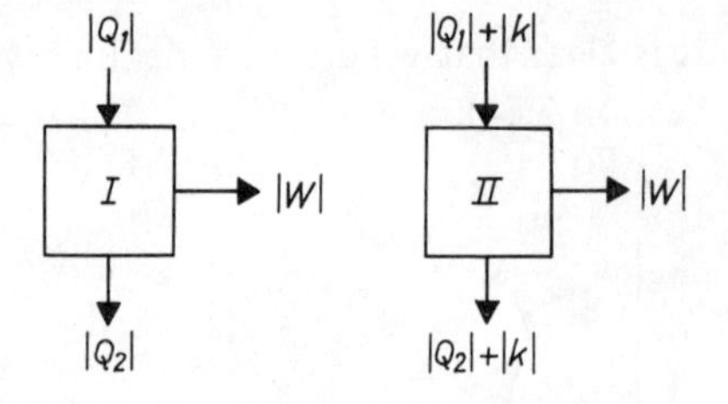

Fig. 6. Symbolic representation of two Carnot cycles, I and II (operating between the same temperatures), the second of which is supposed to have a smaller thermal efficiency than the first.

We can now connect both systems to the same heat reservoirs and use the work supplied by the first process to drive the second, in which the extra quantity of heat occurs, in the reverse direction (see Fig. 7). The total effect of these coupled processes is only that a quantity of heat $|k|$ is transferred from a lower to a higher temperature and this, from the second law, is impossible.

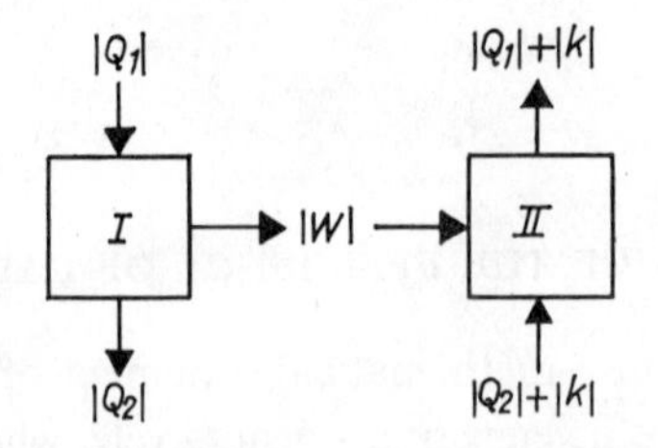

Fig. 7. If cycle II, symbolically indicated in Fig. 6, were reversible, it would be possible to violate the second law by using the work supplied by I to drive II in the reverse direction.

The efficiency of a reversible Carnot cycle is thus independent of the nature of the working substance. It depends only on the two temperatures T_1 and T_2. Thus, if we can calculate the efficiency for one particular substance, we know it for every working substance. If we choose a perfect gas as working substance, the calculation is very simple. If we call the volumes of the gas at points 1, 2, 3 and 4 (Fig. 5) V_1, V_2, V_3 and V_4, then from Equation (1.9.4):

$$|Q_1| = RT_1 \ln \frac{V_2}{V_1} \tag{1.13.2}$$

$$|Q_2| = RT_2 \ln \frac{V_3}{V_4} \tag{1.13.3}$$

Further, for the adiabatic changes of state 1-4 and 2-3, from Equation (1.9.5):

$$T_1 V_1^{(c_p/c_v)-1} = T_2 V_4^{(c_p/c_v)-1} \tag{1.13.4}$$

$$T_1 V_2^{(c_p/c_v)-1} = T_2 V_3^{(c_p/c_v)-1}. \tag{1.13.5}$$

Combining (1.13.4) and (1.13.5):

$$\frac{V_3}{V_4} = \frac{V_2}{V_1}. \tag{1.13.6}$$

From (1.13.2), (1.13.3) and (1.13.6) it then follows that

$$|Q_1| : |Q_2| = T_1 : T_2 \tag{1.13.7}$$

so that, for the efficiency we have

$$\frac{|Q_1| - |Q_2|}{|Q_1|} = \frac{T_1 - T_2}{T_1}. \tag{1.13.8}$$

1.14. ABSOLUTE TEMPERATURE

In the above calculation of the efficiency of a reversible Carnot cycle (Equation (1.13.8)) use was made of the properties of the non-existent perfect gas, the properties of which have already been used as a basis for defining absolute temperature according to the perfect gas scale (Section 1.3). It is fundamentally more satisfying to follow a somewhat different procedure. The fact that the efficiency of a reversible Carnot cycle is independent of the nature of the working substance, may be used as the basis for defining an absolute scale of temperature by (1.13.7) or (1.13.8). It may then be shown that this scale is identical with the perfect gas scale. The reasoning followed is given below.

We start with a temperature scale ϑ, which is quite arbitrary, but which, for the sake of convenience is chosen so that a body A always has a higher value of ϑ than a body B if, on direct contact between the two bodies, heat flows from A to B. Next, we consider a Carnot engine, working reversibly between two heat reservoirs at different temperatures ϑ_1 and ϑ_2 ($\vartheta_1 > \vartheta_2$). The quantity of heat extracted from the high-temperature reservoir during the performance of a cycle is once more called Q_1, while the amount of heat given up to the lower-temperature reservoir is again called Q_2. The work performed is then, as above, given by

$$|W| = |Q_1| - |Q_2|$$

In the same way as in the previous section it is next shown that the efficiency $|W|/|Q_1|$, and therefore also the ratio $|Q_1|/|Q_2|$, has the same value for all bodies, provided that they perform a reversible cycle between the same temperatures. In other words, the ratios mentioned are only functions of the temperatures ϑ_1 and ϑ_2 of the reservoirs. For example, one may write

$$\frac{|Q_1|}{|Q_2|} = \mathrm{f}(\vartheta_1,\vartheta_2). \tag{1.14.1}$$

In order to learn more about this function, a third reservoir at temperature ϑ_3 is brought into the discussion ($\vartheta_1 > \vartheta_2 > \vartheta_3$). It is supposed that one Carnot engine operates between the temperatures ϑ_1 and ϑ_2, a second between the temperatures ϑ_2 and ϑ_3 and a third between ϑ_1 and ϑ_3. If they all operate between the same adiabatics (Fig. 8) we have:

$$\frac{|Q_1|}{|Q_2|} = \mathrm{f}(\vartheta_1,\vartheta_2); \quad \frac{|Q_2|}{|Q_3|} = \mathrm{f}(\vartheta_2,\vartheta_3); \quad \frac{|Q_1|}{|Q_3|} = \mathrm{f}(\vartheta_1,\vartheta_3).$$

From the first two of these three equations it follows that:

$$\frac{|Q_1|}{|Q_3|} = \mathrm{f}\,(\vartheta_1,\vartheta_2)\cdot \mathrm{f}(\vartheta_2,\vartheta_3)$$

The two expressions for $|Q_1|/|Q_3|$ give the relationship

$$\mathrm{f}(\vartheta_1,\vartheta_2)\cdot \mathrm{f}(\vartheta_2,\vartheta_3) = \mathrm{f}(\vartheta_1,\vartheta_3). \tag{1.14.2}$$

i.e. multiplication of $f(\vartheta_1,\vartheta_2)$ by $f(\vartheta_2,\vartheta_3)$ produces a result in which ϑ_2 does not appear at all. This can only be the case if

$$f(\vartheta_a, \vartheta_b) = \frac{g(\vartheta_a)}{g(\vartheta_b)}$$

Equation (1.14.1) can thus be written in the form

$$\frac{|Q_1|}{|Q_2|} = \frac{g(\vartheta_1)}{g(\vartheta_2)} \tag{1.14.3}$$

$g(\vartheta)$ must be a function of the arbitrary temperature ϑ, such that (1.14.3) is satisfied. However, it is not unambiguously defined by (1.14.3), since a function which satisfies this relation still does it after multiplication by a

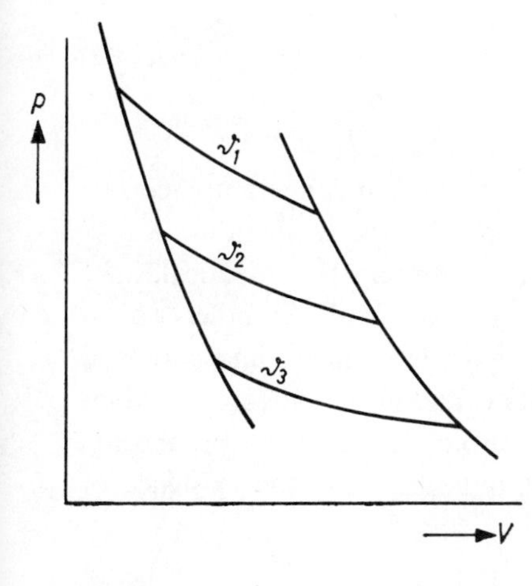

Fig. 8. In the discussions in Section 1.14, three Carnot cycles operate between the same adiabatics.

constant factor. The thermodynamic temperature scale or Kelvin scale T is now so defined that instead of ϑ,

$$g(\vartheta) = T \tag{1.14.4}$$

is chosen as a measure of the temperature and that the arbitrary constant factor is so chosen that the temperature interval between the boiling point T_s and the freezing point T_i of water (at 1 atm pressure) is divided into 100 equal parts, i.e.

$$T_s - T_i = 100° \tag{1.14.5}$$

After this choice, (1.14.4) is unambiguously defined. The form of the function g depends on the scale by which the temperature ϑ is measured, but the value of g(ϑ) is always the same thermodynamic temperature T. From the above it follows that this temperature is absolute, i.e. independent of the properties of particular substances. According to (1.14.3) and (1.14.4) it is given by the defining equation

$$\frac{|Q_1|}{|Q_2|} = \frac{T_1}{T_2} \tag{1.14.6}$$

while the size of the degree is fixed by (1.14.5).

In principle the values of $g(\vartheta_s) = T_s$ and $g(\vartheta_i) = T_i$ can be found by subjecting a body to a reversible Carnot cycle between boiling water (1 atm) and melting ice (1 atm). The ratio of the quantities of heat, Q_s extracted from the boiling water, and Q_i given up to the ice, would be found to be

$$\frac{|Q_s|}{|Q_i|} = \frac{T_s}{T_i} = 1.3661 \tag{1.14.7}$$

From (1.14.5) and (1.14.7), one finds $T_i = 273.1°$ and $T_s = 373.1°$.

Equation (1.14.6), in which T is the thermodynamic temperature, corresponds exactly with Equation (1.13.7), in which T is the temperature according to the perfect gas scale. These two temperatures are thus identical, since the magnitude of the degree is also the same for both. In other words, temperatures measured with the help of a gas thermometer and corrected to values for a perfect gas, have an absolute value. From a theoretical point of view, it is of great importance that the absolute temperature (as appears from this section) can be introduced without supposing the existence of perfect gases.

1.15. ARBITRARY CYCLES

From (1.14.6) it follows that:

$$\frac{|Q_1|}{T_1} = \frac{|Q_2|}{T_2} \tag{1.15.1}$$

If we no longer consider the absolute values of the quantities of heat and work converted, then in accordance with the sign convention of Section 1.6, Equation (1.15.1) becomes

$$\frac{Q_1}{T_1} = -\frac{Q_2}{T_2} \quad \text{or} \quad \frac{Q_1}{T_1} + \frac{Q_2}{T_2} = 0 \tag{1.15.1a}$$

If the ratio of a quantity of heat to the absolute temperature at which it is introduced be called a reduced quantity of heat, then (1.15.1a) can be stated in the following words: For a reversible Carnot cycle, the sum of the reduced quantities of heat supplied to the working substance is zero:

$$\sum \frac{Q}{T} = 0 \tag{1.15.1b}$$

This result can immediately be extended to arbitrary reversible cycles, where these can be represented by a p,V diagram. To do this, we replace the closed curve (see Fig. 9) by a zig-zag line consisting of small pieces of isotherms and adiabatics. If we then proceed to the limit for infinitely small pieces of isotherms and adiabatics, (1.15.1b) becomes

$$\oint \frac{\mathrm{d}Q_{\mathrm{rev}}}{T} = 0 \tag{1.15.2}$$

It may further be shown that (1.15.2) also applies to reversible cycles in which the state of the working substance is determined, not by p and V, but by two other independent variables (cf. Section 1.6). The validity of (1.15.2) can also be demonstrated for reversible cycles in which more than two independent variables are involved.

Finally, irreversible cycles will be included in this discussion. We have seen in Section 1.13 that in two reversible Carnot cycles, in which the same amount

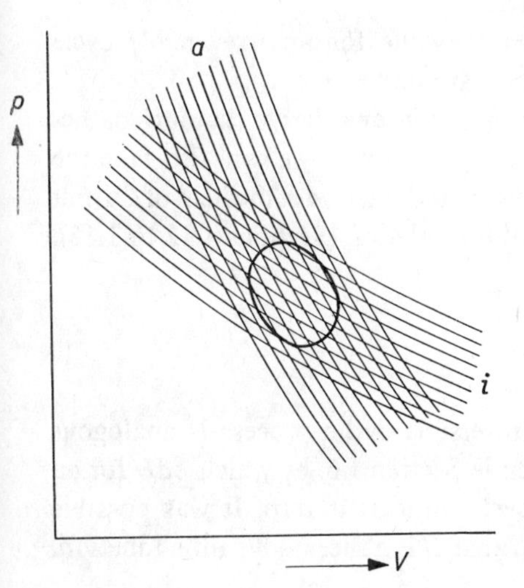

Fig. 9. Arbitrary reversible cycle. Isotherms are indicated by the letter i, adiabatics by a.

of work is done, an equally large quantity of heat flows from the higher to the lower temperature. Indeed, if an extra quantity of heat k should flow in one of these processes, then by using the work supplied by the other process to drive this process in the reverse direction, we should be able to violate the second law (see Fig. 7). However, if one of the processes is irreversible an extra quantity of heat k may flow during its execution, since this process cannot be carried out in the reverse direction. It is even *necessary* that an extra quantity of heat k should be transferred in an irreversible process, since for $k = 0$ the process is reversible (see Section 1.13) and with k negative we should once more be able to violate the second law by using the work obtained from the irreversible process to drive the reversible process in the opposite direction.

Thus, in the irreversible operation of a Carnot cycle, in order to obtain the same quantity of work as in the reversible operation, more heat must be extracted from the hot reservoir. This additional quantity k is transferred to the cold reservoir. A chance to obtain work from heat has here been missed, since in reversible operation of the process, k too could have been partly converted into work. This kind of loss of thermal efficiency is characteristic of the irreversible operation of a process (see Section 3.2).

What effect does this extra heat transfer k have on $\Sigma Q/T$? From (1.15.1b), $\Sigma Q/T$ would be zero if the extra quantity were not involved. For irreversible operation, $|k|/T_1 - |k|/T_2$ must be added, and this amount of reduced heat is negative ($T_2 < T_1$). As was stated above for reversible cycles, this result for irreversible Carnot cycles can be extended directly to arbitrary irreversible cycles. This brings us back to the statement of the second law already introduced as a postulate in Section 1.11:

In every reversible cycle $\oint \mathrm{d}Q/T = 0$, while for an irreversible cycle $\oint \mathrm{d}Q/T < 0$. Cycles in which $\oint \mathrm{d}Q/T > 0$ are impossible.

We may use this axiom as our starting point and derive the non-mathematical formulation from it (Section 1.11). However, we can also take the latter formulation as our axiomatic premise and then, with the help of Carnot cycles, derive from it the mathematical formulation (Sections 1.12 to 1.15).

1.16. ENTROPY

The expression $\oint \mathrm{d}Q_{\mathrm{rev}}/T = 0$ for a reversible cyclic process is analogous to the formulation of the first law given in Section 1.6, by which $\oint \mathrm{d}U$ for an arbitrary (reversible or irreversible) cyclic process is zero. It was possible to deduce from the latter in Section 1.7 that U is a thermodynamic function.

Apparently this must also be so for $\oint \mathrm{d}Q_{\mathrm{rev}}/T$. In fact, if we perform one cycle reversibly as in Fig. 1, then

$$^{\mathrm{I}}\int_1^2 \frac{\mathrm{d}Q}{T} + {}^{\mathrm{II}}\int_2^1 \frac{\mathrm{d}Q}{T} = 0$$

or

$$^{\mathrm{I}}\int_1^2 \frac{\mathrm{d}Q}{T} = {}^{\mathrm{II}}\int_1^2 \frac{\mathrm{d}Q}{T}$$

Thus, if a system follows a *reversible* path from a state 1 to a state 2, the quantity

$$S_{1,2} = \int_1^2 \frac{\mathrm{d}Q}{T} \tag{1.16.1}$$

has a value which is dependent only on the states 1 and 2, and not on the path followed. This is called the *change of entropy* ΔS in the transition from state 1 to state 2. Thus the entropy S, like the internal energy U, is a thermodynamic function which is known with the exception of an additive constant. The differential

$$\mathrm{d}S = \frac{\mathrm{d}Q_{\mathrm{rev}}}{T} \tag{1.16.2}$$

is an exact differential (cf. Section 1.4).

In introducing the entropy function we have roughly followed the historical course, with which well-known names like Carnot, Kelvin and Clausius are associated. Mathematicians often prefer a more abstract mathematical method which was indicated by Carathéodory in 1909. In this method the differential equations of Pfaff play an important part. One starts with the fact that $\mathrm{d}Q$ is not an exact differential and then shows that $\mathrm{d}Q_{\mathrm{rev}}$ possesses integrating factors, i.e. factors which turn this quantity into an exact differential. It is next shown that the reciprocal of each integrating factor has the properties of a temperature and that one of them is identical with the absolute temperature T. The exact differential $\mathrm{d}Q_{\mathrm{rev}}/T$ thus obtained is indicated (as above) by the symbol $\mathrm{d}S$ and S is called the entropy. For a presentation of this method the reader is referred to the literature [1]).

For didactic reasons, we felt obliged to join the many physicists and che-

[1]) P. T. Landsberg, Revs. Mod. Physics **28**, 363 (1956). In this review many important articles are cited.

mists who prefer the historical approach, in which the Carnot cycle plays such an important part. For most readers, untrained in thermodynamics, the entropy concept remains completely incomprehensible when introduced by Carathéodory's method. For these readers, the historical approach is abstract enough. For them, the full significance of the entropy concept will only become apparent with the atomic considerations in the following chapter.

Naturally, with the help of Equation (1.16.1) or (1.16.2) we can only calculate *differences* in entropy and it must always be borne in mind that this can only be done along a reversible path. The real path, from state A to state B, is in general not reversible, so that $\mathrm{d}S = \mathrm{d}Q/T$ does not apply along such a path. In order to be able to calculate the entropy we must consider a reversible path which leads from state A to state B.

For example, if we increase the volume of a perfect gas by connecting the container in which it is enclosed to an evacuated space, this increase in volume will not cause any heat exchange with the surroundings: $\int \mathrm{d}Q/T = 0$. The change in entropy, however, is *not* zero. For, if the volume is increased reversibly, it is found that the gas certainly does take up heat, i.e. the entropy changes when the volume is increased (see the first question in Section 1.17).

Another example is the mutual diffusion of two perfect gases, which occurs when their containers are joined together. Here, too, along the whole (irreversible) path: $\mathrm{d}Q = 0$. One may not conclude from this, however, that no change in entropy occurs during the diffusion, for if the process is carried out reversibly it is found necessary to supply heat (see the second question in Section 1.17).

According to the first law, $\mathrm{d}Q$ in a reversible process is given by

$$\mathrm{d}U = \mathrm{d}Q_{\mathrm{rev}} + \mathrm{d}W_{\mathrm{rev}} \qquad (1.16.3)$$

Assuming that we are dealing only with volume-work (work corresponding to a change in volume of the system) and applying the second law, (1.16.3) becomes

$$\mathrm{d}U = T\mathrm{d}S - p\mathrm{d}V$$

or

$$T\mathrm{d}S = \mathrm{d}U + p\mathrm{d}V \qquad (1.16.4)$$

For a perfect gas we can calculate $\mathrm{d}S$ from this equation, since U and p are known as functions of V and T. U is independent of the volume, so that $\mathrm{d}U$ according to (1.8.1) and (1.8.4) is given for n gram-molecules by

$$dU = nc_v dT. \tag{1.16.5}$$

The pressure p follows from the equation of state $pV = nRT$:

$$p = \frac{nRT}{V} = \frac{nRT}{nv} = \frac{RT}{v}. \tag{1.16.6}$$

Substituting from (1.16.5) and (1.16.6) in (1.16.4) and remembering that $dV = ndv$, we obtain:

$$dS = nc_v \frac{dT}{T} + nR \frac{dv}{v}. \tag{1.16.7}$$

If we regard c_v as a constant, then by integrating we find the following expression for the entropy of n gram-molecules of a perfect gas as a function of T and v:

$$S = n(c_v \ln T + R \ln v) + \text{const.} \tag{1.16.8}$$

It can be seen from this equation that the entropy is proportional to the number of gram-molecules, i.e. proportional to the mass. The entropy per mole or the molar entropy of a perfect gas is given by

$$s = c_v \ln T + R \ln v + \text{const.} \tag{1.16.9}$$

where, as previously, c_v is the specific heat per mole or the molar specific heat and v the volume per mole or molar volume.

1.17. CALCULATION OF ENTROPY CHANGES

A few examples will be given to demonstrate how entropy changes are calculated.

Question: How great is the entropy change in the irreversible expansion of n moles of a perfect gas from volume $V_1 = nv_1$ to volume $V_2 = nv_2$, when the expansion takes place without the addition of heat and without doing work?

Answer: Since no heat or work is exchanged, the internal energy remains unchanged. This means, for a perfect gas, that the temperature remains unchanged. We must therefore find a reversible path from the state (T, V_1) to the state (T, V_2). We already know a suitable path. The gas is put in a cylinder under a movable piston and the cylinder is connected to a heat reservoir at the temperature T. During the expansion, care is taken that the pressure and counter-pressure are always equal. Since $\mathrm{d}U = 0$, from the first law the following equation is valid for every step of the process:

$$0 = \mathrm{d}Q_{\mathrm{rev}} + \mathrm{d}W_{\mathrm{rev}}$$

$$\mathrm{d}Q_{\mathrm{rev}} = -\mathrm{d}W_{\mathrm{rev}} = np\mathrm{d}v$$

$$\mathrm{d}Q_{\mathrm{rev}} = nRT\frac{\mathrm{d}v}{v}$$

The entropy change is thus given by

$$S_2 - S_1 = \int_{v_1}^{v_2} \frac{\mathrm{d}Q_{\mathrm{rev}}}{T} = nR \ln\frac{v_2}{v_1} = nR \ln\frac{V_2}{V_1} \tag{1.17.1}$$

All other reversible paths must produce the same result. To demonstrate this, we imagine that the gas first expands adiabatically-reversibly from volume V_1 to volume V_2 (while the temperature falls from T to T' and the entropy remains constant), we next introduce a quantity of heat in a reversible manner at constant volume, sufficient to raise the temperature again from T' to T. Since $\mathrm{d}V = 0$, the first law gives for every step of this latter process

$$\mathrm{d}U = \mathrm{d}Q_{\mathrm{rev}}$$

or

$$nc_v\mathrm{d}T = \mathrm{d}Q_{\mathrm{rev}}$$

The entropy change is thus given by

$$S_2 - S_1 = \int_{T'}^{T} \frac{\mathrm{d}Q_{\mathrm{rev}}}{T} = nc_v \ln\frac{T}{T'}$$

The ratio T/T' according to Equation (1.9.5) is given by

$$\frac{T}{T'} = \left(\frac{v_2}{v_1}\right)^{(c_p - c_v)/c_v}$$

Substituting this in the equation above, we obtain

$$S_2 - S_1 = n(c_p - c_v) \ln\frac{v_2}{v_1}$$

According to (1.9.2) this latter equation is the same as (1.17.1). It could also have been derived directly from Equation (1.16.8), but the two different reversible paths were chosen for didactic reasons.

Question: What entropy change occurs when a partition is removed to allow two perfect gases at the same temperature to diffuse through one another?

Answer: For this irreversible process also, a reversible path must first be found. Planck allows (in imagination, of course) the reversible mixing of two gases to take place as follows. The first gas (e.g. nitrogen) is contained in a cylinder (see Fig. 10) which is closed at the bottom by a partition, which permits the second gas (e.g. oxygen) to pass but is impermeable to the nitrogen. The oxygen is contained in a cylinder of equal volume, which is closed at the top by a partition which is permeable to nitrogen but not to oxygen.

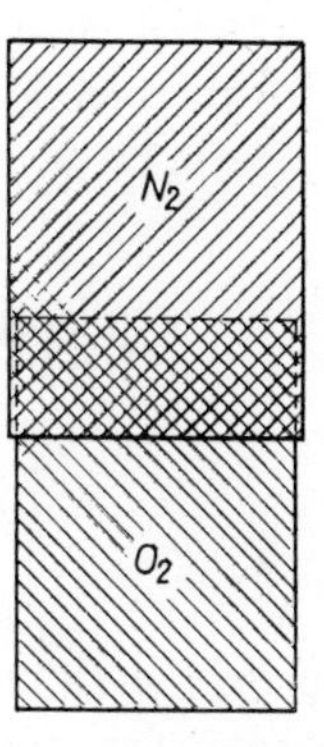

Fig. 10. Arrangement devised by Planck for mixing two gases reversibly.

The cylinders are placed in a vacuum and without friction slowly slid one into the other. In this operation, the two gases continue to fill their own space completely. In the space common to both cylinders, both gases are present, each at its own pressure. It is clear that while the cylinders slip into one another, only equilibrium states are experienced and that finally the two gases will fill one space. By sliding the cylinders out again, the equilibrium states are passed through in the reverse order and the two gases eventually completely separated again. The process is thus entirely reversible and the entropy change can be calculated from $\int dQ/T$. However, no heat is added or extracted, so that there is no change in entropy. Setting up the apparatus in vacuum ensures that no work is performed either, during the process, so that according to the first law the internal energy and with it the temperature remains constant.

What this all amounts to is that the two gases take no notice of one another. Both before and after mixing, the entropy of the system is given by $S = S_1 + S_2$. The entropy of a mixture of perfect gases, like the energy, is thus an additive property. However, this additive property is valid for the entropy only when each constituent occupies the *same volume* before and after mixing. If mixing is effected, as in the question, by removing a partition, the volume of the constituents does not remain constant. The volume of each constituent increases to the sum $V_1 + V_2$ of the original volumes. For this process, a reversible path can be indicated which consists of two steps. (1) First, each of the gases separately is expanded, reversibly and isothermally, to the volume $V_1 + V_2$. (2) Next, the gases are mixed at constant volume by the reversible method of Planck, as described above. Only in the first stage does the entropy change. From Equation (1.17.1), the entropy change for the first gas

$$\Delta S_1 = n_1 R \ln \frac{V_1 + V_2}{V_1}$$

and for the second

$$\Delta S_2 = n_2 R \ln \frac{V_1 + V_2}{V_2}$$

The entropy change which occurs during irreversible mixing, is thus given by

$$\Delta S = R\left(n_1 \ln \frac{V_1 + V_2}{V_1} + n_2 \ln \frac{V_1 + V_2}{V_2}\right). \tag{1.17.2}$$

During the diffusion of two gases, the entropy thus increases. This increase in entropy occurs only because both are permitted to occupy a larger volume and not through the actual mixing. Or in the words of Gibbs: "The entropy of a mixture of perfect gases is equal to the sum of the entropies which each gas separately would have if they each occupied the volume of the mixture."

The result expressed by Equation (1.17.2) is known as Gibbs' paradox. As we can see from the equation, it is independent of the chemical nature of the perfect gases considered. The gases may differ much or little from one another and one might therefore conclude that the same increase in the entropy must occur when dealing with two identical gases. However, it is obvious that macroscopically speaking, nothing happens when a partition is removed between two portions of the same gas, both of which have the same pressure and temperature. The paradox thus lies in the fact that the difference between the two gases may be very small, but not so small that they can no longer be distinguished macroscopically. It should be remarked, however, that the experiment with the two semi-permeable partitions is not conceivable if the two gases are identical. The paradox is resolved by modern atomic theory (see Section 5.7).

Remark: If not only the temperature but also the pressure is equal on both sides of the partition then, since $pV_1 = n_1RT$ and $pV_2 = n_2RT$, we have:

$$V_1 : V_2 = n_1 : n_2 .$$

(1.17.2) may thus be written:

$$\Delta S = R\left(n_1 \ln \frac{n_1 + n_2}{n_1} + n_2 \ln \frac{n_1 + n_2}{n_2}\right). \tag{1.17.2a}$$

If the gases together form just 1 mole, i.e. if $n_1 + n_2 = 1$, we may put $n_1 = x$ (and thus $n_2 = 1 - x$) and the equation becomes:

$$\Delta s = -R\{x \ln x + (1-x) \ln (1-x)\}. \tag{1.17.2b}$$

Question: By what amount is the entropy of 1 mole of water increased when the temperature is raised from 0 °C to 50 °C at a constant pressure of 1 atm?

Answer: The water is heated reversibly from 0 °C to 50 °C at a constant pressure of 1 atm. Hence $dQ_{rev} = c_p dT$ and thus

$$\Delta s = s_{50} - s_0 = \int_{273}^{323} \frac{c_p dT}{T} = c_p \ln \frac{323}{273}$$

where c_p (the specific heat per mole) is taken as constant to a first approximation over the temperature range considered.

Question: A thermostat at −5 °C contains 1 mole of supercooled water at this temperature at a pressure of 1 atm. By seeding with an ice crystal, the water is caused to turn into ice at −5 °C. How large is the entropy change?

Answer: The transition takes place irreversibly. To calculate the change in entropy, we must again find a reversible path. Since water and ice at a pressure of 1 atm can only be in equilibrium with one another at 0 °C, the following reversible path is obvious: The water is first warmed reversibly from −5 °C to 0 °C, the water is then allowed to turn into ice at 0 °C while the latent heat of fusion is slowly extracted and finally the ice is cooled reversibly to −5 °C:

$$\Delta s = \int_{268}^{273} \frac{c_w dT}{T} - \frac{Q}{273} - \int_{268}^{273} \frac{c_i dT}{T}.$$

where c_w is the specific heat per mole of water, c_i that of ice, and Q the latent heat of fusion.

1.18. THE PRINCIPLE OF THE INCREASE OF ENTROPY

Let us consider a cycle in which any arbitrary system is transferred along an irreversible path from a state 1 to a state 2, after which it returns to the initial state by way of a reversible path. The cycle as a whole is irreversible. From Section 1.15 it is governed by the equation

$$\oint_{\text{irr}} \frac{dQ}{T} = \int_{1\,\text{irr}}^{2} \frac{dQ}{T} + \int_{2\,\text{rev}}^{1} \frac{dQ}{T} < 0$$

where, according to Section 1.16, the last integral is given by $S_1 - S_2$. Thus:

$$\int_{1\,\text{irr}}^{2} \frac{dQ}{T} < S_2 - S_1 \qquad (1.18.1)$$

Thus, while $dS = dQ/T$ applies to a reversible change of state, for an irreversible change of state $dS > dQ/T$. Equation (1.18.1) is the mathematical formulation of the warning already stressed in Section 1.16, that the entropy difference between two states can be calculated from the thermal effects only for a *reversible* path.

A very simple example will serve to illustrate the fact that $\int dQ/T$ is smaller than the entropy change in an irreversible process: If ice is melted reversibly, heat must be drawn from a reservoir of which the temperature is only very little higher than 0 °C. If, on the other hand, it is melted irreversibly by heat from a reservoir at 10 °C, for example, the same number of calories will be needed but Q/T will then be smaller than the entropy change.

Equation (1.18.1) becomes particularly important when the irreversible change of state considered is adiabatic. The equation then reads:

$$S_2 - S_1 > 0 \quad \text{or} \quad S_2 > S_1 \qquad (1.18.2)$$

Every change of state can be considered adiabatic by making the system which undergoes the change into an isolated one, which simply means that all bodies concerned in the exchange of energy are considered as part of the system. In this way we come to a conclusion which is so extremely important that we shall express it in three different ways:

a. Processes which occur spontaneously in an isolated system, are always accompanied by an increase in entropy.

b. Changes can take place in an isolated system only so long as it is possible for the entropy to increase.

c. In an isolated system, the entropy tends to a maximum. When this maximum is reached the system is in equilibrium.

Thus, in the magnitude of the entropy we find the criterion needed (cf. Section 1.10) to determine the direction in which natural processes will take place. It must again be stressed, however, that the principle of the increase of entropy only applies to an isolated system. If we place a glass of water at 100 °C on the table, it will cool down, i.e. the entropy of the system considered will decrease. Only when the whole surroundings are taken into consideration, i.e. are included in the system, does it appear that the entropy increases.

We shall demonstrate further the principle of the increase of entropy with a specific example, viz. the chemical reaction

$$CuSO_4aq + Zn \rightarrow ZnSO_4aq + Cu$$

As mentioned in Section 1.5, this reaction may be allowed to proceed either reversibly or irreversibly. It can proceed *reversibly* in a galvanic cell if an opposing voltage $E - dE$ is applied to the poles which is only slightly less than the electromotive force E of the cell. For the reversible process the first law applies in the following form

$$dU = dQ_{rev} + dW_{rev} \tag{1.18.3}$$

According to the second law, dQ_{rev} is given by TdS, while dW_{rev} in this case consists not only of the volume-work $-pdV$, but also of the electrical work $-zFEdn$ (see Section 1.6). For (1.18.3) we may thus write

$$dU = TdS - pdV - zFEdn.$$

The two negative signs are a result of the convention of Section 1.6 that dW shall always represent the work performed *on* the system. The above equation can also be written in the form

$$TdS = dU + pdV + zFEdn \tag{1.18.4}$$

If now the chemical reaction considered is allowed to proceed *irreversibly*, at the same external pressure p, in such a way that no electrical work at all is done on the surroundings (cf. Section 1.5), then (1.18.4) becomes

$$dQ_{irr} = dU' + pdV' \tag{1.18.5}$$

If the process is so performed that for both the reversible and the irreversible cases the initial and final states are identical, then we must have

$$dU = dU', \quad dV = dV' \quad \text{and} \quad dS = dS' \tag{1.18.6}$$

since U, V and S are thermodynamic functions, that is, they depend upon the state of the system. From (1.18.4) it follows that

$$TdS > dU + pdV \tag{1.18.7}$$

or, combining (1.18.5), (1.18.6) and (1.18.7):

$$TdS > dQ_{irr} \tag{1.18.8}$$

If the process is allowed to take place in a closed container of constant volume with walls impermeable to heat, i.e. at constant values of U and V, then (1.18.7) becomes

$$(dS)_{U,V} > 0. \tag{1.18.9}$$

For the special case just considered we thus find once more the same result as found earlier in this section for arbitrary processes, viz. that the entropy increases when a spontaneous change of state takes place in an isolated system.

The principle of the increase of entropy can be regarded as the most general formulation of the second law.

CHAPTER 2

THE STATISTICAL SIGNIFICANCE OF THE ENTROPY CONCEPT

2.1. INTRODUCTION

It has been mentioned in Chapter 1 that the second law of thermodynamics (the principle of the increase of entropy) is based on the experience that all spontaneously-occurring processes have a particular direction, i.e. are irreversible. The second law appeared to be so generally applicable that eventually it (like the first law) was raised to the status of a postulate. The enormous success achieved in the second half of the nineteenth century by the atomic picture of the universe, gave rise to the need to base the second law, too, on this picture.

Boltzmann succeeded in this in 1872 by showing the relation between the concepts of entropy and probability. Before considering this relation in more detail, we shall first discuss the concept of probability.

2.2. THE CONCEPT OF PROBABILITY

If it is certain that a particular event will produce the result X, then by definition the probability $p(X)$ of the occurrence of that result is given the value $p(X) = 1$. For example, if the event is a throw with an abnormal die which bears the number 5 on all its sides, then it is certain that each throw will produce the result 5. In this case thus, $p(5) = 1$. If it is known for certain that a particular result X will *not* occur, then $p(X) = 0$. The probability $p(3)$ of throwing a three with the above-mentioned abnormal die is zero. If there is no certainty about the possible occurrence of the result X, then $p(X)$ will lie somewhere between zero and one.

In many cases the probabilities of the occurrence of certain results are known *a priori*. For example, if a normal die is properly constructed, each of the six symbols will have the same probability of occurrence. One then obtains

$$p(1) = p(2) = \dots = p(6) = \frac{1}{6}$$

The more general formulation of the above reads: If one is dealing with a series of events, each of which produces one of m possible and equally probable results, the probability of occurrence of each individual result X_i is given by:

$$p(X_i) = \frac{1}{m} \tag{2.2.1}$$

If the probability can not be stated *a priori*, the following experimental definition can be employed: The probability $p(X_i)$ that a particular result X_i will occur in a series of events is defined as the ratio between the number of events in which this result is obtained, $\alpha(X_i)$, and the total number of events n or, more exactly, the limit to which this ratio tends as the total number of events increases beyond all bounds:

$$p(X_i) = \lim_{n \to \infty} \frac{\alpha(X_i)}{n} \tag{2.2.2}$$

Thus, if a particular result is obtained an average of 1000 times per 6000 events over a large number of experiments, the probability of that result may be said to be 1/6. The probability defined by (2.2.2) can be determined to any desired degree of accuracy by repeating the experiment often enough under the same conditions.

How great is the probability, when throwing dice, that an even number will occur? This result can be obtained in three different ways, viz. when 2, 4 or 6 comes up. For each of these the probability is 1/6. The required probability is thus 3/6 = 1/2. The more general formulation of this reads: When dealing with a series of events such that each produces one of m possible, equally probable results, the probability $p(X_{\alpha_1}, \ldots, X_{\alpha_g})$, that one of a number g of these results will occur, i.e. that either X_{α_1}, or $X_{\alpha_2}, \ldots$, or X_{α_g} will occur, is given by:

$$p(X_{\alpha_1}, \ldots, X_{\alpha_g}) = \frac{g}{m} \tag{2.2.3}$$

Next we shall consider throwing two dice and discuss the probability that a certain combination of results will occur, e.g. that the first of the dice will show a 2 and the second a 5. Since each of the six symbols on the first die may occur in combination with each of the six symbols on the second, there are, in total, $6^2 = 36$ different possibilities. If the dice have been weighed and measured and found good, it can be assumed *a priori* that the 36 possibilities all have the same probability of occurrence. From Equation (2.2.1) the probability of occurrence of each combination is thus 1/36, i.e.

the product of the probabilities of the two symbols independently. The general formulation thus reads: If a series of events takes place such that in each case two mutually independent results are obtained, the probability $p(X_i \& Y_j)$ that two particular results X_i and Y_j will occur simultaneously is:

$$p(X_i \& Y_j) = p(X_i) \cdot p(Y_j) \tag{2.2.4}$$

How great is the probability of throwing a total of ten with two dice? This result can be obtained in three different ways, viz. with the combinations (4 and 6), (5 and 5), (6 and 4). From the last equation, the probability of each of these combinations is 1/36. According to Equation (2.2.3) the answer to the question is thus 3/36 = 1/12.

2.3. MIXTURE OF PARTICLES OF DIFFERENT KINDS

The relation between entropy and probability appears clearly in the spontaneous mixing of two gases. We shall consider a container divided into two equal compartments by a partition. One contains 0.5 gram-atom of neon, the other 0.5 gram-atom of helium, both under such conditions of pressure and temperature that they can be considered as perfect gases. As described in the previous chapter, rapid mixing of the gases, accompanied by an increase in the entropy, occurs when the partition is removed. The reverse process, the spontaneous separation of a mixture of gases, is never observed. What is the reason for this tendency to form a homogeneous mixture? Can we say that the helium and neon atoms have a "preference" for the homogeneous (the more chaotic) distribution and, if so, what is the reason for this preference?

To answer these questions we mentally replace the helium and neon atoms by a very much smaller number of red and white billiard balls (e.g. 50 + 50). For the initial condition we choose a certain regular arrangement of the balls in the container (Fig. 11) and we imitate the thermal agitation of the atoms by thoroughly shaking the container for a certain length of time. Insofar as there are no specific forces between the balls, this model gives a satisfactory picture of the mixture of helium and neon. Where the filling of the space is concerned, on the other hand, the model shows more resemblance to a crystal.

Experience shows that after the container with the billiard balls has been shaken, the regular initial arrangement is never encountered again, but that the red and white balls always form a disordered distribution. Even so, we must assume that all configurations which are possible must have the same probability, in other words, the initial distribution has the same chance of

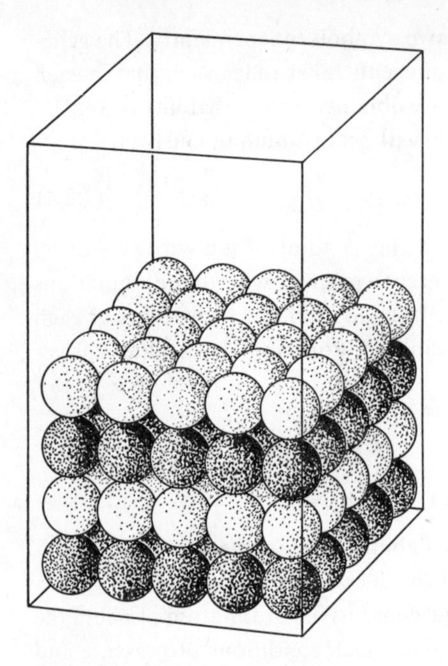

Fig. 11. A regular arrangement of 50 white and 50 red billiard balls.

occurring after shaking as each separate disorderly arrangement. The point is, however — and here we touch the very core of the problem — that there are so many more irregular than regular configurations possible that practically only irregular ones will occur in the shaking experiments. This will be illustrated with the help of a simple calculation.

Suppose that it is possible to distinguish between all of the 100 balls (e.g. by numbering them) and that they are placed one by one in the available places. In placing the first ball we have the choice of 100 possibilities. For *every* position of ball No. 1, ball No. 2 can be placed in *any* of the remaining 99 positions. The first two balls alone can thus be placed in 100×99 different ways. For each of these 9900 possibilities, ball No. 3 can be placed in 98 different positions. There are thus $100 \times 99 \times 98$ different arrangements for three balls. For 100 marked balls and 100 available positions we thus have $100 \times 99 \times 98 \times \ldots \times 2 \times 1$ possibilities for arranging them. This product is indicated by the symbol $100!$, which is called factorial 100.

If the balls are not numbered the number of arrangements which are distinguishable to the eye is much smaller. In fact, both the reversal of any pair of white balls and that of any pair of red balls leave the distribution unchanged. The number of different arrangements m which are now possible is calculated as follows. If only the white balls were numbered, then *each* of the m arrangements would be realisable in 50! different ways because 50 unnumbered white balls can be distributed over 50 positions in only one way, while 50 numbered balls can be distributed in 50! distinguishable ways. If, furthermore, the red balls were now also numbered, the number of possible arrangements would once more be multiplied by 50!. Thus we have

$$m \times 50! \times 50! = 100!$$

or

$$m = \frac{100!}{50!\,50!}$$

Accurate values of the factorials of the numbers from 1 to 100 are to be found in Barlow's Tables [1]). We find that $100! = 9.333 \times 10^{157}$ and $50! = 3.041 \times 10^{64}$. From this follows the value of m

$$m = 1.01 \times 10^{29}.$$

The probability of occurrence of each configuration, according to Equation (2.2.1), is thus only about 10^{-29}.

When the numbers involved are larger than 10, quite an accurate approximation formula for calculating factorials is Stirling's formula:

$$N! = \frac{N^N}{e^N} \sqrt{2\pi N}. \tag{2.3.1}$$

in which e is the base of natural logarithms. It can easily be verified that this equation gives the value of m quoted above.

From the result $m = 10^{29}$ it follows that one must shake an average of 10^{29} times in order to obtain a particular arrangement of the balls. If the balls are shaken each time for one minute, this means that an average of more than 10^{23} years shaking will be needed before a particular distribution occurs. In this context it should be remembered that the age of the universe is estimated as a "mere" 10^{10} years. The above-mentioned experience is thus comprehensible: the chance that after shaking a completely regular arrangement will be encountered, e.g. one in which the white and red balls are found in separate layers, or one in which the white balls have only red and the red balls have only white ones as neighbours (checkboard configuration), is practically nil, since the number of regular arrangements is small compared with the enormous number of irregular arrangements.

The above considerations were concerned with the small number of 100

---

[1]) Barlow's Tables, E. and F.N. Spon Ltd., London, 1947. For the factorials of larger numbers see D. B. Owen and C. M. Williams, Logarithms of factorials from 1 to 2000, Sandia Corporation Monograph, Albuquerque, U.S.A., 1959.

balls. Suppose, to get a better comparison with the mixture of helium and neon, that we are dealing with 0.5 "gram-atom" of white and 0.5 "gram-atom" of red balls, i.e. with a total of $N_0 = 0.6 \times 10^{24}$ balls, then the probability of an orderly arrangement is incomprehensibly smaller. A similar calculation to that carried out above for the 100 "atoms" now gives, using Equation (2.3.1):

$$m = \frac{N_0!}{\{(\frac{1}{2}N_0)!\}^2} = 2^{N_0-40} \tag{2.3.2}$$

This number is so astronomically large that it is quite beyond the grasp of the human imagination. The number of orderly arrangements of the balls is utterly negligible with respect to this total number m.

In section 2.5 we shall discuss more fully the configurations in gases, but we can already consider the questions posed at the beginning of this section as answered: the helium and neon atoms have no preference for particular configurations, but the macroscopically homogeneous distribution corresponds to so many more configurations than do all the more orderly distributions together that the latter never come into consideration. When two gases diffuse through one another the entropy increases (see Equation (1.17.2)) and also the probability of the distribution increases (see Equation (2.2.3)).

2.4. LARGE NUMBERS AND THE MOST PROBABLE STATE

In the 100 positions considered above (Fig. 11) we can place white and red balls in such a way that in each operation white and red each have the same chance. To do this we dip blindfold 100 times into a reservoir containing white and red balls in equal numbers, but otherwise in unlimited quantities. The first ball is put in the first position, the second ball in the second, and so on. The number of different configurations which may be obtained by proceeding in this manner is much greater than was the case in the previous section. Not only is there the possibility of drawing 50 white and 50 red balls, but there is also the chance of drawing 49 white and 51 red or any other colour ratio. Besides the number of configurations in the previous section, given by

$$\frac{100!}{50!\,50!}, \quad \text{there are also} \quad \frac{100!}{49!\,51!}$$

configurations for 49 white and 51 red balls, and so on. The total number of configurations in this case is given by

$$m = \sum_{N_w = 0}^{N_w = 100} \frac{100!}{N_w!\,(100 - N_w)!} \tag{2.4.1}$$

where N_w is the number of white balls. The probability of drawing two white balls in succession, according to Equation (2.2.4) is given by $p = (^1/_2)^2$, the probability of drawing white 100 times in succession by $(^1/_2)^{100}$. Each other particular configuration of 100 balls, e.g. the first white, the second, third and fourth red, the fifth white, etc, naturally has the same probability of occurring. From Equation (2.2.1) the total number of configurations is thus given by

$$m = 2^{100} = 1.26 \times 10^{30} \tag{2.4.2}$$

that is to say, by a number which is more than twelve times as large as that calculated in the previous section for the fixed ratio 50 : 50. That (2.4.1) represents the same number as (2.4.2) is immediately obvious by applying Newton's binomial theorem to the expression $(1 + 1)^{100}$. The terms of the sum given by (2.4.1) are the binomial coefficients for $N = N_w + N_r = 100$, which coefficients can also be read off from the horizontal row for $N = 100$ in Pascal's triangle (Fig. 12). It is easily seen how each row in the Pascal triangle is formed by summation from the preceding row.

For a chosen value of N, the corresponding horizontal row in the Pascal

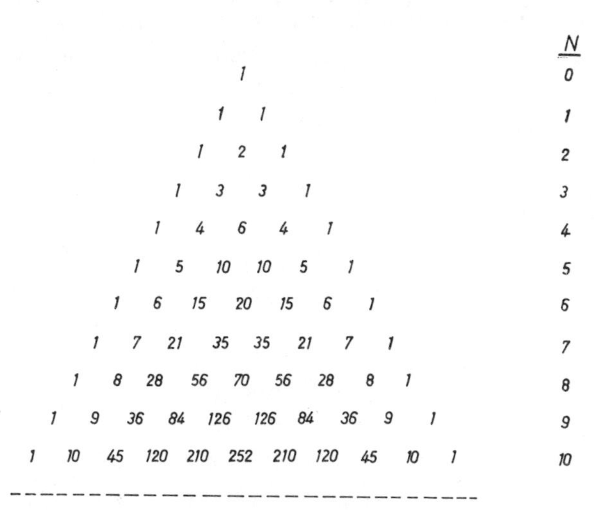

Fig. 12. Pascal's triangle. The numbers in this "triangle" give the coefficients for the expression $(a+b)^N$ for various values of N (at right in the figure). Example: $(a+b)^3 = 1a^3 + 3a^2b + 3ab^2 + 1b^3$.

TABLE 1

N	g_{max}	m	g_{max}/m	$(\ln g_{max})/(\ln m)$
2	2.000	4.000	5.00×10^{-1}	0.500
10	2.520×10^{2}	1.024×10^{3}	2.46×10^{-1}	0.798
100	1.012×10^{29}	1.268×10^{30}	7.98×10^{-2}	0.964
1000	2.704×10^{299}	1.072×10^{301}	2.52×10^{-2}	0.995
10000	1.592×10^{3008}	1.995×10^{3010}	7.98×10^{-3}	0.999

triangle gives the numbers of configurations of the balls for $N_w = 0$, $N_w = 1$, etc. The sum of the numbers in the row gives the total number of configurations $m = 2^N$. The probability of a particular ratio $w : r$ – that is to say, a particular value of N_w – when drawing blindfold (see the beginning of this section) is given according to Equation (2.2.3) by

$$p = \frac{g}{m} = \frac{1}{2^N} \frac{N!}{N_w!(N - N_w)!} \tag{2.4.3}$$

where g is the number of configurations which corresponds to a particular colour ratio. The most probable colour ratio is that which comprises the largest number of configurations, i.e. that for which g is a maximum. As can be seen, this is the colour ratio 1 : 1 ($N_w = N/2$). Table 1 gives for a few values of N the corresponding values of g_{max}, m, g_{max}/m and $(\ln g_{max})/(\ln m)$. It will be noticed that g_{max}/m continues to decrease as the number of balls becomes greater. The chance of drawing exactly equal numbers of white and red balls thus becomes smaller as N increases. On the other hand, the fraction of configurations in which N_w is *almost* equal to $N/2$ becomes larger and larger for increasing values of N. If the number of balls is chosen large enough, a point will be reached where, for example, 99.99% of the configurations deviate by less than 0.001% from $N_w = N/2$. By making N larger and larger these percentages can finally be made to approach 100% and 0% as closely as desired. This tendency for the groups of configurations to gather increasingly closely around $N_w/N = 1/2$ for increasing values of N is demonstrated by Fig. 13, in which g/g_{max} is plotted against N_w/N for various values of N. Strictly speaking, for small values of N, not the curves but only the calculated points have any meaning. These points are given in the figure for $N = 6$, 10 and 20. As N continues to increase, the representative curves approach ever closer to the vertical axis $N_w/N = 1/2$.

In order to compare this with what was dealt with in the previous section, let us consider the total number of configurations

$$m = 2^{N_0}$$

for $N = N_0$ = Avogadro's number (0.6×10^{24}). This number of configurations is 2^{40} times as large as the number calculated in the previous section (Equation (2.3.2)) for the fixed ratio $N_w/N = {}^1/_2$, which was then indicated by m but appears in this section as $g_{\max}$, the number of configurations in the most probable group. The fascinating thing about these extremely large numbers is illustrated in this case by the fact that we can neglect the enormous factor 2^{40} without any objection, whereas this was certainly not justified for the analogous factor 12 in the case of 100 balls (see above). The number 40 in the exponent of 2 (Equation (2.3.2)) is, in fact, completely negligible with respect to $N_0 = 0.6 \times 10^{24}$; Avogadro's number is known to so few decimals that it would actually be physically meaningless to try and make

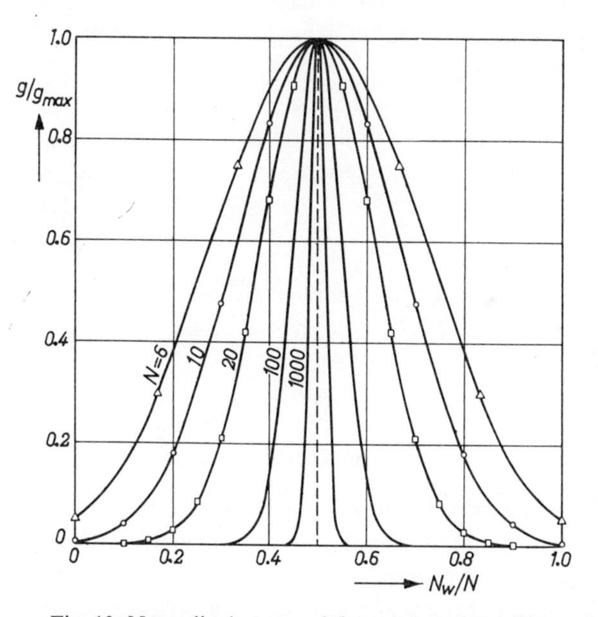

Fig. 13. Normalised curves of the number of possible configurations of N white and red balls as a function of the fraction of white balls, N_w/N. The number of configurations in which N_w/N deviates considerably from 0.5, becomes relatively smaller as N increases.

a distinction between N_0 and $N_0 - 40$. For a sufficiently large value of N there is thus no objection to interchanging $g_{\max}$ and m, although $g_{\max}/m$ is extremely small. The argument is considerably strengthened, as will be seen from the following, by the fact that we are not so much interested in g and m themselves, as in the natural logarithms of these quantities. Although

g_{max}/m becomes smaller and smaller for an increasing number of particles, $(\ln g_{max})/(\ln m)$ approaches closer and closer to 1 (see Table 1).

A physical consequence of the above is that no distinction need be made between the number of configurations which is possible in a closed container holding 0.5 gram-atoms of neon and 0.5 gram-atoms of helium and the number which would be encountered after connecting the container to an infinitely large reservoir holding a gas mixture of the same composition, pressure and temperature. Admittedly after this connection has been made there will be fluctuations around the ratio 1 : 1 in the original container, so that the total number of configurations possible increases, but at the ratio 1 : 1 there is such a sharp maximum in the number of arrangements as a function of the mixing ratio (cf. Fig. 13), that it makes virtually no difference whether one works with the total number of arrangements or with the number corresponding to this maximum.

A mathematical consequence of the above is that for the numbers of atoms with which one normally has to deal in practice, Stirling's Equation (2.3.1) may be used in a rougher approximation, viz. in the approximation

$$\ln N! = N \ln N - N. \tag{2.4.4}$$

It can be easily checked that this equation gives directly the value $m = 2^{N_0}$ for the ratio appearing in (2.3.2).

2.5. CONFIGURATIONS OF THE MOLECULES IN A GAS

The manipulations described with billiard balls were used to illustrate the diffusion of two gases into one another. In fact, a collection of billiard balls is not an adequate model of a gas, since the molecules in a gas only fill a relatively small portion of the available space; in a perfect gas, the molecules occupy no space at all. In a gas there are thus very many more places available than there are molecules, in a perfect gas the number of places would be infinite. If one is dealing with N similar atoms, then in the billiard ball model there is only one possible arrangement; in the gas model, on the other hand, there are many possibilities.

We shall first investigate how the number of possible arrangements or configurations in a pure gas depends on the volume. After that we shall consider the mixing of two gases in more detail.

We imagine that the volume containing a pure gas is divided into very many equal cells which are so small that by far the greater part are empty, while a minute fraction of the cells contain one gas molecule. As a result of the thermal agitation the occupied cells are continually changing. If there are z cells and N molecules, there will be N occupied and $(z - N)$ unoccupied cells. Since the interchange of either two empty or two occupied cells does not alter an arrangement, the total number of configurations is given by

$$m = \frac{z!}{N!(z-N)!} \tag{2.5.1}$$

or, employing Equation (2.4.4):

$$\ln m = z \ln z - N \ln N - (z - N) \ln (z - N)$$

Calculation of this equation by means of the approximation $\ln (1 - N/z) = - N/z$ for $N/z \ll 1$ gives:

$$\ln m = N \ln \frac{z}{N} + N \tag{2.5.2}$$

If we increase the volume by a factor $r = V_2/V_1$, then z must be replaced in the equation by zr. The ratio m_2/m_1 of the number of configurations in the final volume V_2 to that in the original volume V_1, according to (2.5.2) is given by

$$\ln \frac{m_2}{m_1} = N \ln \frac{rz}{N} - N \ln \frac{z}{N} = N \ln r.$$

Since $r = V_2/V_1$:

$$\frac{m_2}{m_1} = \left(\frac{V_2}{V_1}\right)^N \tag{2.5.3}$$

Next we shall consider a container divided by a partition into two compartments of volume V_1 and V_2. The first contains N_1 molecules of a gas 1, the second N_2 molecules of a gas 2. The first volume is imagined as being divided into z cells, the second into $zr = zV_2/V_1$ cells of the same size. Before the partition is removed we have

$$\ln m_1 = N_1 \ln \frac{z}{N_1} + N_1. \tag{2.5.4}$$

$$\ln m_2 = N_2 \ln \frac{rz}{N_2} + N_2. \tag{2.5.5}$$

Since every configuration in V_1 can be combined with every configuration in V_2, the total number of configurations m is given by

$$m = m_1 m_2$$

The natural logarithm of m is thus given by the sum of (2.5.4) and (2.5.5).

After the partition has been removed $N = (N_1 + N_2)$ molecules are distributed over $(r + 1)z$ cells. The number of different configurations m' is now given by

$$m' = \frac{\{(r+1)z\}!}{N_1!N_2!\{(r+1)z - N\}!} \tag{2.5.6}$$

This equation can be worked out in the same way as Equation (2.5.1). Once more, what we are interested in is the ratio m'/m of the number of configurations in the final state to that in the original state. It can be calculated simply that

$$\ln \frac{m'}{m} = N_1 \ln (r+1) + N_2 \ln \left(1 + \frac{1}{r}\right) =$$

$$= N_1 \ln \left(\frac{V_1 + V_2}{V_1}\right) + N_2 \ln \left(\frac{V_1 + V_2}{V_2}\right)$$

If gas 1 consists of n_1 and gas 2 of n_2 gram-molecules we can also write the latter equation as follows

$$\ln \frac{m'}{m} = N_0 \left\{ n_1 \ln \frac{V_1 + V_2}{V_1} + n_2 \ln \frac{V_1 + V_2}{V_2} \right\} \tag{2.5.7}$$

where N_0 is Avogadro's number. In a similar way, Equation (2.5.3) can be written

$$\frac{m_2}{m_1} = \left(\frac{V_2}{V_1}\right)^{nN_0} \tag{2.5.8}$$

2.6. MICRO-STATES AND MACRO-STATES

In the experiments with the billiard balls, it had to be assumed that all possible arrangements of the balls (all "micro-states") had the same probability.

In the case of the fixed colour ratio 1 : 1 (see Section 2.3) the various micro-states can be brought together into groups, each characterized by a particular degree of disorder. A regular distribution, like that illustrated in Fig. 11, can only be achieved in one way or, if one wishes, in two ways (horizontal layers *w-r-w-r* or *r-w-r-w*). The same applies to a chequered distribution in which each red ball has only white and each white ball only red as nearest neighbours. In an orderly distribution a "defect" can be introduced by interchanging a white and a red ball. Since each of the 50 white balls can be interchanged with each of the 50 red balls, the group characterized by one defect contains as many as 2500 micro-states. If one increases the number of defects to two or more, one obtains groups which contain considerably more micro-states. The probability of each group is determined according to Equation (2.2.3) by the number of micro-states or possibilities of realization g which correspond to this state.

In the case where the colour ratio was not constant (see Section 2.4) the division into groups of micro-states was obvious, each being characterized by a particular colour ratio. The probability of each group is once more determined by its magnitude.

Atomic configurations in a gas (see Section 2.5) can also be collected into groups. As in the example of the balls, it is assumed *a priori* that all configurations (micro-states) which are possible in the available space, have the same probability. The probability of a group of micro-states is again given by Equation (2.2.3). Since no single micro-state is favoured above another, one can imagine that the gas, as a result of the movements of the molecules, in the course of time will pass through every imaginable micro-state, and so, for example, also that in which all the molecules are concentrated in only 10% of the available space. And yet one always encounters the state in which the gases fill the available space evenly, as far as the limits of our sensory observations reach. This is explained by the fact that the uniform distribution comprises so very many more micro-states than all the other distributions together that it is always observed to the exclusion of all the others.

A group of micro-states, corresponding to a thermodynamic state (see Section 1.2), will be referred to as a macro-state. Macro-states are characterized by a small number of macroscopic quantities, such as temperature, pressure, volume, etc. They can be physically distinguished from one another, whereas arbitrary collections of micro-states in general can not.

To show just how improbable a non-uniform distribution of a gas is, we shall consider the imaginary case where 10^{24} gas molecules fill 99% of the available volume V, while a certain portion amounting to 1% of V is empty. The probability of this state is given according to Equation (2.2.3) by the number of micro-states g corresponding to the volume 0.99 V, divided by the number of micro-states m belonging to the total volume V. The required probability is given directly by Equation (2.5.3):

$$p = \frac{g}{m} = (0.99)^{10^{24}} = 10^{-44.10^{20}}$$

From this result it follows that the condition where 1% of the volume is empty will be found, on an average, only once in $10^{44.10^{20}}$ observations, i.e. if we make one observation per year, an average of once in $10^{44.10^{20}}$ years. If we make extremely rapid observations and make ten thousand observations per second, i.e. $10^4 \times 3600 \times 24 \times 365 = 10^{11.5}$ observations per year, then the state considered will be found, on an average, once in $10^{44.10^{20}-11.5}$ years. More rapid observation, thus, is not the slightest help, for the figure 11.5 in the exponent can be neglected with respect to 44.10^{20}.

2.7. THE STATISTICAL DEFINITION OF ENTROPY

It is particularly interesting to compare Equations (2.5.7) and (2.5.8) for the ratio of the numbers of micro-states before and after mixing of two gases and before and after occupation of a larger volume respectively, with Equations (1.17.2) and (1.17.1), which give the increase in entropy for these processes:

$$\Delta S = R\left\{n_1 \ln \frac{V_1 + V_2}{V_1} + n_2 \ln \frac{V_1 + V_2}{V_2}\right\} \qquad (1.17.2)$$

$$\Delta S = nR \ln \frac{V_2}{V_1} \qquad (1.17.1)$$

From (2.5.7) and (1.17.2) the increase in entropy due to the mixing of two gases is given by

$$\Delta S = \frac{R}{N_0} \ln \frac{m'}{m} \qquad (2.7.1)$$

and for an increase in volume of a gas, from (2.5.8) and (1.17.1) by

$$\Delta S = \frac{R}{N_0} \ln \frac{m_2}{m_1} \qquad (2.7.2)$$

in both cases, therefore, by the value of $k \ln m$ for the final state, less the value of $k \ln m$ for the initial state:

$$S_f - S_i = k \ln m_f - k \ln m_i \qquad (2.7.3)$$

k being Boltzmann's constant (the gas constant per molecule):

$$k = \frac{R}{N_0} \tag{2.7.4}$$

If we choose some standard reference state as zero point of the entropy then, on the basis of Equation (2.7.3), we may simply write for the entropy

$$S = k \ln m \tag{2.7.5}$$

just as we normally speak of the height of a mountain, choosing the average level of the sea as reference state. Nernst's heat theorem (the "third law"), to be discussed in Section 2.17, will show that the most natural reference state is that of substances in internal equilibrium at 0 °K. These substances at 0 °K are in a state of complete order, so that for them we find $m = 1$ and consequently $S = 0$.

The statistical entropy Equation (2.7.3) or (2.7.5) was derived above only for the number of possible arrangements in a gas or mixture of gases. However, its validity is so general that in all cases as yet investigated it leads to the same results as the thermodynamic definition (1.16.1). The tendency of entropy to a maximum value, according to (2.7.5), means nothing else than a tendency to a state with as great as possible a number of micro-states m, i.e. nothing else than a tendency to a more probable state.

From Section 2.4, m in Equation (2.7.5) may be replaced by $g_{\max}$, the number of micro-states in the most probable distribution. One can therefore write

$$S = k \ln g_{\max} \tag{2.7.6}$$

In applying this equation to a gas one only counts those micro-states in which the gas, macroscopically considered, is distributed *uniformly* over the available space. When using Equation (2.7.5), on the other hand, one also counts all the micro-states which correspond to non-uniform distributions. Although the non-uniform distributions comprise very many micro-states, these can be neglected in the logarithm with respect to the enormous number represented by $g_{\max}$. Equations (2.7.5) and (2.7.6) thus lead to practically the same result which, however, is often easier to arrive at via (2.7.5) than via (2.7.6).

The statistical definition of entropy is often introduced in a rather different manner than that given above. Starting from the connection noted in Section 2.3 between entropy and probability, one postulates that entropy is a function of the number of different ways m in which a thermodynamic state can be realized:

$$S = \mathrm{f}(m) \tag{2.7.7}$$

If one now considers two separated systems A and B as one system AB, then we have for the entropy

$$S_{AB} = S_A + S_B \tag{2.7.8}$$

since the quantity of heat needed to heat AB in a reversible manner from T to $T + \mathrm{d}T$ is the sum of the quantities of heat needed to produce this rise in temperature in A and B separately (cf. Chapter 1). Further, since every micro-state of A can be combined with every micro-state of B, we have:

$$m_{AB} = m_A m_B. \tag{2.7.9}$$

From (2.7.8) and (2.7.9) it follows that the relation between S and m postulated in (2.7.7) must be given by an equation of the form of (2.7.5), in which k is still undetermined. The magnitude of k is then determined by applying the equation to a perfect gas.

Boltzmann's equation $S = k \ln m$ or $S = k \ln g_{\max}$ forms the bridge between the second law of classical thermodynamics and the atomic picture. According to classical thermodynamics an unchangeable state is reached in an isolated system as soon as the entropy is at a maximum. According to the atomic conception, on the other hand, even in an equilibrium state a system still passes continually through one micro-state after another, but in general these are not macroscopically distinguishable from each other. Only under very special conditions can the variations about the state of maximum entropy be observed. The blue colour of the sky, for example, reveals the existence of local fluctuations in the density of the air, while also the well-known Brownian motion in colloidal suspensions is based on the irregular thermal agitation of the molecules. An analogous fluctuation phenomenon occurs through the thermal agitation of electrons in a conductor. This causes extremely small, alternating voltages to appear between the extremities of a resistor. These voltages can, with sufficient amplification, be made audible as noise in a loudspeaker ("thermal noise").

The above atomic considerations and the experimentally observed fluctuation phenomena show us that the second law of thermodynamics does not apply rigorously, but only by approximation. It is not absolutely certain, but only extremely probable that the entropy increases in every spontaneous process. The entropy may sometimes decrease and, if one waits long enough, even the most improbable states will occur. From the calculations in the previous section however, it can be seen that it is almost certain that a local vacuum will not occur suddenly in a room, for however short a space of time. There is therefore no objection to the continued use of the thermodynamic laws.

2.8. QUANTUM STATES

In the above, we have only considered the micro-states corresponding to the spatial distribution of one or two kinds of particle. To describe the micro-state of a gas completely in the classical manner, however, we must specify not only the positions of the molecules, but also their velocities. To do this, we can proceed in the same way as we did in specifying the spatial distribution. Just as the space with co-ordinates x, y and z was divided into a large number of cells $\Delta x \Delta y \Delta z$, after which the distribution of the molecules in the cells was specified, so we can construct a "velocity space" with velocity co-ordinates u, v and w, divide this into a large number of cells $\Delta u \Delta v \Delta w$ and indicate how the velocities of the molecules are distributed in these cells. It is now an obvious step in some special cases not to consider the position and velocity separately, but to deal at once with a six-dimensional position-velocity space or, even better, a six-dimensional position-momentum space (momentum = mass × velocity) and to divide this into very many, equal cells.

Although the division of the position space and the momentum space into tiny cells $\Delta x \Delta y \Delta z$ and $\Delta p_x \Delta p_y \Delta p_z$ respectively (p = momentum) is a mathematical artifice which leaves the size of the cells undetermined, there is nothing arbitrary in the division of the position-momentum space or "phase space" into equal cells $\Delta x \Delta p_x \Delta y \Delta p_y \Delta z \Delta p_z$. Modern atomic theory (see Chapter 4) not only shows us that this division is a physical necessity, it gives, at the same time, a decision on the size of the cells. This is given by

$$\Delta x \Delta p_x \Delta y \Delta p_y \Delta z \Delta p_z = h^3.$$

where h is Planck's constant. A particle enclosed in a limited space can, in fact, only be in definite quantum states and each h^3 cell in the phase space corresponds with one of these quantum states (see Chapter 4).

The number of cells in the phase space is much larger than the number of particles. The molecules can be distributed over the cells in very many ways. Each separate distribution represents a micro-state of the whole system. If the system considered is isolated, not only the number of particles N and the volume V have a constant value, but also the internal energy U is constant. One starts from the hypothesis that all micro-states which are compatible with these conditions have the same probability. Considered over a very long period of time, the system will therefore pass equal lengths of time in each permissible micro-state. The condition of constant energy makes many other micro-states inaccessible for the system.

Apart from fluctuation phenomena the great majority of micro-states are macroscopically indistinguishable from one another. Together they form the most probable macro-state, the state with the largest number of possibilities of realization. One of the central problems in statistical thermodynamics is the determination of this most probable distribution for a system of N identical particles and a given value of the total energy U. For a gas, this problem will not be dealt with until Chapter 4. In this chapter we shall discuss the micro-states and the grouping of these states for a greatly simplified model of a solid, a so-called Einstein solid.

2.9. STATES IN A SCHEMATIC SOLID

In the model of a solid according to Einstein the atoms perform their oscillations virtually independently of one another. Each atom, as in an actual crystal, is only alotted a volume of the order of magnitude of 10^{-23} cm^3. Since a certain interaction is necessary in order to achieve thermal equilibrium, a very loose coupling is assumed to exist between the atoms, making a mutual exchange of energy possible but having no effect on the manner of oscillation.

We shall not, for the present, consider an atom in a solid as having *three* degrees of freedom of oscillation. We reduce this number in our model to one and are thus dealing with a schematic solid in which the atoms behave as linear, harmonic oscillators which perform their oscillations about fixed centres. These centres are arranged in space like the points of a crystal lattice. The energy levels of these localized oscillators, according to quantum mechanics, are equidistant, i.e. above their lowest energy ε_0 they can absorb amounts of energy $\varepsilon_1 = h\nu$, $\varepsilon_2 = 2h\nu$ and so on, where ν is the frequency of their characteristic oscillation and h is Planck's constant. Each energy level corresponds to one particular quantum state of a particle.

At first we take a very small number of oscillators and consider 25 of them, represented by one of the four horizontal layers of balls of the same colour in Fig. 11. At absolute zero temperature the oscillators are all in the lowest energy state ε_0, i.e. in the lowest energy level. In order to raise the temperature, energy must be introduced. This causes particles to move from the lowest to higher quantum states, i.e. higher energy levels. The important point in this discussion is, once more, the number of different ways in which it is possible to distribute the energy. If we introduce 25 energy quanta $h\nu$ to the 25 oscillators, either by the supply of heat $\mathrm{d}Q$ or by performing work $\mathrm{d}W$ on the system, we must look for the number of ways in which the energy $\mathrm{d}U = 25\,h\nu$ can be divided between the oscillators. As with the shaking experiments with billiard balls (Section 2.3), we are not bothered now by the fact that the system considered is really too small for the effective application of statistical-thermodynamical considerations. For the present we merely wish to demonstrate the counting method.

If each of the oscillators receives one quantum, then no new state will be created by interchanging any two atoms. The regular distribution of energy, in other words, can only be achieved in one way: it represents one micro-state and will thus only occur extremely rarely. A less regular distribution is quite a different matter. Consider the following example of an irregular distribution, shown in Fig. 14. The six oscillators in the positions *C*3, *C*4,

	A	B	C	D	E
1	0	0	0	1	2
2	0	0	0	1	2
3	0	0	1	1	3
4	0	0	1	2	3
5	0	0	1	2	5

Fig. 14. Schematic representation of a particular distribution of 25 energy quanta among 25 oscillators. Each square in the figure corresponds with one oscillator. The number appearing in each square indicates the number of quanta per oscillator. These numbers can be distributed in the diagram in about 10^{12} different ways. The total number of possible distributions of 25 quanta among 25 oscillators is considerably larger, viz. about 6×10^{13}.

*C*5, *D*1, *D*2, *D*3, have each absorbed one quantum, the four oscillators at *D*4, *D*5, *E*1, *E*2 two quanta each, the two oscillators at *E*3, *E*4, three quanta each and the oscillator at *E*5 five quanta. The other twelve oscillators have absorbed no energy, Since each oscillator has its own place in the "crystal", a new micro-state will be created when two oscillators in *different* quantum states are exchanged. (In a gas in which each particle has the whole gas volume at its disposal, this is not the case).

The number of distributions when six *arbitrary* oscillators have absorbed one quantum each, four two each, two three each and one oscillator five quanta, is thus given by

$$g = \frac{25!}{12!\,6!\,4!\,2!\,1!} = 9.4 \times 10^{11}$$

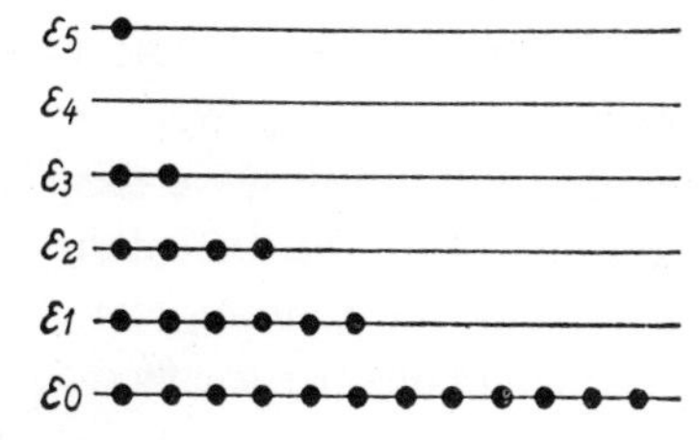

Fig. 15. Distribution of 25 oscillators among equidistant energy levels in such a way that a total of 25 quanta are absorbed. This diagram comprises the circa 10^{12} distributions referred to in the caption to Fig. 14.

Instead of speaking of a distribution of 25 quanta among 25 oscillators as we have just done, we can equally well say that the 25 oscillators are so distributed in the available energy levels that 12 are in the lowest level, 6 in level 1 and so on. We then obtain the diagram in Fig. 15. Since the positions $A1$, $A2$ etc. of the oscillators cannot be deduced from the figure, Fig. 15 does not correspond to one but to a whole group of about 10^{12} micro-states, one of which is indicated by Fig. 14.

2.10. THE MOST PROBABLE DISTRIBUTION

Besides the distribution of the oscillators in the available energy levels given in Fig. 15 there are many other distributions possible. According to Equation (2.2.3) the most probable distribution is that which comprises the greatest number of micro-states. In the present case, the most probable distribution is characterized by the populations $n_0 = 11$, $n_1 = 7$, $n_2 = 4$, $n_3 = 2$, $n_4 = 1$. The reader will note that this distribution does, in fact, correspond to 25 oscillators and 25 quanta and that it comprises 1.6×10^{12} micro-states.

The general expression for the number of micro-states corresponding to a particular distribution over the energy levels in our schematic solid is seen from the above to be given by

$$g = \frac{N!}{n_0!\, n_1!\, n_2!\, n_3! \dots} \tag{2.10.1}$$

where N is the total number of oscillators, while the series $n_0, n_1, n_2, n_3 \dots$ gives the populations of the energy levels, i.e. the numbers of oscillators with energies $\varepsilon_0, \varepsilon_1, \varepsilon_2, \varepsilon_3 \dots$ (cf. Fig. 15, in which $n_0 = 12$, $n_1 = 6$, etc). Only those series are permissible which fulfil the restrictive conditions

$$\sum n_i = N. \tag{2.10.2}$$

$$\sum n_i \varepsilon_i = U. \tag{2.10.3}$$

where $U = qh\nu$ is the total energy introduced (q = number of quanta).

We shall now show that a maximum occurs in g (complying with the restrictive conditions (2.10.2) and (2.10.3)) when the occupation numbers $n_0, n_1, n_2, n_3, \dots$ decrease according to a geometrical progression. When g has a maximum value, a small variation in the distribution will leave this value unchanged. The variation must be such that, from (2.10.2) and (2.10.3), N and U remain unchanged. The variation may, for example, consist of transferring one particle from level 1 to level 0 and one particle from level 1 to level 2. The value of g as given by (2.10.1) then changes to

$$g' = \frac{N!}{(n_0 + 1)!\,(n_1 - 2)!\,(n_2 + 1)!\,n_3!\,\ldots}$$

The ratio g'/g is given by

$$\frac{g'}{g} = \frac{n_0!\,n_1!\,n_2!}{(n_0 + 1)!\,(n_1 - 2)!\,(n_2 + 1)!} = \frac{n_1(n_1 - 1)}{(n_0 + 1)\,(n_2 + 1)}$$

For very large numbers of oscillators this expression may be replaced by

$$\frac{g'}{g} = \frac{n_1^2}{n_0 n_2}$$

If g is not noticeably changed by the rearrangement, then $g'/g = 1$ and thus

$$\frac{n_0}{n_1} = \frac{n_1}{n_2}$$

A similar relationship naturally applies also for levels 1, 2, and 3; for levels 2, 3 and 4, etc. The condition for a maximum value of g is thus

$$\frac{n_0}{n_1} = \frac{n_1}{n_2} = \frac{n_2}{n_3} = \cdots \qquad (2.10.4)$$

This is satisfied in a special way by $n_0 = n_1 = n_2 = n_3 = \ldots$. The restrictive conditions (2.10.2) and (2.10.3), however, exclude this uniform occupation of the energy levels: the higher the energy levels are, the less densely populated they will be (cf. Fig. 15). To obtain a *maximum* value of g while satisfying the restrictive conditions, $n_0, n_1, n_2, n_3, \ldots$ must form a diminishing geometrical progression:

$$n_0,\; n_1 = n_0 c^{-1},\; n_2 = n_0 c^{-2},\; \ldots,\; n_i = n_0 c^{-i},\; \ldots$$

where c is given by (2.10.4). The energies corresponding to these numbers are

$$\varepsilon_0,\; \varepsilon_1 = h\nu \;,\; \varepsilon_2 = 2h\nu \;,\; \ldots,\; \varepsilon_i = ih\nu \;,\; \ldots$$

It follows that

$$n_i = n_0 c^{-\varepsilon_i/h\nu} = n_0 e^{-\varepsilon_i (\ln c)/h\nu}$$

or

$$n_i = n_0 e^{-\beta \varepsilon_i}, \qquad (2.10.5)$$

where $\beta = (h\nu)^{-1} \ln c$ is a positive quantity.

2.11. STATISTICAL DEFINITION OF THE TEMPERATURE AND STATE SUM

Since all the oscillators are in their lowest quantum state at absolute zero temperature, β must become infinite for $T=0$. As the temperature rises (introduction of more and more quanta) the oscillators spread out over more and more levels; from (2.10.5) β will thus become smaller as the temperature rises. In accordance with this, the statistical definition of temperature reads

$$T = \frac{1}{k\beta} \tag{2.11.1}$$

where k is Boltzmann's constant. In Section 2.16 it will be shown that this temperature is identical with the absolute temperature according to Kelvin which was discussed in Chapter 1.

By summation of all the numbers given by Equation (2.10.5) and by making use of (2.10.2) and (2.11.1), we obtain:

$$n_0 = \frac{N}{\sum e^{-\varepsilon_i/kT}} \tag{2.11.2}$$

The sum appearing in the denominator is called the *state sum* or partition function of the harmonic oscillator. For a particle with non-equidistant energy levels the expression

$$Z = \sum e^{-\varepsilon_i/kT} \tag{2.11.3}$$

is also called the state sum of the particle. The summation must be carried out over all the stationary states which are open to the particle. The state sum of a *system* of particles is also frequently employed. The total energy of the whole system then appears in the exponent and one must sum over all the states possible for the system (see Section 5.8).

Employing (2.11.2) the distribution numbers (2.10.5) for the most probable macro-state become

$$n_i = \frac{Ne^{-\varepsilon_i/kT}}{\sum e^{-\varepsilon_i/kT}} \tag{2.11.4}$$

summing over all values of i. In the present case of a system of linear harmonic oscillators the sum over the states is given directly by the geometrical progression

$$\sum e^{-\varepsilon_i/kT} = 1 + e^{-h\nu/kT} + e^{-2h\nu/kT} + \ldots = \frac{1}{1 - e^{-h\nu/kT}} \tag{2.11.5}$$

To show how the oscillators distribute themselves over the equidistant levels at various temperatures, we shall calculate the distribution for the temperatures $T_1 = h\nu/k$, $T_2 = 2h\nu/k$ and $T_3 = 4h\nu/k$. From (2.11.4) and (2.11.5), and since $\varepsilon_i = ih\nu$, the distribution numbers at these temperatures are given by

$$n_i(T_1) = Ne^{-i}(1 - e^{-1}) = 0.632\ Ne^{-i}$$
$$n_i(T_2) = Ne^{-i/2}(1 - e^{-1/2}) = 0.393\ Ne^{-i/2}$$
$$n_i(T_3) = Ne^{-i/4}(1 - e^{-1/4}) = 0.221\ Ne^{-i/4}$$

where i must be substituted in turn by all the integers from zero to infinity. Graphical representations of the distribution numbers thus calculated can be found in Fig. 16. The figure demonstrates clearly that the particles

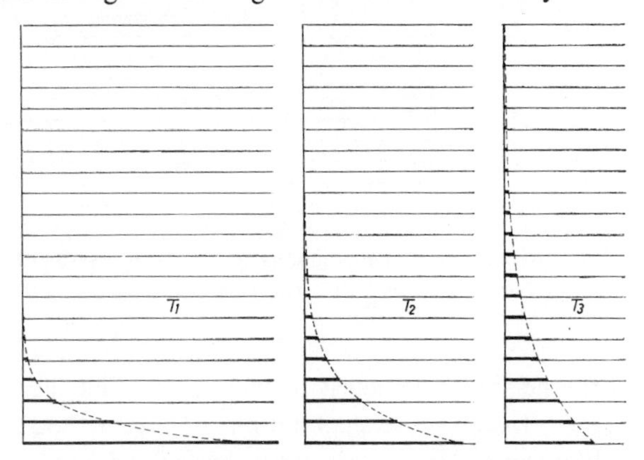

Fig. 16. Distribution of the particles among the energy levels of a system of linear harmonic oscillators at three temperatures $T_1 = h\nu/k$, $T_2 = 2h\nu/k$ and $T_3 = 4h\nu/k$. The total number of particles is constant, which is shown by the constant total length of the thick horizontal lines which indicate the population of each separate level.

spread out over more and more levels as the temperature rises. The relative values of the energy of the particles in the various levels as derived from the distribution numbers are shown in Fig. 17. The maximum value of the energy shifts to higher levels at increasing temperature.

For a discussion of "negative absolute temperatures" and lasers the reader is referred to Section 3 of the Appendix.

2.12. THE METHOD OF UNDETERMINED MULTIPLIERS

The important Equation (2.11.4) was derived in the preceding section only for a series of equidistant levels. It is also valid, however, for the most probable distribution over an arbitrary series of levels. It is not difficult to demonstrate this in a manner similar to that employed above for equidistant levels, by removing a number of particles from an arbitrary level and distributing

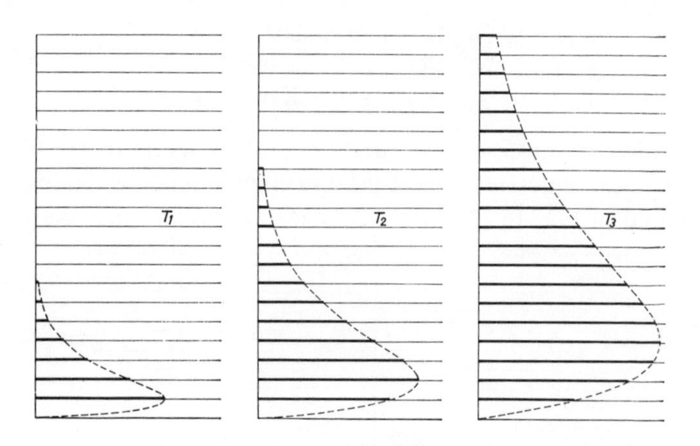

Fig. 17. Distribution of energy among the levels of a system of linear harmonic oscillators for the three temperatures in Fig. 16.

them between the neighbouring levels in such a way that the total energy remains unchanged. However, we shall not pursue this method of treatment further but shall proceed at once to the most general and most usual method of proof: the method of Lagrange's undetermined multipliers.

We again consider a system of N identical, localized particles which vibrate virtually independently of each other. The energy levels of the particles, however, are now not equidistant; the particles do not behave as harmonic oscillators. Each particle may be in one of a number of stationary states with energies $\varepsilon_0, \varepsilon_1, \varepsilon_2, \ldots, \varepsilon_i, \ldots$ The distribution numbers for these states are again indicated by $n_0, n_1, n_2, \ldots, n_i, \ldots$, which are restricted by the conditions (2.10.2) and (2.10.3). The number of micro-states corresponding to a particular distribution of the particles over the levels is once more given by (2.10.1). We wish to know which distribution gives a maximum value for this number of micro-states, i.e. we wish to know the most probable state. Once g has reached a maximum, small changes in the distribution numbers will leave this quantity unchanged, provided that the restrictive conditions are satisfied.

If the number of particles in the level with energy ε_i changes by the small number δn_i, then the energy of the particles in that level will change by the amount $\varepsilon_i \delta n_i$, the sign of which is determined by that of δn_i. The condition that for any arbitrary rearrangement of the particles in the available levels their total number remains unchanged reads

$$\Sigma \delta n_i = 0.$$

The condition that the total energy must remain unchanged reads

$$\Sigma \varepsilon_i \delta n_i = 0$$

Note that these relationships are merely the differential forms of (2.10.2) and (2.10.3):

$$\delta N = \Sigma \delta n_i = 0, \tag{2.12.1}$$

$$\delta U = \Sigma \varepsilon_i \delta n_i = 0. \tag{2.12.2}$$

From Equations (2.10.1) and (2.4.4) the number of micro-states comprised in a particular distribution is given by

$$\ln g = \ln N! - \Sigma \ln n_i! = \text{constant} - \Sigma(n_i \ln n_i - n_i) \tag{2.12.3}$$

If changes δn_i occur in the distribution numbers, then in general $\ln g$ will also change by an amount $\delta \ln g$ which, from (2.12.3), is given by

$$\delta \ln g = -\Sigma \ln n_i \,.\, \delta n_i \,. \tag{2.12.4}$$

When g has the maximum attainable value, this value will remain unchanged by the small variations δn_i in the distribution numbers. In other words, in that case δg (and also $\delta \ln g$) is zero, provided that the accessory conditions (2.12.1) and (2.12.2) are fulfilled. The latter implies that the most probable distribution cannot be found simply by making (2.12.4) equal to zero, i.e. by writing

$$\Sigma \ln n_i \,.\, \delta n_i = 0, \tag{2.12.5}$$

because this expression does not comprise the two constraining relations (2.12.1) and (2.12.2). These expressions are taken into account by multiplying them, according to Lagrange's method of undetermined multipliers, by the constants α and β, the values of which are not at once specified, and subsequently adding them to (2.12.5):

$$\Sigma (\ln n_i + \alpha + \beta \varepsilon_i) \delta n_i = 0. \tag{2.12.6}$$

It is clear that *one* way in which this equation can be satisfied is that the bracketed expression is equal to zero for each value of i. It is shown in the

Appendix that this is the *only* way in which (2.12.6) can be satisfied. Thus, in general,

$$\ln n_i + \alpha + \beta\varepsilon_i = 0 \tag{2.12.7}$$

From (2.12.7) it follows that

$$n_i = e^{-\alpha}e^{-\beta\varepsilon_i}. \tag{2.12.8}$$

By comparing this equation with (2.10.5) we see that the quantity β which appears here is identical with the quantity β in the two preceding sections.

Making use of Equation (2.10.2), we obtain:

$$N = e^{-\alpha} \sum e^{-\beta\varepsilon_i}. \tag{2.12.9}$$

With the definition (2.11.1) of the temperature T it follows from (2.12.8) and (2.12.9) that

$$n_i = \frac{Ne^{-\varepsilon_i/kT}}{\sum e^{-\varepsilon_i/kT}} \tag{2.12.10}$$

This is the same equation as was derived in Section 2.11 for a system of particles with equidistant energy levels. From the above it can be seen that it holds for any arbitrary system of localized and virtually independent particles, whatever the mutual separation of the levels may be. From (2.10.3) and (2.12.10) the total energy of the system is given by

$$U = \frac{N\sum\varepsilon_i e^{-\varepsilon_i/kT}}{\sum e^{-\varepsilon_i/kT}} \tag{2.12.11}$$

For the average energy per particle we have:

$$\bar{\varepsilon} = \frac{U}{N} = \frac{\sum\varepsilon_i e^{-\varepsilon_i/kT}}{\sum e^{-\varepsilon_i/kT}} \tag{2.12.12}$$

The entropy of the system of localized and virtually independent particles can be found with the help of Equation (2.7.6),

$$S = k \ln g_{\max}$$

(cf. the following section). From Equation (2.12.3) we have

$$\ln g = N \ln N - N - \sum(n_i \ln n_i - n_i)$$

or, combining with (2.10.2):

$$\ln g = N \ln N - \sum n_i \ln n_i \tag{2.12.13}$$

The maximum value of g is reached at the values of n_i given by (2.12.8):

$$\ln g_{\max} = N \ln N - \sum n_i(-\alpha - \beta\varepsilon_i).$$

or, using (2.10.2) and (2.10.3):

$$\ln g_{\max} = N \ln N + \alpha N + \beta U.$$

Using Equation (2.12.9) to calculate α and the definition (2.11.1) for the temperature, we finally obtain:

$$S = k \ln g_{\max} = kN \ln \sum e^{-\varepsilon_i/kT} + \frac{U}{T} \tag{2.12.14}$$

In the next chapter we shall meet a new thermodynamic function, the free energy $F = U - TS$. According to (2.12.14) this is given for the solids discussed by

$$F = -kTN \ln \sum e^{-\varepsilon_i/kT} \tag{2.12.15}$$

2.13. THE TOTAL NUMBER OF MICRO-STATES

The number of micro-states corresponding to a certain distribution n_0, n_1, $n_2, \ldots, n_i, \ldots$ over the available levels is given by Equation (2.10.1):

$$g = \frac{N!}{n_0! n_1! n_2! \ldots n_i! \ldots} \tag{2.13.1}$$

The *total* number of micro-states m is given by the sum of all the expressions of this form which satisfy the restrictive conditions (2.10.2) and (2.10.3):

$$m = \sum \frac{N!}{n_0! n_1! n_2! \ldots n_i! \ldots} \tag{2.13.2}$$

In the special case where the energy levels are equidistant, we can calculate this total number of micro-states for N oscillators and an energy $U = qh\nu$ (q=number of quanta absorbed) in a very simple way without making use of (2.13.2). In imagination we place the q quanta and $N-1$ of the N oscillators in arbitrary succession and independently of one another along a straight line. At the extreme right we place the Nth oscillator. Let us agree that the quanta placed between two oscillators belong to the oscillator on their right; then each succession represents a complete distribution, since the last position at the right is always occupied by an oscillator. The required total number of micro-states is thus given by the number of different ways in which the oscillators and quanta can be arranged along the line. This number can be written down at once by remembering that the last oscillator does not participate in the possible rearrangements and that neither the interchange of two oscillators nor that of two quanta have any effect on the arrangement:

$$m = \frac{(q + N - 1)!}{q!(N - 1)!} \tag{2.13.3}$$

If we apply this equation to the numerical example discussed in Section 2.9 (25 oscillators and 25 quanta), we find:

$$m = \frac{(25 + 24)!}{25!24!} = 6.3 \times 10^{13}$$

a number which is nearly 40 times as large as that of the most probable distribution (see Section 2.10). If the number of oscillators and quanta becomes very large, it is no longer necessary to discriminate between the number of micro-states in the most probable distribution and the total number of micro-states. In other words, the sum in (2.13.2) may then be replaced by its greatest term, in the same manner as was discussed in Section 2.4 with respect to the micro-states corresponding to the mixing of two sorts of particles. This applies, not only when the energy levels of the particles are equidistant, but also when this is not the case.

If we consider the above further we note that the remarks made in Section 2.4 with respect to the various configurations of white and red billiard balls, can be applied almost word for word to the present problem. As the number of particles increases $g_{\max}/m$ becomes smaller. The chance of encountering *exactly* the most probable distribution as given by (2.12.10) becomes smaller as N increases. On the other hand, the fraction of the number of micro-states belonging to *nearly* exponential distributions continues to increase as N becomes larger. These almost exponential distributions cannot be distinguished macroscopically from the pure exponential distribution. For example, if a particular energy level in the exponential distribution has a population of 10^{19} particles, no change will be observed macroscopically if this number increases or decreases by 10^6.

The exponential and nearly exponential distributions together comprise the overwhelming majority of all the micro-states. Instead of the macro-state, determined by the number of oscillators N and the energy U, being regarded as a collection of all permissible micro-states, it can therefore also be regarded as a collection of those micro-states belonging to the exponential and nearly exponential distributions. One can even go a step further and take the macro-state to include exclusively the pure exponential distribution, although this comprises only a small fraction of all the micro-states. The justification for this is the same as for the replacement in Section 2.4 of the number 2^{N_0-40} by the number 2^{N_0}, although one differs from the other by a factor 2^{40}. As in Section 2.4, the argument is considerably strengthened by the fact that, in connection with the statistical definition of entropy (Section 2.7), we are not so much interested in g and m themselves as in the natural logarithms of these quantities. The continual decrease of g_{max}/m for an increasing number of particles does not prevent $(\ln g_{max})/(\ln m)$ from approaching unity under these circumstances. We shall demonstrate this for equidistant levels by calculating the above ratios for three different populations of the three lowest levels and for the following numbers of oscillators N and quanta q:

TABLE 2

	1	2	3
N	111	1110	11100
q	12	120	1200

The most probable distributions are given in these three cases by the populations:

TABLE 3

	1	2	3
n_0	100	1000	10000
n_1	10	100	1000
n_2	1	10	100

With the help of Equations (2.13.1), (2.13.3) and (2.3.1) we calculate:

TABLE 4

N	g_{max}	m	g_{max}/m	$(\ln g_{max})/(\ln m)$
111	$5.2 \cdot 10^{15}$	$1.3 \cdot 10^{16}$	$4 \cdot 10^{-1}$	0.975
1110	$1.0 \cdot 10^{168}$	$2.5 \cdot 10^{169}$	$4 \cdot 10^{-2}$	0.992
11100	$3.6 \cdot 10^{1699}$	$8.7 \cdot 10^{1705}$	$4 \cdot 10^{-7}$	0.996

We see that $(\ln g_{max})/(\ln m)$ really does approach unity as the number of particles increases (cf. Table 1 where the same occurs). Real crystals usually contain at least 10^{20}, more often 10^{21} to 10^{24} atoms. For such numbers of oscillators m is vastly greater than g_{max}, while at the same time the difference between $\ln g_{max}$ and $\ln m$ is completely negligible. Therefore, as in Section 2.7, we come to the conclusion that entropy can be calculated either by means of Equation (2.7.5) or Equation (2.7.6).

The discussions in this and preceding sections all deal with a system which, in the course of time, passes through one micro-state after another. An alternative procedure is the following (cf. the General Introduction). One supposes that the system of N particles to be investigated exists in very many independent, but identical samples. The whole collection of similar samples is called an *ensemble* of systems. Instead of considering the behaviour of one sample in the course of time (Boltzmann, Einstein), one considers the state of the whole ensemble at any moment (Gibbs). All micro-states are accorded the same *a priori* probability and one investigates the relative populations of the levels in the whole ensemble. This treatment leads to the same results as were obtained above.

2.14. THE SPECIFIC HEAT OF AN EINSTEIN SOLID

In the idealized solid discussed in Sections 2.9 and 2.10, the identical atoms behave as linear harmonic oscillators which vibrate about fixed centres and virtually independently of each other. That part of the internal energy of this solid which varies with the temperature is given by Equation (2.12.11), which is valid for any system of localized and virtually independent particles.

In Section 2.9 *et seq.*, for the sake of simplicity it was not taken into account that an atom in an Einstein solid must be attributed *three* degrees of vibrational freedom. An actual crystal of N atoms in this model can be regarded as a system of $3N$ linear oscillators. For an Einstein solid we thus have:

$$U = 3N \frac{\sum \varepsilon_i e^{-\varepsilon_i/kT}}{\sum e^{-\varepsilon_i/kT}} \qquad (2.14.1)$$

or, in another form:

$$U = 3NkT^2 \frac{\partial}{\partial T} \ln \sum e^{-\varepsilon_i/kT} \qquad (2.14.2)$$

where, again, the summation must be carried out from $i = 0$ to $i = \infty$. Making use of (2.11.5) it is easy to obtain:

$$U = 3N \frac{h\nu}{e^{h\nu/kT} - 1} \qquad (2.14.3)$$

According to (1.8.4) the heat capacity at constant volume is given by

$$C_V = \left(\frac{\partial U}{\partial T}\right)_V$$

hence, from (2.14.3), by

$$C_V = 3kN\left(\frac{h\nu}{kT}\right)^2 \frac{e^{h\nu/kT}}{(e^{h\nu/kT} - 1)^2} \tag{2.14.4}$$

In deriving this equation, the starting point was (2.12.11), an equation based on the number of micro-states in the most probable distribution. After discussing free energy in the following chapter we shall show that the derivation of (2.14.4) is very simple when it is based on the *total number* of micro-states.

Equation (2.14.4) was derived by Einstein as early as 1907. If kT is very much larger than $h\nu$, then:

$$e^{h\nu/kT} \cong 1 + h\nu/kT$$

and thus, substituting in (2.14.3):

$$U \cong 3\,NkT$$
$$C_V \cong 3\,Nk$$

At relatively high temperatures, thus, the heat capacity of an Einstein solid reaches a constant value which is not only independent of the temperature, but also independent of ν. If this idealized solid were representative of the behaviour of real (elementary) solids, then the latter should all have the same specific heat per gram-atom at all not too low temperatures. This would amount to $c_v = 3N_0k = 3R$, where N_0 is Avogadro's number and R the gas constant. According to the experimentally established law of Dulong and Petit, many elementary solids do, in fact, have a constant specific heat of about 6 cal per degree per gram-atom (R = approx. 2 cal per degree) at not too low temperatures. Equation (2.14.4) requires that c_v shall finally approach zero as the temperature decreases; this is also in agreement with observation, cf. Fig. 18 which represents the relationship between c_v/R and $kT/h\nu$ given by Equation (2.14.4).

To a first approximation, thus, this equation provides a good expression for the thermal behaviour of the solids considered here. It is no exaggeration to say that Einstein's theory which we have just discussed, was one of the first great successes of the quantum theory, bearing in mind that the drop in specific heat at low temperatures was completely inexplicable by classical theories. The form of the experimental $c_v(T)$ curves does not agree in detail with the equation, however. The largest deviations occur at very low temperatures. The reason is that, in reality, a crystal cannot be regarded as an assembly of nearly independent oscillators of equal frequency. A crystal has a great number of modes of vibration at widely divergent frequencies; they may not be regarded as the vibrations of individual atoms but are intrinsic to

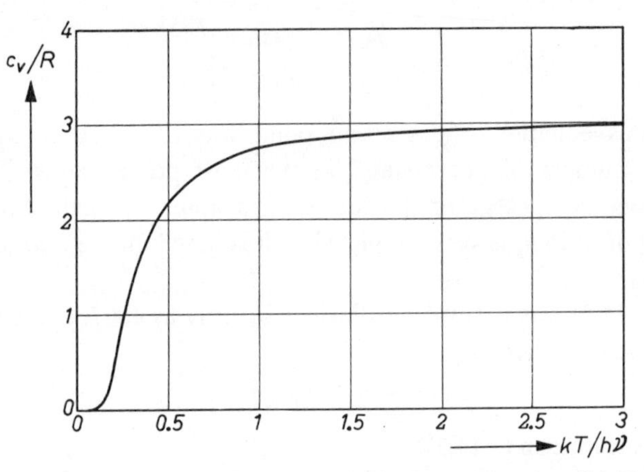

Fig. 18. c_v/R as a function of $kT/h\nu$ for an Einstein solid (Equation (2.14.4), applied to one gram-atom).

the lattice as a whole: they can be pictured as standing waves. A theory based on this view and in better agreement with experiment has been formulated by Debye.

In Debye's theory a crystal is regarded as a continuum for which the number of natural frequencies in the range from ν to $\nu + d\nu$ is proportional to ν^2. Strictly speaking, this is not justifiable since a continuum possesses an infinite number of natural frequencies, whilst the number of frequencies in a crystal is limited by the number of atoms N (it amounts to $3N$). Debye resolved this inconsistency in a manner reminiscent of the cutting of the Gordian knot, namely by terminating the frequency spectrum abruptly at the frequency ν_{max} at which the total number of frequencies has reached $3N$. We shall not enter into Debye's theory since, from a purely thermodynamic point of view, it covers no new ground. One of its consequences, which is in good agreement with experiment, is worthy of mention: at low temperatures the energy corresponding to the vibrations in a solid is proportional to T^4, so that the specific heat is proportional to T^3. This is the same relationship as that valid for radiation within an enclosure at all temperatures (Stefan-Boltzmann Law). It is true that an enclosure possesses an infinite number of modes of vibration and the solid only $3N$, but at low temperatures this difference is irrelevant, since according to quantum theory the high frequency vibrations then play no part.

It should also be noted that, according to Einstein's theory (Equation (2.14.4)), the curve in Fig. 18 should apply to all elementary solids, since not T but $kT/h\nu$ is plotted as abscissa. In Einstein's theory, for a given substance ν has a constant value ν_E. If we consider a series of substances then, for a given value of $kT/h\nu_E$, T will be proportional to ν_E. In other words: as ν_E becomes larger, the Dulong and Petit specific heat will only be reached at a higher temperature.

By definition a characteristic temperature (or Einstein temperature) $\theta_E = h\nu_E/k$ has been introduced, which can be determined by comparing the experimental curve with the standard curve in Fig. 18. The value of θ_E gives that of ν_E directly. For an extremely hard substance like diamond the value of θ_E is found to lie in the region of 2000 °K (ν_E approx. 4.10^{13} sec^{-1}), on the other hand, for a fairly soft metal like silver its value is only about 200 °K (ν_E approx. 4.10^{12} sec^{-1}).

In Debye's theory a characteristic temperature (Debye temperature) is also introduced and defined as

$$\theta_D = \frac{h\nu_{\max}}{k}$$

The Debye temperatures do not differ greatly from the Einstein temperatures (at most a few tens percent).

2.15. THE VIBRATIONAL SPECIFIC HEAT OF GASES

We have seen in the previous section that only to a rough approximation can a solid be regarded as a collection of independent oscillators of equal frequency. A diatomic or polyatomic gas, on the other hand, conforms most satisfactorily to this model. In a diatomic gas each molecule can vibrate in such a way that the atoms oscillate along the straight line through their centres. Energy is exchanged when gas molecules collide with one another. Apart from this, the vibrations proceed independently. Therefore Equation (2.14.4), after deletion of factor 3, can be applied with considerably greater accuracy to the vibrations of a diatomic gas than to those of a solid. If we once more introduce a characteristic temperature of the vibration, defined as $\theta_{\text{vi}} = h\nu/k$, we then have for one mole of gas:

$$c_{\text{vi}} = R\left(\frac{\theta_{\text{vi}}}{T}\right)^2 \frac{e^{\theta_{\text{vi}}/T}}{(e^{\theta_{\text{vi}}/T}-1)^2} \qquad (2.15.1)$$

Three points should immediately be raised. (*a*) As will be discussed in Chapter 4, the calculation of the number of micro-states of a system of non-localized particles does not correspond to the calculation discussed earlier (Equation (2.10.1)). This, however, does not impair the validity of Equation (2.15.1). (*b*) Besides vibrational motion, the gas molecules also possess rotational and translational motion. Equation (2.15.1) thus only gives a portion of the specific heat of a diatomic gas. (*c*) The characteristic temperatures of the vibration lie much higher for most diatomic gases than for solids. For the latter the constant final value of the specific heat is virtually reached at room temperature in nearly every case; on the other hand, for most diatomic gases, vibration plays no appreciable part in the specific heat at room temperature.

Table 5 shows the characteristic temperatures of the vibration for a few diatomic gases. They have been obtained by deducing the frequencies of vibration ν from the spectra (cf. Chapter 6).

TABLE 5

CHARACTERISTIC TEMPERATURES OF VIBRATION FOR SOME DIATOMIC GASES

Gas	θ_{vi} (°K)	Gas	θ_{vi} (°K)
I_2	310	NO	2740
Br_2	465	CO	3120
Cl_2	810	HI	3320
O_2	2270	HBr	3810
N_2	3380	HCl	4300
D_2	4490	HF	5960
H_2	6340	HD	5500

It can be calculated from (2.15.1) that at a temperature $T = \theta_{vi}$ the vibrational specific heat will have reached 92% of its final value R, at $T = 0.1\ \theta_{vi}$, on the other hand, barely 0.5%. Therefore, only for the first three gases mentioned in the table will the vibration make an appreciable contribution to the specific heat at room temperature.

In order to demonstrate the value of Equation (2.15.1) for calculating the specific heat of diatomic gases, we shall apply it to nitrogen, N_2. That part of the specific heat which depends on the translational and rotational motion of the N_2 molecules is given, at all temperatures at which N_2 is gaseous, by the law of the equipartition of energy. Per mole this portion amounts to $3R/2$ for translation and R for rotation, together $5R/2$. The total specific heat at constant volume is found by adding to this the quantity given by Equation (2.15.1). For $R = 1.987$ cal per degree per mole and $\theta_{vi}\ (N_2) = 3380$ K, we find the following values for four different temperatures (Table 6). For the sake of comparison the last column contains the "exact" values of c_v, calculated by a very accurate method discussed in Chapter 6.

TABLE 6

CALCULATED SPECIFIC HEAT OF N_2 AT CONSTANT VOLUME (IN CAL/DEG.MOLE)

Temp. (°K)	$c_{tr}+c_{ro}$	c_{vi}	c_v (tot.)	c_v ("exact") [1]
100	4.97	0.00	4.97	4.97
1000	4.97	0.83	5.80	5.83
2000	4.97	1.57	6.54	6.62
3000	4.97	1.79	6.76	6.88

[1]) H. L. Johnston and C. O. Davis, J. Amer. Chem. Soc. **56**, 271 (1934).

It can be seen that the agreement between the values we have calculated and the "exact" values is satisfactory at not very high temperatures. The discrepancies become greater as the temperature increases. They are due partly to the fact that the oscillations are not purely harmonic and to interaction between the vibration and rotation, and partly to an electronic contribution to the specific heat which has been neglected. The electrons in a N_2 molecule can exist in various stationary states of widely-differing energy and therefore make a contribution to the specific heat at very high temperatures (see Chapter 6).

For polyatomic gases with non-linear molecules $c_{tr}+c_{ro} = 3R$, since the number of degrees of freedom of rotation is one more than for linear molecules. Furthermore, the number of modes of vibration increases rapidly as the number of atoms per molecule increases. A triatomic non-linear molecule has three modes of vibration, which are shown in Fig. 19 for a water mole-

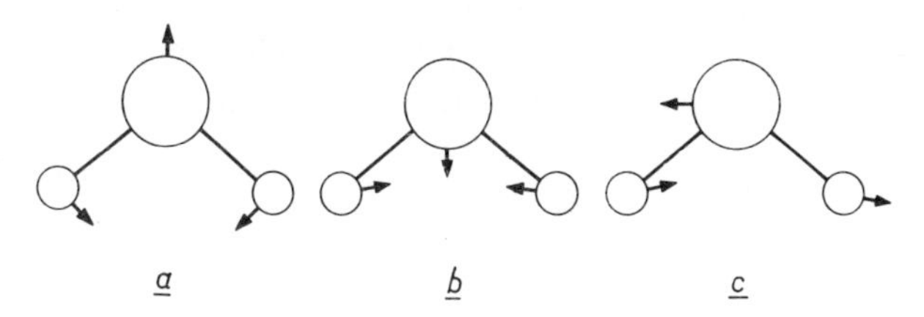

Fig. 19. An H_2O molecule can vibrate in three different ways.

cule. The arrows in this figure show the directions in which the separate atoms move. The characteristic temperatures of the three vibrations can be derived from the spectrum of water vapour and amount to (*a*) 2290 °K, (*b*) 5170 °K, (*c*) 5400 °K. The specific heat of gaseous water is found by adding three Einstein terms to the above-mentioned quantity, $3R$. The terms are obtained by substituting the characteristic temperatures above in Equation (2.15.1).

Table 7 gives the results for four different temperatures. The last column of the table once more gives the "exact" values of c_v for comparison.

TABLE 7

CALCULATED SPECIFIC HEAT OF GASEOUS H_2O AT CONSTANT VOLUME (IN CAL/DEG.MOLE)

Temp.(°K)	$c_{tr} + c_{ro}$	c_{vi} (*a*)	c_{vi} (*b*)	c_{vi} (*c*)	c_v (tot)	c_v("exact") [1]
300	5.96	0.06	0.00	0.00	6.02	6.02
500	5.96	0.44	0.01	0.00	6.41	6.39
1000	5.96	1.31	0.31	0.26	7.84	7.81
1500	5.96	1.64	0.80	0.74	9.14	9.17

[1]) A. R. Gordon, J. Chem. Phys. **2,** 65 (1934).

Here, too, it will be seen that the agreement between calculated and "exact" values is satisfactory. At higher temperatures than those given in Tables 6 and 7, the dissociation of N_2 and H_2O begins to play an important part.

The specific heats of gases at constant pressure, as far as they can be considered perfect, are derived from the calculated specific heats at constant volume by making use of the relationship $c_p = c_v + R$ (Equation (1.9.2)).

2.16. THE IDENTITY OF "STATISTICAL" AND ABSOLUTE TEMPERATURE

In Section 2.11 we introduced a "statistical" temperature T, defined by (2.11.1). We shall now show that it is identical with the absolute temperature in Chapter 1, which (to distinguish it from the "statistical" temperature) we shall refer to in this section by the symbol T'.

Let us consider a system of localized and virtually independent particles, the entropy per gram-atom of which is given, according to Equation(2.12.14), by

$$s = R \ln \sum e^{-\varepsilon_i/kT} + \frac{u}{T} \tag{2.16.1}$$

From this we wish to calculate $\partial s/\partial u$ for a constant separation of the energy levels, i.e. at constant volume. The result can then be compared with that obtained by classical thermodynamics. The latter can be written down immediately since, from the first and second laws, the following equation is valid for a reversible change of state of a gram-atom of an element (see Section 1.16):

$$T'\mathrm{d}s = \mathrm{d}u + p\mathrm{d}v$$

and thus

$$\left(\frac{\partial s}{\partial u}\right)_v = \frac{1}{T'} \tag{2.16.2}$$

The "statistical" value of this derivative can be deduced from (2.16.1) with the help of the following relationship, based on (1.4.4):

$$\left(\frac{\partial s}{\partial u}\right)_v = \left(\frac{\partial s}{\partial u}\right)_T + \left(\frac{\partial s}{\partial T}\right)_u \left(\frac{\partial T}{\partial u}\right)_v \tag{2.16.3}$$

Using Equation (2.12.11) it is easy to show that (2.16.1) yields:

$$\left(\frac{\partial s}{\partial T}\right)_u = 0 \tag{2.16.4}$$

Combining (2.16.3) and (2.16.4) we further derive from (2.16.1):

$$\left(\frac{\partial s}{\partial u}\right)_v = \frac{1}{T} \tag{2.16.5}$$

From (2.16.2) and (2.16.5) it follows that T and T' are identical. In the following chapter we shall be able to demonstrate this identity by another method.

2.17. ZERO-POINT ENTROPY AND THE "THIRD LAW"

According to Nernst's heat theorem, the entropy of all systems in stable or metastable equilibrium tends to zero as the absolute zero of temperature is approached. This means, from Equations (2.7.5) and (2.7.6), that at absolute zero they can only exist in one micro-state. From the point of view of quantum theory this is immediately understandable: at $T = 0$ all the particles are in their lowest quantum state. In the diagram in Fig. 16 all the higher energy levels are gradually vacated as the temperature falls, until finally all the particles occupy the level ε_0. Nernst's theorem would *not* apply, however, if the lowest energy level corresponded to two or more quantum states, in other words, if this level were "degenerate". In the equilibrium state the particles would then be distributed evenly among these states even at 0 °K. Nernst's theorem thus implies the hypothesis that the lowest energy level is not degenerate.

In many cases the validity of the theorem can be tested experimentally. A classic example is the transition of white tin into grey tin. Grey tin is stable below 13 °C (286 °K), white tin is stable above this temperature. Since white tin can be supercooled to the lowest attainable temperatures, it has been possible to measure the specific heat c_p of both modifications at low temperatures. If Nernst's theorem is correct, then from Equation (1.16.1) the entropy of white tin at 286 °K can be found in two ways, either directly from c_p measurements on white tin or from those on grey tin and from the entropy change which occurs with the transition from grey to white tin. Both methods should lead to the same result, i.e. the following should be valid:

$$\int_0^{286} \frac{c_p(w)\mathrm{d}T}{T} = \int_0^{286} \frac{c_p(g)\mathrm{d}T}{T} + \frac{Q}{286} \tag{2.17.1}$$

where $c_p(w)$ and $c_p(g)$ are the specific heats per gram-atom of white and grey tin and Q the heat of transition, i.e. the quantity of heat absorbed during the isothermal and reversible transition of one gram-atom of tin from the grey to the white modification. It has, indeed, been found that (2.17.1) is satisfied within the limits of experimental accuracy. Unfortunately, the heat of transition Q is not known with such accuracy that very great value can be attached to this agreement. Moreover, even if complete agreement were established this would only prove that the *difference* in entropy between the two modifications at 0 °K is zero. The heat theorem is therefore often worded in the following rather more cautious form: at the absolute zero-point all entropy *differences* between the thermodynamical states of a system in internal equilibrium vanish. This formulation has the same practical significance as that in which the separate entropies are said to approach zero, since one can now usefully define the zero point of the entropy of all substances in stable or metastable equilibrium as lying at absolute zero temperature. If, for example, in every chemical reaction $A + B \rightleftarrows AB$, the entropy change is zero at 0 °K, then it is logical to assign a zero-point entropy of zero to A and B as well as to AB. This is not possible for the energy, as extrapolations to absolute zero show that there is no question of the heats of reaction disappearing at the absolute zero-point.

Stronger evidence of the validity of Nernst's theorem is derived from measurements on gases. As we shall see in Chapter 6, the statistical entropy formula $S = k \ln m$ (or $S = k \ln g_{\mathrm{max}}$) can be used to calculate the entropy of many gases, with the help of information derived from spectra on the rotational and vibrational states of their molecules. On the other hand, if Nernst's theorem is valid, the entropy of gases may be calculated with the help of the classical formula $\mathrm{d}S = \mathrm{d}Q_{\mathrm{rev}}/T$, making use of the known values of the specific heats c of these substances in the solid, liquid and gaseous states, and of the heats of transformation, fusion and vaporization (cf.

Chapter 1). The entropy of a substance in the gaseous state at temperature T, assuming there are no transition points in the solid state, can be written directly as:

$$S = \int_0^{T_f} \frac{c_{\mathrm{sol}}}{T}\, \mathrm{d}T + \frac{Q_f}{T_f} + \int_{T_f}^{T_v} \frac{c_{\mathrm{liq}}}{T}\, \mathrm{d}T + \frac{Q_v}{T_v} + \int_{T_v}^{T} \frac{c_{\mathrm{gas}}}{T}\, \mathrm{d}T. \quad (2.17.2)$$

where T_f and T_v are the melting and boiling points and Q_f and Q_v the heats of fusion and vaporization.

This "calorimetric entropy" is thus obtained entirely without reference to the existence of atoms; it is based solely on the results of calorimetric measurements. The "statistical entropy" on the other hand is found by a method which requires no knowledge of the existence of the liquid and solid states. The beauty of it is that both ways generally lead to the same result, while the exceptions which have been found can be satisfactorily explained. These exceptions are of two kinds. Some are only apparent exceptions, caused by the fact that measurements of specific heats were not extended to sufficiently low temperatures, other exceptions occur because equilibrium is not maintained as the temperature falls. Neither type of exception contradicts Nernst's theorem, which only claims validity at the absolute zero-point and then only for systems which are in internal equilibrium.

The above apparent exceptions in the agreement between calorimetric and statistical entropy occur when the "lowest energy level" of the particles appears, on closer inspection, to consist of a group of energy levels whose separation $\Delta\varepsilon$ is so small that even at the lowest temperatures of measurement, kT is still large with respect to $\Delta\varepsilon$. When this is the case, even at the lowest temperature of measurement the particles will still be evenly distributed, according to Equation (2.12.10), over the above-mentioned group of levels. The states corresponding to these levels will, therefore, not manifest themselves in the specific heat. A gradual emptying of the higher levels in the group will only start when the temperature reaches a value for which kT is of the order of magnitude of $\Delta\varepsilon$, and only at temperatures for which kT is much smaller than $\Delta\varepsilon$ will all the particles be found in the lowest level ε_0. This redistribution will manifest itself by a maximum in the curve of specific heat as a function of the temperature (see Fig. 20). If a peak of this sort exists, but is not discovered because measurements are not carried to sufficiently low temperatures, the value obtained for the calorimetric entropy will be too low. An interesting example of this will be found in Chapter 6 in the discussion of the entropy of hydrogen.

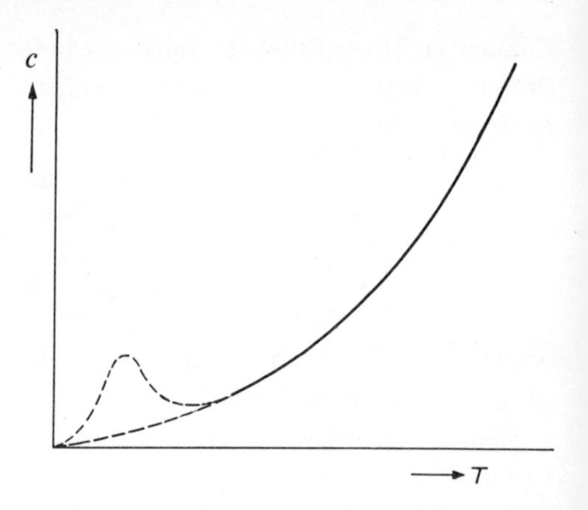

Fig. 20. The calorimetric entropy can be calculated when the specific heat c is known as a function of the temperature T. One is forced to extrapolate from the lowest temperatures of measurement to absolute zero temperature, although this involves the risk of overlooking a possible peak in the specific heat curve below the range of measurement.

In the above, the value found for the calorimetric entropy was too small due to the extrapolation of the specific heat from too high a temperature. The cause, however, may well be a different one, viz. the failure to attain equilibrium, already mentioned. An example of a "frozen" distribution of the molecules over two energy levels is solid carbon monoxide CO, at low temperatures (cf. Chapter 6). A value is found for the calorimetric entropy which is smaller by an amount $R \ln 2 = k \ln 2^{N_0}$ cal/mole.deg. than the statistical entropy derived from the spectrum of CO. This discrepancy corresponds to a number of micro-states $m = 2^{N_0}$. This immediately suggests two possible orientations of the molecules in the crystal. It is assumed that the CO molecules have such a small electric moment and are so very nearly symmetrical that the crystal lattice shows no strong preference for either of the two orientations, CO or OC. As a result, the two opposed directions of orientation would remain irregularly distributed between the lattice positions down to the lowest possible temperature of measurement. It is highly improbable that the rotation of 180°, required to produce equilibrium, is possible in solid CO. In other words: it is not to be expected that an extension of the measurements to lower temperatures will eliminate the discrepancy. If this assumption is correct, then the difference observed here between the calorimetric and statistical entropy merely means that the energy difference between the two CO positions at the freezing point is still too small with respect to kT_f to cause a particular orientation.

Other examples of systems which possess a zero-point entropy because internal equilibrium is not attained at low temperatures are the numerous disordered solid solutions of metals. To attain the equilibrium state, these solid solutions would either have to be transformed into the ordered state or have to split up into two or more pure or ordered phases (cf. Chapter 3). At sufficiently low temperatures this applies even to mixtures of isotopes. The diffusion required to accomplish this phase separation or ordering does not occur, however, at a noticeable rate at low temperatures.

Finally, a good example of a system with a zero-point entropy because equilibrium is not established is glass, which can be regarded as a rigid, supercooled liquid. As long as the temperature remains above the congealing temperature, the supercooled liquid will be in internal equilibrium, below that temperature, however, this will no longer be the case. If crystallization takes place, the system assumes one particular configuration; in the glass-

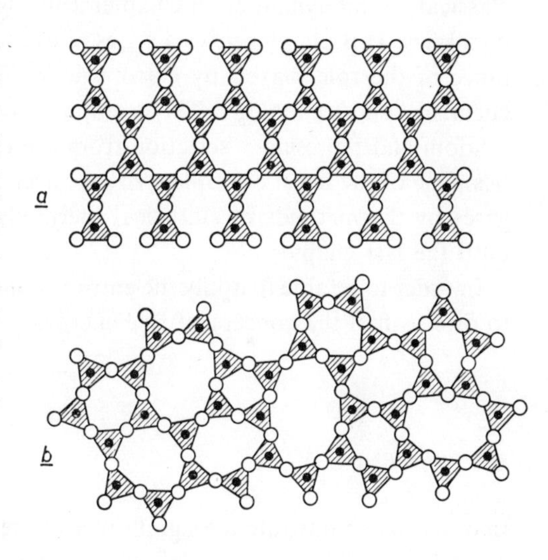

Fig. 21. Various oxides may occur in either a crystalline or a glass-like state. According to Zachariasen, these oxides in either state are built up of the same bricks, namely, polyhedra of oxygen ions (tetrahedra or triangles); at the centre of each polyhedron is a higher-charged positive ion (B^{3+}, Si^{4+}, P^{5+}). The figure illustrates schematically that (*a*) in the crystalline state the polyhedra are arranged regularly, (*b*) in the glass-like state, on the other hand, they are not.

like condition, on the other hand, innumerable configurations are possible (see Fig. 21).

Nernst's theorem cannot be derived from the first and second laws of thermodynamics and is therefore often called the "third law of thermodynamics" [1]).

[1]) cf. F. Simon, Year Book of the Physical Society, London, 1956, and S. J. Glass and M. J. Klein, Investigations on the Third Law of Thermodynamics, Atomic Energy Commission Techn. Report No. 2, Case Inst. of Techn., Cleveland, Ohio, 1958.

CHAPTER 3

APPLICATIONS OF THE CONCEPT OF ENTROPY

3.1. INTRODUCTION

In Chapter 1 we discussed the significance of the concept of entropy in classical thermodynamics, in Chapter 2 its significance in statistical thermodynamics. This chapter gives a series of examples demonstrating the importance of the role played by entropy in widely divergent fields of physics, chemistry and technology. The examples have been chosen more or less at random and represent a selection from a much larger number of possibilities. One of the finest examples, viz. the accurate calculation of equilibria in gases by the methods of statistical thermodynamics, cannot be dealt with until the last chapter.

In order to be able to apply the entropy concept successfully, it is desirable to discuss first the concept of free energy.

3.2. FREE ENERGY

In order to demonstrate the significance of the concept of entropy, our attention was turned in the foregoing to the perfect gas state, i.e. a state of matter in which the molecules exert no attractive forces upon one another. In this state, equilibrium always corresponds to a disordered distribution of the different sorts of molecules. This is not necessarily the case when there are attractive forces at work between the molecules or when the distribution of the molecules is influenced by external forces or fields. If, in the case of the 100 billiard balls illustrated in Fig. 11, the 50 white balls attract each other strongly (e.g. due to magnets incorporated in each of the balls), then after each shaking experiment the system would usually be found to be separated into two "phases", one containing almost exclusively white balls and one almost exclusively red balls. In this case, a disordered distribution of the balls corresponds to a state of higher energy. Apparently, thus, the striving towards a maximum value of the entropy can in some cases be counteracted by another tendency, viz. the striving towards a minimum value of the energy. A quantitative comparison of these tendencies can be made by using the two laws of thermodynamics in the form

$$dU = dQ + dW \qquad \text{(first law)}$$
$$dQ \leqslant TdS. \qquad \text{(second law)}$$

Combining these equations, we obtain the relation

$$dU - TdS \leqslant dW \tag{3.2.1}$$

or, for a constant value of T:

$$d(U - TS)_T \leqslant dW \tag{3.2.2}$$

where the equality sign applies to reversible changes of state, the inequality sign to irreversible changes. In (3.2.2) the quantity $(U - TS)$ is a thermodynamic function since U, T and S are thermodynamic functions. This thermodynamic function is called the *Helmholtz free energy* or the free energy at constant volume and is indicated by the symbol F. According to (3.2.2), the work which must be performed on a system in order to bring about an isothermal change of state, is greater when the change takes place irreversibly than when it is reversible. Conversely, the work supplied (performed on the surroundings) by a system during an isothermal change of state is smaller in an irreversible than in a reversible process, since

$$-dW \leqslant -dF$$

When a process takes place irreversibly there is therefore always a loss of work in the sense that more work must be performed on the system or less work is supplied by the system. The reader is referred to Section 1.15, in which this loss of efficiency has already been mentioned for a special case.

From (3.2.2) the work performed on a system in an isothermal change of state is equal to the change in the free energy of the system if the process takes place reversibly. This justifies the assertion made in Section 1.7, viz. that the work performed is equal to the change in a thermodynamic function only in the special case of a reversible and isothermal change of state.

We shall now consider the case of an isothermal change of state in which no work at all is performed, the case in which not only the temperature but also the volume of the system remains constant, the same being true for every other parameter the variation of which can give rise to the performance of external work. In that case, from (3.2.2), we have for an *irreversible* process:

$$d(U - TS)_{T,V} < 0$$

Thus, if we are not dealing with an isolated system but with a system of which the temperature T and the volume V are kept constant, the "striving of the entropy to a maximum" is replaced by another tendency, viz. that of the free energy $F = U - TS$ towards a minimum.

If, instead of the temperature and the volume, the temperature and the pressure are kept constant, then even for an irreversible process the system will still supply some work, viz. the volume-work $-\mathrm{d}W = p\mathrm{d}V$ (in which p this time represents the external, not the internal, pressure), and it is immediately obvious from (3.2.2) that now another thermodynamic function, the function $U - TS + pV$, tends towards a minimum value. This function is called the *Gibbs free energy*, but also the free energy at constant pressure or thermodynamic potential. It is commonly indicated by the symbol G. Another name often used for this function is *free enthalpy*, which is derived from the name enthalpy used for the thermodynamic function $H = U + pV$ (cf. Section 1.8). If, in agreement with fairly general practice, the name Helmholtz free energy, $F = U - TS$, is abbreviated to free energy, then it is logical to call the function $G = H - TS$ the free enthalpy. For the sake of clarity, the three thermodynamic functions are listed once more below [1]:

$F = U - TS$, (Helmholtz) free energy,
$H = U + pV$, enthalpy,
$G = H - TS$, free enthalpy or Gibbs free energy.

The relationships found for irreversible processes

$$\mathrm{d}F_{T,V} = \mathrm{d}(U - TS)_{T,V} < 0 \tag{3.2.3}$$

and

$$\mathrm{d}G_{T,p} = \mathrm{d}(H - TS)_{T,p} < 0 \tag{3.2.4}$$

can be split, somewhat artificially, into two parts, $\mathrm{d}U_{T,V} < 0$ or $\mathrm{d}H_{T,p} < 0$; and $\mathrm{d}S_{T,V} > 0$ or $\mathrm{d}S_{T,p} > 0$. They can be regarded as the mathematical formulation of the two opposed tendencies which were indicated at the beginning of this section. If only the latter tendency ($\mathrm{d}S > 0$) existed, one would expect only those processes and reactions to occur spontaneously for which the number of realization possibilities (the "disorder") increases. On the other hand, if only the first tendency were active ($\mathrm{d}U < 0$ or $\mathrm{d}H < 0$), one would only observe processes for which the opposite was the case, i.e. in which heat is liberated and in which, in general, the order increases. Thus, in order to determine the direction in which a process will take place we must consider the free energy which involves both thermodynamic functions U and S, or H and S. If we do this we see from (3.2.3) and (3.2.4) that at low

[1] Warning: In the literature the symbol F is sometimes used instead of G for the Gibbs free energy. It is therefore always necessary to check the context carefully to be sure which of the two free energies is meant.

temperatures the tendency towards minimum energy (enthalpy) and the corresponding order predominates, while at high temperatures (violent shaking of the billiard balls) the tendency to maximum entropy and the accompanying disorder is predominant.

In accordance with the above it is found that at low temperatures the atoms and molecules, under the influence of their mutual attraction, form the orderly periodic structures we know as crystals. As the temperature rises, the order in the crystals decreases gradually due to the increasingly violent vibration of the particles. Finally, at a sufficiently high temperature, all matter is transformed into the chaotic gas state, often by way of the liquid state. The temperature range in which this entropy effect begins to predominate depends on the magnitude of the attractive forces. Even in gases a certain order is often present in the form of orderly groups of atoms (molecules). A further rise in the temperature, however, causes this order also to disappear and at the surface of the sun (temperature about 6000 °C) matter only exists as a gaseous mixture of the atoms of the various elements. But even this is not complete chaos and as the temperature rises higher and higher even the order in the electron shells is finally completely destroyed by thermal ionization. At a few million degrees, ionization is complete. This complete splitting of atoms into nuclei and electrons occurs in the interior of our sun and of innumerable other stars. The only order remaining there is that of the nuclei. Complete dissociation of the nuclei into protons and neutrons would require higher temperatures than appear to occur in the hottest stars.

3.3. GENERAL EXPRESSION FOR FREE ENTHALPY

In the discussions in the previous section the work dW appearing in Equations (3.2.1) and (3.2.2) was assumed to be exclusively volume-work. In fact, from the considerations in Section 1.6 we know that this does not necessarily have to be so. We shall start by considering the rather more general case of a deformable and magnetisable rod, which is subjected to varying pressures, tensions and magnetic fields. For small changes in the volume V, the length l and the magnetic moment M, the work performed on the rod will be

$$dW = -p dV + K dl + \mathrm{H} dM \tag{3.3.1}$$

where p represents the hydrostatic pressure, K the tensile force and H the magnetic field strength (not to be confused with the enthalpy H). According to Equation (3.2.1) the combination of the first and second laws now takes the form

$$\mathrm{d}U - T\mathrm{d}S + p\mathrm{d}V - K\mathrm{d}l - \mathrm{H}\mathrm{d}M \leqslant 0 \tag{3.3.2}$$

If T, p, K and H are constant during the change, this equation becomes, for an irreversible process:

$$\mathrm{d}(U - TS + pV - Kl - \mathrm{H}M) < 0 \tag{3.3.3}$$

In this case, thus, the thermodynamic function

$$G = U - TS + pV - Kl - \mathrm{H}M \tag{3.3.4}$$

tends to a minimum value. It is once more referred to as free enthalpy, while the function

$$H = U + pV - Kl - \mathrm{H}M \tag{3.3.5}$$

is indicated by the name enthalpy. It is easily seen from the following considerations that this does not conflict with Equation (1.8.6), in which the enthalpy made its first appearance. The internal energy in the present case is not only a function of p and T, but also of K and H:

$$U = \mathrm{f}(p, T, K, \mathrm{H}).$$

For the differential of U we can write:

$$\mathrm{d}U = \left(\frac{\partial U}{\partial p}\right)\mathrm{d}p + \left(\frac{\partial U}{\partial T}\right)\mathrm{d}T + \left(\frac{\partial U}{\partial K}\right)\mathrm{d}K + \left(\frac{\partial U}{\partial \mathrm{H}}\right)\mathrm{d}\mathrm{H} \tag{3.3.6}$$

Furthermore, from the first law:

$$\mathrm{d}Q = \mathrm{d}U + p\mathrm{d}V - K\mathrm{d}l - \mathrm{H}\mathrm{d}M. \tag{3.3.7}$$

From (3.3.6) and (3.3.7) it follows that:

$$C_{p,K,\mathrm{H}} = \left(\frac{\mathrm{d}Q}{\mathrm{d}T}\right)_{p,K,\mathrm{H}} = \left(\frac{\partial U}{\partial T}\right) + p\left(\frac{\partial V}{\partial T}\right) - K\left(\frac{\partial l}{\partial T}\right) - \mathrm{H}\left(\frac{\partial M}{\partial T}\right)$$

From (3.3.5), in accordance with (1.8.6), the above expression can indeed be written

$$C_{p,K,\mathrm{H}} = \left(\frac{\partial H}{\partial T}\right)_{p,K,\mathrm{H}} \tag{3.3.8}$$

The general expressions for enthalpy and free enthalpy as found above are:

$$H = U - \Sigma xy, \tag{3.3.9}$$

$$G = H - TS = U - TS - \Sigma xy, \tag{3.3.10}$$

where x may be, for example, a negative pressure $-p$, a tensile force K, a magnetic field strength H or an electric field strength E, while y then represents a volume V, a length l, a magnetic moment M_m or an electric moment M_e, respectively.

The zero-point of the enthalpy and the free enthalpy is undetermined. If we take as an example an elastically deformable rod subjected to a tensile force K, we can define G as

$$G = U - TS - Kl$$

but, if desired, also as

$$G' = U - TS - K(l - l_0) \tag{3.3.11}$$

where l_0 is the length of the rod when $K = 0$. We can write:

$$K(l - l_0) = \sigma A l_0 \frac{l - l_0}{l_0} = \sigma \varepsilon V_0 \tag{3.3.12}$$

where σ is the tensile stress, A the area of the perpendicular cross-section, V_0 the volume and $\varepsilon = (l - l_0)/l_0$ the tensile strain of the rod (cf. Section 1.6). Substituting (3.3.12) in (3.3.11) we obtain

$$G' = U - TS - \sigma \varepsilon V_0. \tag{3.3.13}$$

The free enthalpy per unit volume is given by

$$G'' = U' - TS' - \sigma\varepsilon. \tag{3.3.14}$$

where U' and S' are the energy and entropy per unit volume.

3.4. APPLICATION TO CHEMICAL EQUILIBRIUM

Chemical affinity

If a chemical reaction takes place in some system, volume-work is usually the only form of work which is exchanged with the surroundings. Naturally, this volume-work will be zero if the volume of the system is kept constant. In chemistry and metallurgy one is usually more interested, for practical reasons, in processes which take place at constant pressure than in those at constant volume. This implies a greater interest in the free enthalpy or Gibbs free energy G than in the (Helmholtz) free energy F, since G is the quantity which tends to a minimum value at constant pressure and temperature. In the foregoing discussions we saw that this tendency could be regarded as a compromise between two opposite tendencies: the striving of the enthalpy H to a minimum and that of the entropy S to a maximum value.

At constant temperature and pressure, according to Equation (3.2.4), the only chemical reactions which will proceed spontaneously will be those in which the change in the free enthalpy

$$\Delta G = \Delta H - T\Delta S \tag{3.4.1}$$

is negative. In this equation ΔH is the heat of reaction at constant pressure, i.e. the quantity of heat absorbed by the system when the chemical reaction proceeds irreversibly at constant values of p and T, while ΔS is the increase in entropy caused by the reaction. It may be useful to recall here that the heat introduced into a system is regarded as positive (see Section 1.6), so that an exothermal reaction has a negative heat of reaction.

It follows directly from the first law which here takes the form

$$Q_{\mathrm{irr}} = \Delta U + p\Delta V = \Delta H \tag{3.4.2}$$

that ΔH is indeed the normal (irreversible) heat of reaction at constant pressure. ΔU and ΔH, the heats of reaction at constant volume and constant pressure, are virtually equal to one another when no gaseous substances appear in the formula for the reaction (e.g. $PbS + Fe \rightarrow Pb + FeS$) or when the number of gas molecules remains unchanged by the reaction (e.g. $H_2 + Cl_2 \rightarrow 2\ HCl$). If, however, the number of gas molecules formed during the reaction is not equal to the number which disappear (as in the reaction $2\ NH_3 \rightarrow N_2 + 3\ H_2$), then ΔU and ΔH will differ, according to (3.4.2), by an amount

$$p\Delta V = \Delta(pV) \simeq nRT.$$

where n is the number of moles formed less the number which disappears.

From the thermodynamic point of view, a chemical reaction will have a greater tendency to occur as ΔG for this reaction becomes more negative. Therefore, in thermodynamics $-\Delta G = -\Delta H + T\Delta S$ is often called the *affinity* of the reaction. It increases as ΔH becomes more strongly negative, i.e. the reaction more exothermic, and as ΔS becomes more strongly positive, i.e. the greater the increase in entropy during the reaction.

Chemical reactions in which only solids are involved are usually accompanied by a comparatively small change in entropy and are thus generally exothermic when they occur spontaneously. When gases appear in the reaction formula, ΔS will generally be small if the number of gas molecules remains unchanged during the reaction. If, on the other hand, the number of gas molecules increases in the course of the reaction, the entropy will generally increase considerably, since the number of available micro-states m is much greater in the gaseous state than in the condensed state (see Chapter 2, where we saw that $S = k \ln m$).

As an example, let us consider the oxidation of carbon according to the two equations

$$C + O_2 \rightarrow CO_2 \quad (3.4.3)$$

$$2\,C + O_2 \rightarrow 2\,CO \quad (3.4.4)$$

During the course of (3.4.3) much more heat is evolved (the enthalpy decreases much more) than during the course of (3.4.4). On the other hand the entropy hardly changes during (3.4.3), while in (3.4.4), due to the doubling of the number of gas molecules, it increases considerably. If only a tendency towards minimum enthalpy existed, all carbon would oxidize to CO_2. If there were only the tendency towards maximum entropy, combustion would only produce CO. In reality, a compromise is reached by the formation of a mixture of the two gases. At relatively low temperatures the energy (or enthalpy) effect predominates and carbon is chiefly converted into CO_2. At high temperatures, multiplication by T (see Equation (3.4.1)) makes the entropy term more important and CO is principally produced.

In the latter case (ΔS positive), according to (3.4.1), ΔG will become more negative as the temperature rises. In other words: the affinity of carbon for oxygen, as regards the formation of CO, increases continuously with a rise of temperature. Conversely, in the case of the oxidation of solid or liquid metals into solid or liquid oxides, e.g.

$$2\,Fe + O_2 \rightarrow 2\,FeO \quad (3.4.5)$$

$$\tfrac{4}{3}\,Al + O_2 \rightarrow \tfrac{2}{3}\,Al_2O_3 \quad (3.4.6)$$

the affinity decreases with rising temperature, since ΔS is negative due to the decrease in the number of gas molecules. The fact that rising temperature causes a decline in the affinity of oxidation reactions of the types (3.4.5) and (3.4.6) and a simultaneous increase in the affinity of reaction (3.4.4) means that for every metal there is some temperature above which oxygen at a certain pressure has a smaller tendency to combine with it than with carbon. We have thus here the entropy effect which is so important in extractive metallurgy: at a sufficiently high temperature all liquid or solid metallic oxides can be reduced by carbon.

For a more quantitative treatment of the above, we must know the dependence of ΔG on the gas pressure. This dependence can most easily be found by introducing *molar quantities*, each of which relates to 1 gram-molecule (1 mole) of a particular substance.

Molar quantities

Let us consider a chemical reaction taking place at constant values of p and T, e.g. the reaction

$$PbO_2 + 2\,Fe \rightarrow Pb + 2\,FeO.$$

The system embracing the substances participating in the reaction suffers a change of volume ΔV, a change of entropy ΔS, a change of enthalpy ΔH and so on. These changes are given by the molar quantities. This is most easily demonstrated by first examining the volume; it is clear that the initial volume of the system is that of 1 mole of PbO_2 and 2 moles of Fe, the final volume is that of 1 mole Pb and 2 moles FeO. The total change of volume is thus given by

$$\Delta V = v_{Pb} + 2\,v_{FeO} - v_{PbO_2} - 2\,v_{Fe}$$

In general:

$$\Delta V = \Sigma \nu_i v_i. \qquad (3.4.7)$$

where the symbol v_i refers to the molar quantities under consideration, i.e. the molar volumes of the various substances, and ν_i to the coefficients in the equation of the reaction. The latter are always regarded as positive for the substances produced during the reaction (the products) and negative for those which vanish (the reactants).

What has here been said about volume changes which occur in the course of a reaction can be reiterated in almost exactly the same words for changes in H, S, G, etc. One obtains relationships completely analogous to (3.4.7), e.g.:

$$\Delta H = \Sigma \nu_i h_i. \qquad (3.4.8)$$

$$\Delta S = \Sigma \nu_i s_i. \qquad (3.4.9)$$

where h is the symbol for the molar enthalpy and s for the molar entropy. For historical reasons the molar free enthalpies are often indicated by the symbol μ, rather than by g:

$$\Delta G = \Sigma \nu_i \mu_i. \qquad (3.4.10)$$

If the substances taking part in a reaction are in solution, the molar quantities in (3.4.7) to (3.4.10) must be replaced by so-called *partial* molar quantities. By far the most important of these is the partial molar free enthalpy of a substance in solution, which is often referred to as the chemical potential and, like the molar free enthalpy, indicated by the symbol μ. We need not consider the partial molar quantities here since we only wish to discuss reactions between substances which are pure when in the solid or liquid state and which can be regarded as perfect when they are gaseous.

There is no interaction between the molecules in a mixture of perfect gases, so that the properties of one component of a mixture are the same as if it alone were present in the available space. The partial molar free enthalpy of a gas A in a mixture of perfect gases is therefore equal to the molar free enthalpy of the pure gas A, provided that its pressure is equal to the partial pressure of A in the mixture. Gases at a pressure of about 1 atm at room temperature and higher temperatures can usually be regarded as approximately perfect.

Since only those reactions for which ΔG is negative can occur spontaneously, reactions for which ΔG is positive have a tendency to proceed in the reverse direction. When $\Delta G = 0$ holds for a chemical reaction, it cannot proceed in either direction. In such a case, the chemical system in question is in equilibrium. Proceeding from (3.4.10), we can thus write as a condition of chemical equilibrium:

$$\Delta G = \sum \nu_i \mu_i = 0 \tag{3.4.11}$$

Dependence of the molar free enthalpy on the pressure

It follows from the above that the molar free enthalpy of an element or a compound is given by

$$\mu = u - Ts + pv \tag{3.4.12}$$

where u represents the molar energy, s the molar entropy and v the molar volume of the pure substance. From (3.4.12) we have

$$\mathrm{d}\mu = \mathrm{d}u - T\mathrm{d}s - s\mathrm{d}T + p\mathrm{d}v + v\mathrm{d}p \tag{3.4.13}$$

According to the first and second laws,

$$\mathrm{d}u = T\mathrm{d}s - p\mathrm{d}v$$

applies to the substance considered under equilibrium conditions. Substitution in (3.4.13) gives:

$$\mathrm{d}\mu = -s\mathrm{d}T + v\mathrm{d}p \tag{3.4.14}$$

At constant temperature we thus have

$$\mathrm{d}\mu = v\mathrm{d}p$$

or, integrating:

$$\mu = \mu^0 + \int_{p=1}^{p} v\mathrm{d}p \tag{3.4.15}$$

where μ^0 is the molar free enthalpy of the substance at the chosen temperature and at unit pressure. The unit of pressure is chosen as 1 atmosphere; this choice will be maintained in all our further thermodynamic discussions. If we apply Equation (3.4.15) to a perfect gas, then v may be replaced by RT/p. We then obtain:

$$\mu = \mu^0 + RT \ln p \tag{3.4.16}$$

As we have already explained, this relationship also applies for the *partial* molar free enthalpy (the chemical potential) of a gas in a mixture of perfect gases, provided that the partial pressure is inserted for p.

For a condensed phase (liquid or solid), the integral appearing in (3.4.15) will have a negligibly small value if pressure changes are limited between zero and a few atmospheres. For example, if the molar volume of the solid or liquid phase is 20 cm³ then for $p = 0.01$ atm, the integral will have an absolute value of 0.99×20 cm³·atm, which is equivalent to only about 0.5 cal. In contrast to this, for a perfect gas the integral in this case would have an absolute value of $RT \ln 100$, which corresponds at 300 °K to about 2700 cal. Thus if the pressure changes occurring do not exceed a few atmospheres, the chemical potential of a liquid or solid substance can be equated, to a good approximation, to the chemical potential at a pressure of 1 atmosphere:

$$\mu = \mu^0 \tag{3.4.17}$$

The reason for this can be found in the small molar volume of a condensed phase.

Standard affinity and reaction constant

For the reactions we are considering, in which only gases and pure condensed substances are involved, substitution of (3.4.16) for the participating gases and (3.4.17) for the pure substances gives Equation (3.4.10) the following form:

$$\Delta G = \Sigma \nu_i \mu_i^0 + \Sigma \nu_j RT \ln p_j \tag{3.4.18}$$

As follows from the above, the terms $\nu_j RT \ln p_j$ only apply to the *gases* taking part in the reaction, while the terms of the form $\nu_i \mu_i^0$ apply to *all* the substances participating. It is therefore convenient to write

$$\Sigma \nu_i \mu_i^0 = \Delta G^0. \tag{3.4.19}$$

where ΔG^0 is the change in the free enthalpy caused by the reaction in question in the special case when all the substances involved are in their "standard state". In the cases which interest us here, the standard state is the pure state for liquids and solids, and the perfect state at a pressure of 1 atm for gases. The quantity $-\Delta G^0$ is called the "*standard affinity*" of the reaction.

Combination of (3.4.18) and (3.4.19) gives:

$$\Delta G = \Delta G^0 + \sum RT \ln p_j^{\nu_j} \tag{3.4.20}$$

where the ν's are once again positive for the gaseous products and negative for the gaseous reactants. In the equilibrium state $\Delta G = 0$ and thus, for this state we have:

$$\Delta G^0 = -RT \ln \Pi\, p_j^{\nu_j} \tag{3.4.21}$$

Here the sum of a number of logarithms is replaced by the logarithm of a product (Π is the mathematical symbol for a product).

Since the quantities μ^0 are pure functions of the temperature, the right-hand side of (3.4.21) will vary only with the temperature. The product occurring there must therefore have a constant value at a given temperature. It is called the equilibrium constant or reaction constant K_p:

$$K_p = \Pi p_j^{\nu_j} \tag{3.4.22}$$

As seen in (3.4.21), the reaction constant is given by the standard affinity:

$$\Delta G^0 = -RT \ln K_p \tag{3.4.23}$$

Equation (3.4.23) is one of the most frequently used equations in "chemical thermodynamics".

For a simple oxidation reaction, e.g. (3.4.5) or (3.4.6), the value of K_p is found from (3.4.22) to be

$$K_p = p_{O_2}^{-1}$$

In this case, from (3.4.23) we have

$$\Delta G^0 = RT \ln p_{O_2} \tag{3.4.24}$$

where p_{O_2} is the dissociation pressure of FeO or Al_2O_3. For reaction (3.4.4) it follows from (3.4.22) that

$$K_p = p_{CO}^2 \cdot p_{O_2}^{-1}$$

so that in this case ΔG^0 is given by

$$\Delta G^0 = - RT \ln \frac{p_{CO}^2}{p_{O_2}}. \tag{3.4.25}$$

The equilibrium

$$2\,C + O_2 \rightleftarrows 2\,CO$$

at any temperature can thus be calculated directly by means of (3.4.25) if the value of ΔG^0 for the reaction is known as a function of temperature.

Determination of ΔG_0 values [1])

There are three methods by which the standard affinities of chemical reactions can be determined quantitatively. Sometimes they are determined directly from equilibrium measurements, i.e. from measurements of gas pressures in the case of reactions between gases and pure condensed substances (see above). Use is then made of the relationship, given above,

$$\Delta G_T^0 = -RT \ln K_p. \tag{3.4.23}$$

In many cases ΔG^0 values can be obtained from calorimetric data by means of the relationship (3.4.1), applied to a reaction occurring under standard conditions:

$$\Delta G_T^0 = \Delta H_T^0 - T\Delta S_T^0 \tag{3.4.26}$$

ΔS_T^0, the entropy change which occurs when the reaction takes place under standard conditions at a temperature T, can be calculated by way of the "third law" if the molar specific heats of the reacting substances are known over the range $0° - T$ °K and, furthermore, if one has knowledge of the latent heats of all physical changes of state undergone in this range by the substances participating in the reaction. For the necessary calculations, the reader is referred to Chapter 2 and especially to Equation (2.17.2), which may need to be extended to account for any crystallographic transformations which may occur. Molar entropies, calculated with the aid of (2.17.2), and possibly supplemented with molar entropies of gases derived from spectroscopic data (see Chapters 5 and 6) can be used to calculate ΔS_T^0 from the relationship (3.4.9). ΔH^0, the heat of reaction at constant pressure under standard conditions, is known, in many cases, at *one* temperature from calorimetric measurements. With the help of molar specific heats and latent heats it can then be calculated for the required temperature T.

There are some cases where ΔG^0 can be measured directly as an electromotive force. If a chemical reaction, e.g. the reaction

$$CuSO_4aq + Zn \rightarrow ZnSO_4aq + Cu$$

[1]) See also Section 2 of the Appendix, where the most simple kind of chemical reaction is discussed.

can proceed in a galvanic cell, supplying electrical energy, then from Section 1.18, for a small reversible transformation we have

$$dU = TdS - pdV - zFEdn \quad (3.4.27)$$

where E is the electromotive force of the cell, F the charge on one gram-ion of monovalent positive ions (F = 1 faraday = 96500 coulomb) and z is the number of faradays transported during the transformation of the quantity of substance indicated in the reaction formula. In the case under consideration (3.4.27) corresponds to the transformation of only dn moles $CuSO_4$ and $z = 2$. Since $G = U - TS + pV$, it follows from (3.4.27) for constant values of p and T that

$$dG = -zFEdn \quad (3.4.28)$$

or, for the conversion of a quantity of matter as indicated in the reaction equation:

$$\Delta G = -zFE$$

For specially chosen standard conditions, this formula takes the form

$$\Delta G_T^0 = -zFE_T^0 \quad (3.4.29)$$

Values of ΔG are usually given in kcal. In order to convert the right-hand side of the equation into kcal, use must be made of the relationship

$$1 \text{ faraday volt} = 23.06 \text{ kcal.}$$

i.e. the value

$$F = 23.06 \text{ kcal/volt}$$

must be substituted for F in Equation (3.4.29):

$$\Delta G_T^0 = -23.06\, zE_T^0 \text{ kcal.} \quad (3.4.30)$$

ΔG^0 values of a great number of chemical reactions determined with the help of the relationships (3.4.23), (3.4.26) and (3.4.29) are compiled in tables[1]).

Standard affinities of oxidation reactions

Fig. 22 gives the standard affinities of a number of oxidation reactions as a function of the temperature. It will be seen that, apart from a few kinks, the curves diverge but little from straight lines. The reason for this is that ΔH^0 and ΔS^0 change very little with temperature, while those changes which do occur roughly compensate each other in most cases. Use is often made, therefore, of the approximation formula

[1]) See e.g. F. D. Rossini *et al.*, Selected Values of Chemical Thermodynamic Properties, Nat. Bureau of Standards, Washington, 1952, and O. Kubaschewski, E. LL. Evans and C. B. Alcock, Metallurgical Thermochemistry, Pergamon Press, London, 1967.

$$\Delta G^0_T \simeq \Delta H^0_{T_a} - T\Delta S^0_{T_a}. \tag{3.4.31}$$

where T_a represents a constant temperature.

Diagrams of the type shown in Fig. 22 were introduced in 1944 by Ellingham [1]). He demonstrated their usefulness for oxides and sulphides. These are of great importance, particularly to metallurgists, because many metals occur in their natural state as oxides or sulphides, or as compounds which can easily be converted into oxides.

All the lines in Fig. 22 relate to a reaction with 1 mole of oxygen. The

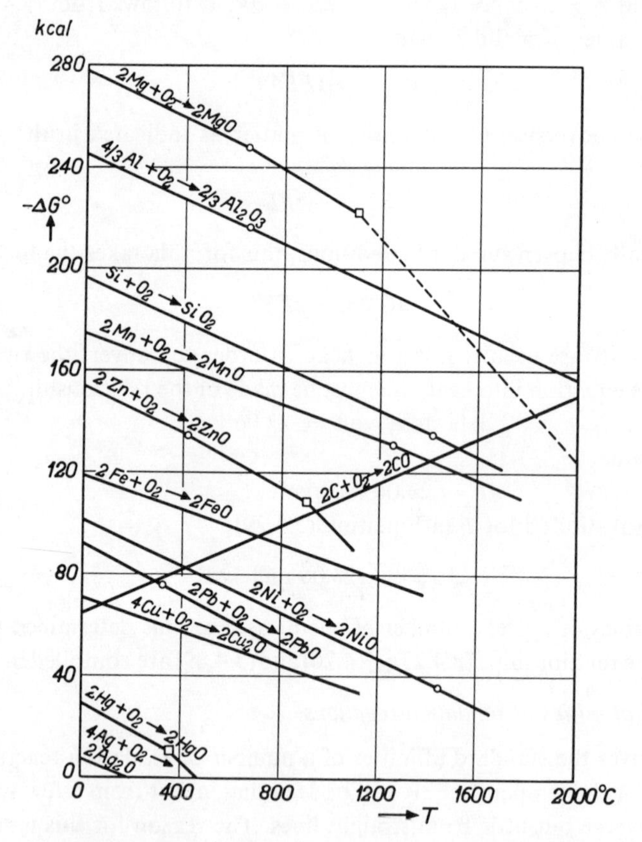

Fig. 22. Standard affinity in kcal of some oxidation reactions as a function of the temperature. Each circle in the figure corresponds to the melting point of a metal and each square to its boiling point. The affinity lines have not been produced beyond the melting and boiling points of the oxides.

---

1) H. J. T. Ellingham, Trans. Soc. Chem. Ind. **63**, 125 (1944).

higher the position of a line in the figure, the greater is the affinity of the element concerned for oxygen. In principle, it is possible to reduce oxides from the lower lines with metals from the higher lines. For example, Al can be used to reduce FeO, and Fe to reduce Cu_2O. This conclusion, in all its generality, is justified because, as already stated, all the reactions relate to 1 mole O_2, so that O_2 cancels out when two reactions are subtracted from one another. For instance, if (3.4.5) is subtracted from (3.4.6) one obtains the reaction

$$\tfrac{4}{3}\,\mathrm{Al} + 2\,\mathrm{FeO} \rightarrow \tfrac{2}{3}\,\mathrm{Al_2O_3} + 2\,\mathrm{Fe}.$$

For this reaction ΔG^0 is the difference between the ΔG^0 values for (3.4.6) and (3.4.5). The figure shows that this difference has a negative value at all temperatures, so that the reaction will, indeed, have the tendency to occur spontaneously.

Relating all the reactions to the same quantity of oxygen has a second advantage, viz. that when the reactions are brought about electrochemically, the quantity of electricity transported is always the same. For the chosen 1 mole O_2, this quantity is four faradays for every reaction. From the $(-\Delta G^0)$ values read off from the figure, Equation (3.4.30) enables us to find E^0 values directly in volts by dividing the values found by $4 \times 23.06 = 92.2$. Each value of E^0 gives the reversible dissociation voltage (at a particular temperature) of an oxide in pure state or in saturated solution in any molten salt. It is the minimum voltage required to dissociate the oxide by electrolysis into metal and oxygen at 1 atm.

In the extreme lower left-hand corner of the figure is the line for the reaction

$$4\,\mathrm{Ag} + \mathrm{O_2} \rightarrow 2\,\mathrm{Ag_2O}. \tag{3.4.32}$$

It is seen that ΔG^0 is slightly negative below about 200 °C ($-\Delta G^0$ slightly positive). The slope of the "straight" line (and the same applies to the other lines) is determined according to (3.4.9), (3.4.10) and (3.4.14) by the relationship

$$\frac{\partial(-\Delta G^0)}{\partial T} = \Delta S^0. \tag{3.4.33}$$

i.e. by the entropy change which takes place during the reaction. It has already been shown that this entropy change is strongly negative, so that the affinity of the reaction decreases as the temperature rises. The figure shows that at about 200 °C the standard affinity reaches zero. At that temperature Ag and Ag_2O are thus in equilibrium with O_2 at 1 atm; in other words: at that temperature the dissociation pressure of Ag_2O is 1 atm.

Above 200 °C Ag_2O decomposes into silver and oxygen even at an oxygen pressure of 1 atm in the surroundings.

If the metal or the oxide undergoes a change of state, the affinity line will show a discontinuity. This can be seen in the second line from the bottom (Fig. 22) relating to the reaction

$$2\,Hg + O_2 \rightarrow 2\,HgO \tag{3.4.34}$$

A discontinuity occurs at 357 °C, the boiling point of mercury. Above this temperature mercury is gaseous at a pressure of 1 atm; its entropy is then much greater than in the liquid state. ΔS^0 (34) is thus more strongly negative above 357 °C than below this temperature and the curve, in agreement with (3.4.33), slopes more steeply downwards. Melting points and crystallographic transformation points, due to the comparatively small entropy changes involved, produce only small changes in the direction of the affinity curves. Roughly speaking, for the solid and liquid states of both metal and oxide, all the curves have about the same slope. This depends on the fact that the entropy change in all these oxidation reactions is chiefly determined by the disappearance of 1 mole O_2.

An exception which is of the greatest importance in metallurgy is the curve for the reaction $2\,C + O_2 \rightarrow 2\,CO$ which, for reasons already discussed, rises with rising temperature. Where it intersects another line, the standard affinities of the reactions corresponding to the two lines are equal. At temperatures below that of the intersection point, the affinity of oxygen at 1 atm for the metal concerned is greater than for carbon. At temperatures above that of the intersection, the reverse is true. As an example we shall consider the intersection with the curve

$$Si + O_2 \rightarrow SiO_2 \tag{3.4.35}$$

Subtracting (3.4.35) from (3.4.4) we obtain the reaction

$$SiO_2 + 2\,C \rightarrow Si + 2\,CO \tag{3.4.36}$$

The standard affinity of this reaction at the temperature of the intersection point will be zero. From (3.4.23) K_p then has the value 1. Since $K_p = p_{CO}^2$, the CO equilibrium pressure of reaction (3.4.36) at the intersection temperature will be just 1 atm. In fact, this pressure is reached at a lower temperature because of the rather large affinity which exists between Si and C and which causes the reaction with carbon to proceed further, according to the equation:

$$SiO_2 + 3\,C \rightarrow SiC + 2\,CO$$

This reaction corresponds to greater CO pressures than reaction (3.4.36). Finally, one must take into account the formation of gaseous SiO, which results in an appreciable interaction between C and SiO_2 at lower temperatures than would be expected from (3.4.36).

The above reasoning was followed as if only CO (and no CO_2) is formed in the reaction between carbon and a metal oxide. In agreement with what was discussed previously, this can be done at high temperatures (above about 900 °C) without introducing great errors (cf. sub-section "Chemical affinity").

Rate and maximum yield of chemical reactions

In connection with the above it should be noted that the affinity values do not give any information about the *rate* of reactions. Cases exist in which ΔG is strongly negative and yet the rate of reaction is negligibly small. For example, the standard affinity of reaction (3.4.6) at room temperature is very large (see Fig. 22), but from experience we know that aluminium objects do not react appreciably with oxygen at this temperature. Further examination reveals that a very thin film of oxide is formed on the surface of the metal, protecting it from further corrosion, since neither aluminium nor oxygen atoms (or ions) can permeate through this skin.

Investigation of the rate of chemical reactions falls outside the scope of thermodynamics, which is only concerned with equilibrium states. The great value of thermodynamics in chemistry becomes clear, however, when one considers that it must first be known whether ΔG is negative for a desired reaction, before any useful purpose can be served by looking for suitable catalysts to accelerate the reaction. It is of even greater importance that from ΔG^0, as shown in preceding sections, one can calculate the chemical equilibrium, i.e. the *maximum yield* of the reaction products. Just how useful this can be to the metallurgist appears, for instance, from the striking example noted in the last century by the great French chemist Le Chatelier. He described it approximately as follows.

It is known that the reduction of iron oxides in blast-furnaces is mainly effected by their reaction with CO, forming CO_2. The gas leaving the blast-furnace at the top, however, contains a considerable percentage of CO, which was naturally considered wasteful and undesirable. Since the incompleteness of the reaction was ascribed to an insufficiently prolonged contact between CO and the iron ore, blast-furnaces of abnormal height were built. The percentage of CO in the escaping gases did not, however, diminish. These costly experiments showed that there is an upper limit to the power of CO to reduce iron ore, which cannot be exceeded. Familiarity with the laws of chemical equilibrium would have enabled the same conclusion to be reached much more quickly and at far less cost.

We might add to this that it is only necessary to know the reaction enthalpy ΔH^0 and the reaction entropy ΔS^0 of the reaction

$$FeO + CO \rightarrow Fe + CO_2$$

for the standard conditions and the temperatures of interest in order to be able to calculate the maximum fraction of the CO which can be converted into CO_2. This calculation can be carried out directly with the help of the equations

$$\Delta G^0 = \Delta H^0 - T\Delta S^0$$

and

$$\ln \frac{p_{CO_2}}{p_{CO}} = -\frac{\Delta G^0}{RT}$$

From Equation (3.4.31) the position of the equilibrium at various temperatures can also be calculated approximately if ΔH^0 and ΔS^0 are known at only one temperature.

3.5. PARAMAGNETISM AND LOW TEMPERATURES

Paramagnetism

The molecules of paramagnetic materials possess a magnetic moment, i.e. they behave to a certain extent like small magnets. Paramagnetism may result from both the orbital motions of the electrons in the molecule and from the intrinsic angular momentum (spin) of the electrons. In most molecules spin and orbital moments of the electrons are so coupled that the resultant magnetic moment is zero.

An example of a gaseous paramagnetic substance is oxygen. The magnetic properties of O_2 are based on the fact that two of the electrons in each oxygen molecule are "unpaired", i.e. have parallel angular momenta (spins) and consequently also parallel magnetic moments (see Chapter 6). On the other hand, all the other electrons in the O_2 molecule are paired. A gaseous alkali metal presents an even simpler example, since the paramagnetic properties in this case depend on the presence of only one non-compensated electron spin per atom. Various hydrated salts show a strong analogy with these paramagnetic gases since the carriers of the magnetic moment (groups of electrons with parallel spins) in these salts are so far apart that their interaction, as in the gases, can be neglected to a first approximation. Throughout salt or gas there is complete disorder of the atomic or molecular magnets. In other words: the axes of the little magnets are distributed at random over the various directions.

If a quantity of oxygen, gaseous sodium or paramagnetic salt is introduced into a magnetic field, the tendency towards a minimum value of the enthalpy ($U-\mathrm{H}M$; see Section 3.3) corresponds to the tendency of the permanent atomic magnets to orientate themselves parallel to the field. The striving towards maximum entropy, however, which becomes stronger as the temperature rises, resists any preferred orientation. At room temperature the thermal agitation causes the orientation of each elementary magnet to change continually, but an average over all the molecules produces a small resultant moment in the direction of the applied field.

Taking into account these two opposed tendencies – the enthalpy and entropy effects – it is not surprising that the resultant moment (the magnetization) is found to be directly proportional to the field strength H and inversely proportional to the absolute temperature T. This temperature-dependence for "ideal" paramagnetic substances is known as Curie's Law.

If the field strength could be continually increased, the magnetization would, in the long run, fail to increase proportionately to H. A state of saturation would finally occur in which all the magnetic axes were similarly orientated. Fields strong enough to produce this saturation at room temperature, however, cannot be generated in the laboratory. By applying a combination of strong fields and very low temperatures, i.e. by making the enthalpy effect large and the entropy effect small, it has been possible to approach saturation very closely for some paramagnetic salts. It has also been possible to employ this principle to attain considerably lower temperatures than was previously possible. Before describing this in the second part of this section, we shall first calculate the dependence of the magnetization on H and T for a simple case.

We shall consider the case where the paramagnetic properties depend on the presence of one uncompensated electron spin per molecule, atom or ion and where the directions of the magnetic moments are virtually independent of each other. The quantum theory states that the component of the spin in the direction of a magnetic field can only take the values $-\frac{1}{2}$ and $+\frac{1}{2}$ (expressed in units of $h/2\pi$). This corresponds to a magnetic moment of $+\mu_B$ or $-\mu_B$ in the direction of the field, the symbol μ_B representing the value of a Bohr magneton. The energies (actually enthalpies) of these two states coincide as the field strength tends to zero. At a field strength H, on the other hand, the energies have the values $-\mu_B\mathrm{H}$ and $+\mu_B\mathrm{H}$ with respect to the field-free state. We are thus dealing with the case of only two energy levels. The distribution of the elementary magnets between these two levels can be calculated as a function of H and T in the following manner.

In a system of N molecules, like that described, at some value of H there will be a greater fraction at the lower than at the higher energy level. If the difference in the populations is denoted by n, then $\frac{1}{2}(N + n)$ molecules will have their magnetic moment in the direction of the field and $\frac{1}{2}(N - n)$ in the opposite direction. The system then possesses a moment of $n\mu_B$ in the direction of the field and an enthalpy which, from Section 3.3, is given by

$$U - \mathrm{H}M = U - n\mu_B\mathrm{H} \tag{3.5.1}$$

The entropy of this system consists of a portion S_0 which is independent and a portion S_{sp} which is dependent on the orientation of the spins:

$$S = S_0 + S_{\mathrm{sp}}$$

Molecules with different directions of spin are distinguishable, molecules with the same direction are not. According to Chapter 2, the "entropy of mixing" S_{sp} is therefore given by the number of ways in which $\frac{1}{2}(N + n)$ "white" and $\frac{1}{2}(N - n)$ "red" molecules can be distributed between N positions:

$$S_{\mathrm{sp}} = k \ln \frac{N!}{\left\{\frac{1}{2}(N + n)\right\}!\left\{\frac{1}{2}(N - n)\right\}!} \tag{3.5.2}$$

The free enthalpy

$$G = U - T(S_0 + S_{\mathrm{sp}}) - n\mu_B\mathrm{H}$$

consists of a part $U - TS_0$ which is independent of n and another part

$$G(n) = -TS_{\mathrm{sp}} - n\mu_B\mathrm{H}$$

which is dependent on n. Using (3.5.2) and Stirling's formula in its approximate form (2.4.4) we can write for this:

$$G(n) = kT\left\{\frac{N+n}{2}\ln\frac{N+n}{2N} + \frac{N-n}{2}\ln\frac{N-n}{2N}\right\} - n\mu_B\mathrm{H} \tag{3.5.3}$$

If the system could submit entirely to its tendency towards a minimum value of the enthalpy, then all the molecules would be found in the lowest energy level ($n = N$). If, on the other hand, it could submit entirely to the tendency towards maximum entropy, then the molecules would be equally divided between the two levels ($n = 0$). The compromise (the equilibrium state) lies at the point where the free enthalpy is at a minimum, i.e. where $dG(n)/dn = 0$. From this it follows that:

$$\frac{kT}{2} \ln \frac{N+n}{N-n} = \mu_B \mathrm{H}$$

or, rearranging:

$$\frac{n}{N} = \frac{e^{\mu_B \mathrm{H}/kT} - e^{-\mu_B \mathrm{H}/kT}}{e^{\mu_B \mathrm{H}/kT} + e^{-\mu_B \mathrm{H}/kT}} = \tanh(\mu_B \mathrm{H}/kT). \qquad (3.5.4)$$

The derivation of this formula has been given in order to illustrate more clearly the "competition" between enthalpy and entropy. It can also be derived much more simply by making use of Equation (2.12.10), which gives the populations of the two levels directly. According to (2.12.10), we have for the lower level:

$$\tfrac{1}{2}(N+n) = \frac{Ne^{\mu_B \mathrm{H}/kT}}{e^{\mu_B \mathrm{H}/kT} + e^{-\mu_B \mathrm{H}/kT}}$$

For the upper level (2.12.10) gives:

$$\tfrac{1}{2}(N-n) = \frac{Ne^{-\mu_B \mathrm{H}/kT}}{e^{\mu_B \mathrm{H}/kT} + e^{-\mu_B \mathrm{H}/kT}}$$

Equation (3.5.4) follows directly from these two equations. The magnetic moment of the system is $n\mu_B$ and thus from (3.5.4):

$$n\mu_B = N\mu_B \tanh(\mu_B \mathrm{H}/kT).$$

The moment per unit volume is

$$I = N'\mu_B \tanh(\mu_B \mathrm{H}/kT) \qquad (3.5.5)$$

where N' is the number of molecules per unit volume. For $\mu_B \mathrm{H}/kT \ll 1$, i.e. for comparatively weak fields or high temperatures, $\tanh(\mu_B \mathrm{H}/kT) \simeq \mu_B \mathrm{H}/kT$ and consequently

$$I \simeq \frac{N'\mu_B^2}{kT} \mathrm{H} \qquad (3.5.6)$$

For this limiting case we have thus derived the proportionality of I to H and $1/T$ previously mentioned. The susceptibility χ is given in this region by

$$\chi = \frac{I}{\mathrm{H}} = \frac{N'\mu_B^2}{kT} \qquad (3.5.7)$$

If the molecule has more than one uncompensated electron spin, these spins are often mutually coupled to one resultant spin for the molecule. A resultant spin of this sort has more than two possible orientations in an applied magnetic field. In general, for a spin quantum number S ($S = \frac{1}{2}$ for one uncompensated electron spin) the component in the direction of the magnetic field can only assume the values $S, S-1, S-2, \ldots, -S+1, -S$. In the case discussed ($S = \frac{1}{2}$) the components $+\frac{1}{2}$ and $-\frac{1}{2}$ were possible. When $S = 1$ (two parallel spins per molecule) we find the three possibilities $+1$, 0 and -1. In general there are $2S+1$ different ways in which the resultant spin can be orientated (see Chapter 5).

A calculation completely analogous to that given above for $S = \frac{1}{2}$, shows that in the more general case (if $\mu_B \mathrm{H}/kT \ll 1$):

$$\chi = \frac{N' 4S(S+1)\mu_B^2}{3kT} \tag{3.5.8}$$

a formula which reduces to (3.5.7) when $S = \frac{1}{2}$.

Equation (3.5.8) is sometimes written in the form

$$\chi = \frac{N' \mu_{\mathrm{eff}}^2}{3kT} \tag{3.5.9}$$

where μ_{eff} represents an effective magnetic moment defined by

$$\mu_{\mathrm{eff}}^2 = 4S(S+1)\mu_B^2$$

Equation (3.5.9) is identical in form to the equation for the dipole contribution to the electrical susceptibility, χ_{dip}, for a rarefied gas of molecules such as H_2O, NH_3, HCl, which possess a permanent electric moment μ_{el}:

$$\chi_{\mathrm{dip}} = \frac{N' \mu_{\mathrm{el}}^2}{3kT} . \tag{3.5.10}$$

The theoretical background of the Equations (3.5.9) and (3.5.10), however, is different. In very strong fields or at very low temperatures the electrical polarization is given by $N'\mu_{\mathrm{el}}$, while the magnetization is *not* given by the analogous expression $N'\mu_{\mathrm{eff}} = N'2\sqrt{S(S+1)}\,\mu_B$ but by $N'2S\mu_B$.

Fig. 23 gives the curves of the magnetic moment as a function of H/T for three paramagnetic salts. Let us consider the middle curve, which relates to $FeNH_4(SO_4)_2 . 12H_2O$, i.e. ferric ammonium alum. In this compound it is the ferric ion which is responsible for the magnetic properties. In its outer electron shell there are $2 + 6 + 5$ electrons ($3s^2\ 3p^6\ 3d^5$). Of these, the five 3d electrons are unpaired, i.e. their spins are parallel. All the other electrons are paired and cannot be affected by a magnetic field. The spin quantum number of this ion is thus $S = 5/2$. In agreement with this, the figure shows

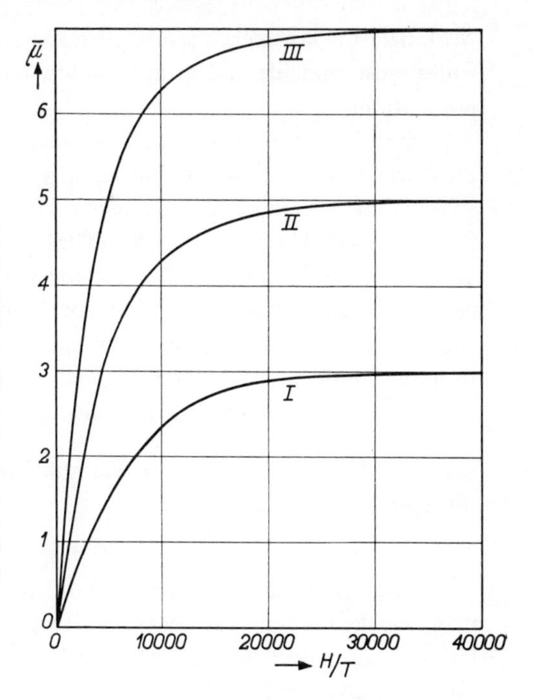

Fig. 23. Average magnetic moment $\bar{\mu}$ in the direction of the field in Bohr magnetons per molecule as a function of H/T in oersteds per degree for (I) chromium potassium alum (S = 3/2), (II) ferric ammonium alum (S = 5/2) and (III) gadolinium sulphate octahydrate (S = 7/2). From measurements by C. J. Gorter, W. J. de Haas and J. van den Handel, Proc. Kon. Ned. Akad. Wetensch. **36,** 158 (1933), later extended by W. E. Henry, Phys. Rev. **88,** 559 (1952).

that the average magnetic moment in the direction of the field at large values of H/T reaches a saturation value of 5 Bohr magnetons per ion. It is also clear from the figure that the mean moment in the direction of the field for small values of H/T, is proportional to H/T. The two other curves refer to chromium potassium alum (S = 3/2) and gadolinium sulphate octahydrate (S = 7/2).

Attaining very low temperatures

By allowing liquid helium to boil under reduced pressure, a temperature not much lower than about 1 °K can be reached. Considerably lower temperatures can be attained by making use of the paramagnetic properties of a salt such as the ferric ammonium alum mentioned above. The effect depends on entropy changes of the system of the electron spins of the ferric ions [1]).

We have already seen that the resultant of five parallel spins, and hence also the resultant spin of a ferric ion, can only assume six different orientations with respect to a magnetic field. If the field is very weak, there is virtually no energetic preference for any one of the six orientations. In other words: in a

[1]) See, e.g. W. J. de Haas, E. C. Wiersma and H. A. Kramers, Physica **1,** 1 (1934).

weak field the six energy levels, corresponding to the possible orientations, will almost coincide and at not too low temperatures they will have equal populations.

Each molecule of the ferric ammonium alum, at small values of H/T, can thus occur in six different, but equally probable, paramagnetic states. A second molecule may occur in one of six states for each of the six states of the first molecule. There are thus 6^2 different states for two molecules and 6^{N_0} different states for a mole (N_0 molecules). The part of the entropy dependent on the paramagnetic properties of the salt, also called the paramagnetic entropy or, more specifically, spin entropy, according to the well-known formula $S = k \ln m$, has the value

$$S_{sp} = k \ln 6^{N_0} = R \ln 6$$

per mole. In general: if S is the quantum number of the resultant spin, then the spin entropy per mole is

$$S_{sp} = R \ln (2S + 1) \qquad (3.5.11)$$

The energy levels of the $(2S + 1)$ different states lie further apart as the magnetic field becomes stronger. In a very strong field and with continuously falling temperature, the higher energy levels will gradually empty and finally, the previously-mentioned saturation condition will be approached, in which nearly all the molecules will be found at the lowest level. In this extreme case, the paramagnetic entropy per mole will be

$$S_{sp} \sim R \ln 1 = 0.$$

A considerable lowering of the temperature can now be attained by demagnetizing, adiabatically and reversibly, the magnetically saturated substance. The effect obtained is completely analogous to the temperature drop which occurs during the adiabatic-reversible expansion of a gas. In both cases a temperature drop can be achieved in two steps. First step: the volume of the gas is reduced in an isothermal-reversible manner, or, the magnetization of the salt is increased in an isothermal-reversible manner. In both cases this leads to a decrease in the entropy (for the case of the gas, see preceding chapters) and thus, since $TdS = dQ_{rev}$, to a flow of heat out of the gas or salt. This is carried off to a heat-sink of constant temperature (in the paramagnetic case, a quantity of liquid helium boiling at reduced pressure). Second step: thermal contact with the surroundings is broken, after which the pressure of the gas is reduced reversibly or the magnetic field is permitted to drop reversibly to zero. In this adiabatic-reversible process, the total entropy of the gas or salt, since $dS = dQ_{rev}/T$, remains constant. In the case of the gas it can therefore be said directly that the increase in entropy due to increasing the volume must be compensated by a reduction of the thermal entropy. This implies a decrease in the temperature.

The case of the paramagnetic salt is not quite so simple. An analogous reasoning would be as follows. When the magnetic field is reduced to zero, the paramagnetic entropy $S_{sp} = R \ln (2S + 1)$ is restored at the cost of the entropy of the atomic vibrations, since the total entropy must remain constant. In fact, the vibrational entropy at an initial temperature of about 1 °K is extremely small, much smaller than the paramagnetic entropy. At first sight it could therefore be expected that during the adiabatic demagnetization the temperature would drop to 0 °K, while furthermore a large portion of the magnetic order would be retained. The latter is indeed the case, but the temperature does not drop to 0 °K. The lowest temperatures which are reached lie between 0.01 and 0.001 °K.

The fact that the temperature does not fall even further is the result of a small, previously neglected, interaction between the magnetic moments and between the moments and the crystal lattice. These interactions have the effect that the energy levels of the $(2S + 1)$ possible states of the paramagnetic ions do not completely coincide even in the absence of an external magnetic field. In accordance with the third law of thermodynamics (see Chapter 2) all the spins therefore have a fixed orientation at absolute zero temperature. In other words: at very low temperatures the spin-disorder disappears spontaneously in a field-free space under the influence of the forces of

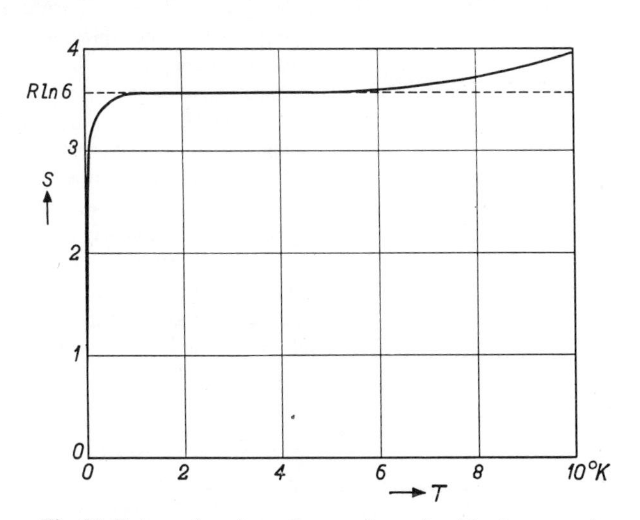

Fig. 24. Entropy in cal per degree of a mole of ferric ammonium alum (in the absence of an external magnetic field) as a function of the temperature.

interaction, so that the entropy is reduced to zero. Fig. 24 shows the curve of the entropy of a mole of ferric ammonium alum (in the absence of an external magnetic field) as a function of the temperature. Between 0.5 °K and 5 °K the entropy has an almost constant value of $R \ln 6$ as mentioned earlier. The vibrational entropy only begins to play a noticeable part above 5 °K.

We know from the beginning of this section that the two energy levels of an electron in a field H are separated by an amount 2 μ_BH. To a first approximation we can assume that the $(2S + 1)$ levels of the paramagnetic ions are separated by equal amounts 2 μ_BH in a strong field H and that there is a much smaller separation $k\theta_0$ in a zero field (see Fig. 25). Furthermore, (see above) we may disregard the entropy of the atomic vibrations at temperatures of 1 °K or lower with respect to the paramagnetic entropy. Consequently, the latter remains virtually constant during the adiabatic "demagnetization" and this means, in the statistical concept of entropy, that the electron spins at the final temperature T_2 (after removal of the field) are distributed between the right-hand group of levels in Fig. 25 in the same way as they were between the middle group at the initial temperature T_1 in the

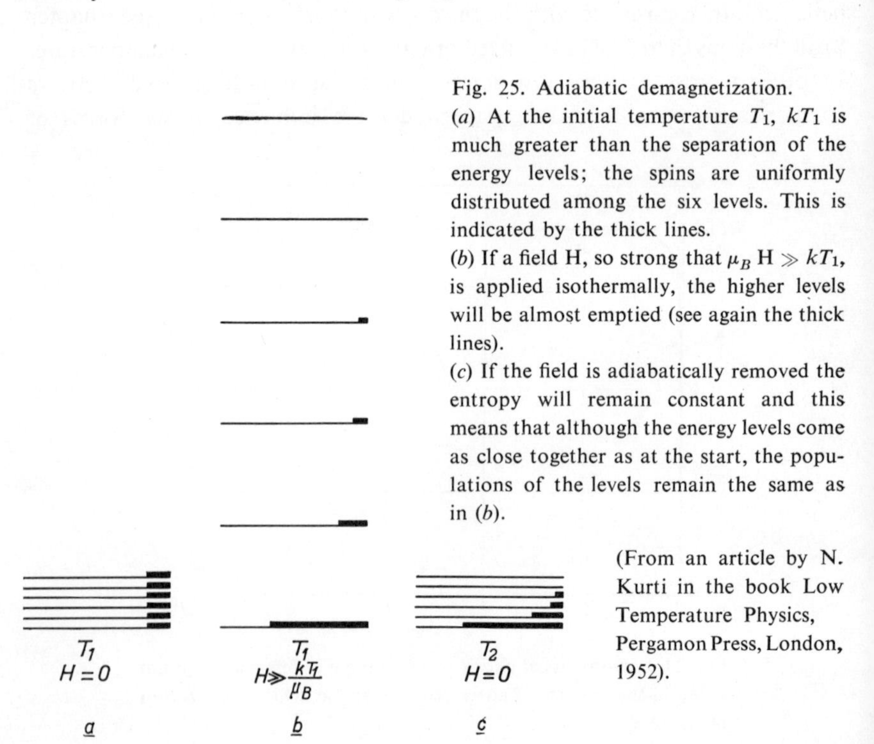

Fig. 25. Adiabatic demagnetization.
(*a*) At the initial temperature T_1, kT_1 is much greater than the separation of the energy levels; the spins are uniformly distributed among the six levels. This is indicated by the thick lines.
(*b*) If a field H, so strong that μ_B H $\gg kT_1$, is applied isothermally, the higher levels will be almost emptied (see again the thick lines).
(*c*) If the field is adiabatically removed the entropy will remain constant and this means that although the energy levels come as close together as at the start, the populations of the levels remain the same as in (*b*).

(From an article by N. Kurti in the book Low Temperature Physics, Pergamon Press, London, 1952).

field H. According to Chapter 2, this distribution is determined by the ratio between the separation of the levels and kT, i.e. by $2\mu_B H/kT_1$ and $k\theta_0/kT_2$. To a rough approximation, thus:

$$\frac{2\mu_B H}{kT_1} \simeq \frac{k\theta_0}{kT_2} \quad \text{or} \quad T_2 \simeq \frac{k\theta_0}{2\mu_B H} T_1 \tag{3.5.12}$$

The final temperature T_2 will thus be lower as the interaction between the magnetic moments (and thus $k\theta_0$) is less, the field H stronger and the initial temperature T_1 lower. The first condition is satisfied by using salts such as $FeNH_4(SO_4)_2.12H_2O$, in which the magnetic ions (Fe^{+++}) are widely separated.

From the foregoing it is evident that below 1 °K the concept of temperature is not so much coupled to the vibrational disorder but to the disorder of the spins in the systems considered. The higher energy levels are already beginning to empty noticeably at a temperature for which kT is equal to the separation of the levels, i.e. at a temperature $T = \theta_0$. This "characteristic temperature" can be regarded as a Curie temperature. It can be determined by approximation from measurements of the specific heat.

From the foregoing it follows that still lower temperatures could be

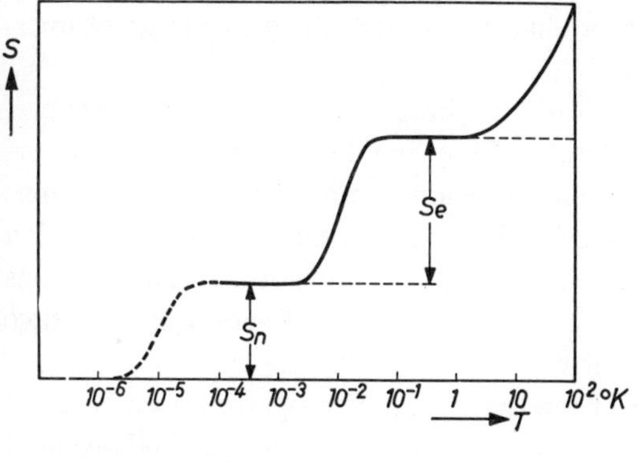

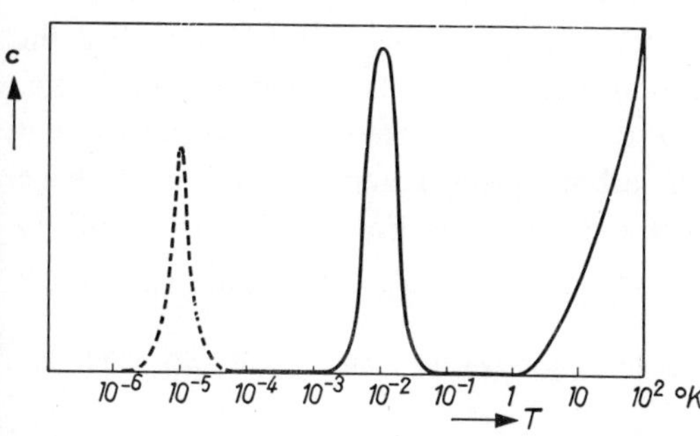

Fig. 26. Schematic representation of the entropy and specific heat as a function of the temperature for a paramagnetic substance, the nuclei of which possess a magnetic moment. S_e is the maximum entropy of the electron spins, S_n that of the nuclear spins. (From an article by F. Simon in the Year Book of the Physical Society, London, 1956).

reached if one had a substance in which the interaction between the elementary magnetic moments was yet weaker than in the substances discussed above. A substance whose atoms have a *nuclear* magnetic moment fulfils this requirement. Nuclear magnetic moments are so much smaller than the electronic magnetic moment, that it is only at temperatures of 10^{-4} to 10^{-6} °K that the weak interaction can give rise to the spontaneous ordering of these moments. This temperature region has indeed recently been reached by the application of nuclear demagnetization [1]).

Fig. 26 shows schematically the temperature dependence of the entropy and the specific heat of a paramagnetic salt whose atoms possess a nuclear magnetic moment.

3.6. INTERSTITIAL ATOMS IN BODY-CENTRED CUBIC METALS

The Snoek effect

In the preceding section we saw that an external *magnetic* field can produce ordering of the directions of the molecular magnetic moments in a paramagnetic salt. In this section we shall discuss the conditions under which an external *mechanical stress* can produce ordering in the distribution of interstitial atoms in a solid metal.

If relatively small atoms, e.g. carbon, nitrogen or oxygen atoms, are present in solution in a solid metal, they will be found in the spaces between the metal atoms. In this case each unit cell retains the normal number of atoms of the metal, but here and there one will contain an extra atom of the foreign element. Solutions of this kind are referred to as interstitial (solid) solutions as distinct from substitutional solutions in which a fraction of the lattice atoms is replaced by the foreign atoms.

During the second world war the Eindhoven investigator Snoek [2]) found that the distribution of foreign atoms in the interstices of a body-centred cubic metal was affected by forces exerted on the metal. He studied this phenomenon mainly in solutions of carbon in the body-centred cubic (b.c.c.) modification of iron (α iron), but the same holds for solutions of nitrogen in α iron and also for solutions of carbon, nitrogen and oxygen in other b.c.c. metals, e.g. vanadium, niobium and tantalum. For the sake of simplicity,

[1]) N. Kurti *et al.*, Nature **178,** 450 (1956); M. V. Hobden and N. Kurti, Phil. Mag. **4,** 1092 (1959).

[2]) J. L. Snoek, New Developments in Ferromagnetic Materials, Elsevier Publ. Co., Amsterdam, 1947.

we shall discuss further only solutions of carbon in α iron.

The carbon atoms dissolved in α iron are to be found in the octahedral spaces in the lattice, the centres of which coincide with the centres of the edges and of the faces of the elementary b.c.c. cell. The nuclei of the six atoms which surround an octahedral space in α iron are situated at the corners of an octahedron, one body diagonal of which is shorter than the other two (Fig. 27). It is immediately obvious from the figure that the distance AB is equal to the lattice constant, while the distances CE and DF are $\sqrt{2}$ times as large. The diameter of a carbon atom is considerably greater than the space available on the line AB. Consequently, each dissolved carbon atom is squeezed between two iron atoms and causes a local uniaxial strain in the direction of one of the three axes of the elementary cell, which we refer to as the x, y and z axes. The interstices can correspondingly be divided into three groups: x, y and z interstices. When the iron crystal is not deformed, the carbon atoms are distributed regularly among these sites. This, however, by no means indicates that they have fixed positions. At not too low temperatures the C atoms jump continually from one to another of the interstices. At room temperature the mean time spent in one site is about one second.

Due to the regular distribution of the carbon atoms, an iron crystal containing carbon is equally deformed in all directions. It is still cubic, but the lattice constant is slightly greater than that of pure iron. The cubic symmetry is destroyed when the carbon atoms are irregularly distributed among the three groups of sites. Later we shall discuss the existence of "martensite", i.e. α iron rich in carbon atoms, all of which are found in similar sites and which consequently produce a tetragonal deformation (dilation parallel to one of the axes, contraction parallel to the other two axes).

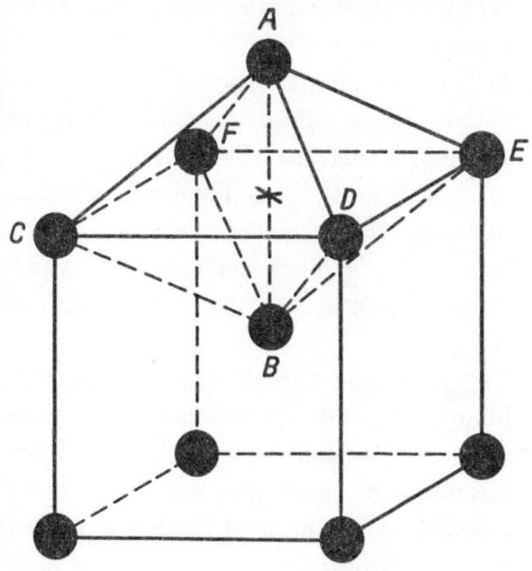

Fig. 27. The cross in the figure indicates the centre of an octahedral space in a cubic body-centred metal. This type of space is not equi-axial: the distances CE and DF are $\sqrt{2}$ times as large as the distance AB.

Since the manner in which the carbon atoms are distributed in an α iron crystal influences the shape of the crystal, it may be expected that a change in this shape as a result of external forces will affect the distribution of the carbon atoms, at least at temperatures at which the carbon atoms are sufficiently mobile. We shall now consider this in more detail.

Let us consider an α iron crystal containing carbon atoms distributed at random among the octahedral spaces and let us suppose that a constant tensile stress σ is suddenly applied in the z direction of the crystal. This tensile stress should be much smaller than that required to produce plastic deformation. The iron crystal will respond immediately to the applied force by an elastic strain ε_1, given by:

$$\varepsilon_1 = \frac{\sigma}{E}, \tag{3.6.1}$$

where E is the modulus of elasticity in a $\langle 100 \rangle$ direction. The asymmetrical deformation caused by a dissolved carbon atom has the effect that, for a constant tensile stress σ, an extra strain ε_2 will occur if carbon atoms move from x and y sites to z sites. Since this extra strain causes the enthalpy $(U - \sigma\varepsilon)$ to decrease, the distribution equilibrium will, indeed, be shifted in this way. This redistribution exhibits a strong analogy to the reorientation of the molecular magnetic moments in a paramagnetic salt discussed in the preceding section. Here, too, complete order is not reached by a long way,

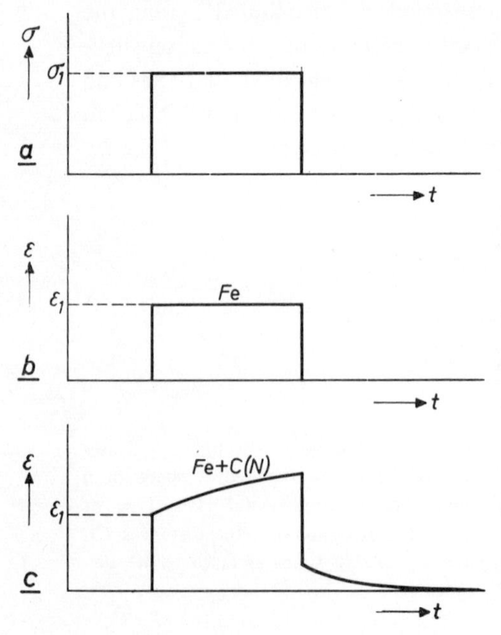

Fig. 28. (*a*) At a given moment a constant tensile stress σ_1 (much smaller than the yield strength) is applied to an iron crystal in the direction of one of the cube axes.
(*b*) The immediate reaction of the metal is a certain elastic strain ε_1, which remains constant in the case of pure iron.
(*c*) If the iron contains dissolved carbon or nitrogen, the instantaneous strain ε_1 is followed by a much smaller strain which gradually tends to a final value (after-effect).

If, later, the tensile stress is removed, the total strain disappears instantly in pure iron; with iron containing C or N, the same instantaneous shortening will occur, to be followed by the gradual disappearance of the extra strain.

since the tendency of the entropy to a maximum value resists it. Just as the magnetic moment of a paramagnetic substance is found, in general, to be proportional to H/T, the degree of order amongst the carbon atoms will be found to be proportional to σ/T (see below).

The adjustment of the distribution equilibrium requires a certain time which depends on the frequency with which the carbon atoms jump and therefore on the temperature. *After* the elastic extension has occurred, thus, the deformation will be seen to increase gradually (see Fig. 28). This is known as *elastic after-effect.*

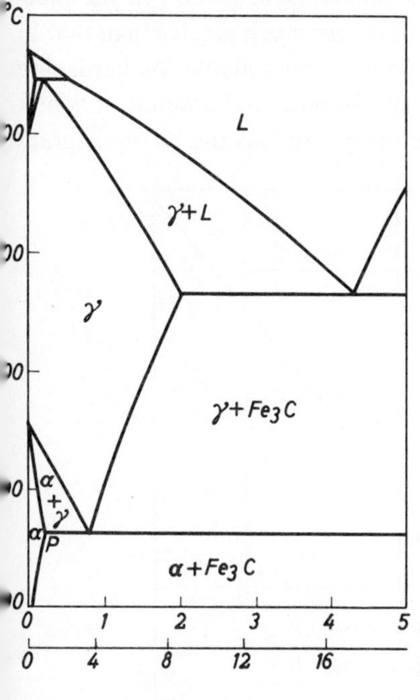

Fig. 29. Part of the iron-carbon phase diagram. For the sake of clarity the point P has been drawn too far to the right: in reality the maximum solubility of C in α Fe, corresponding to this point, only amounts to 0.02% by weight. L is the liquid phase. The phase which is stable above 900 °C is not called β but γ phase; there is a historical reason for this, since originally it was believed, erroneously, that the transition at ca. 780 °C of iron from the ferromagnetic to the paramagnetic state (see Section 3.8) was accompanied by a transition to another phase, which was then called the β phase.

The total strain produced by a tensile stress σ is given, from the above, by

$$\varepsilon = \varepsilon_1 + \varepsilon_2. \tag{3.6.2}$$

in which ε_1 is independent of the time and is called the ideal elastic strain, while ε_2 is dependent on the time and is referred to as the anelastic strain. For the final value of ε_2 one can write:

$$\varepsilon_{2e} = \Delta_r \frac{\sigma}{E}. \tag{3.6.3}$$

where $\Delta_r = \varepsilon_{2e}/\varepsilon_1$, the relaxation strength, is a measure of the magnitude of the anelastic effect. Before showing how the relaxation strength can be calculated thermodynamically, we shall first discuss the martensite already mentioned.

Martensite

Martensite is a tetragonal solid solution of carbon in iron, which plays an essential part in the hardening of steel. This phase is metastable and consequently does not appear in the equilibrium diagram for carbon and iron. In discussing its properties, however, it is convenient to start from the equilibrium diagram (Fig. 29).

As can be seen from the diagram, the solubility of carbon in the face-centred cubic modification of iron (γ iron) is very much greater than that in the body-centred cubic modification (α iron). Steel suitable for hardening contains a lot of carbon, e.g. 1% by weight. In order to harden it, it is first heated to a temperature at which it is in the γ state. As the phase diagram

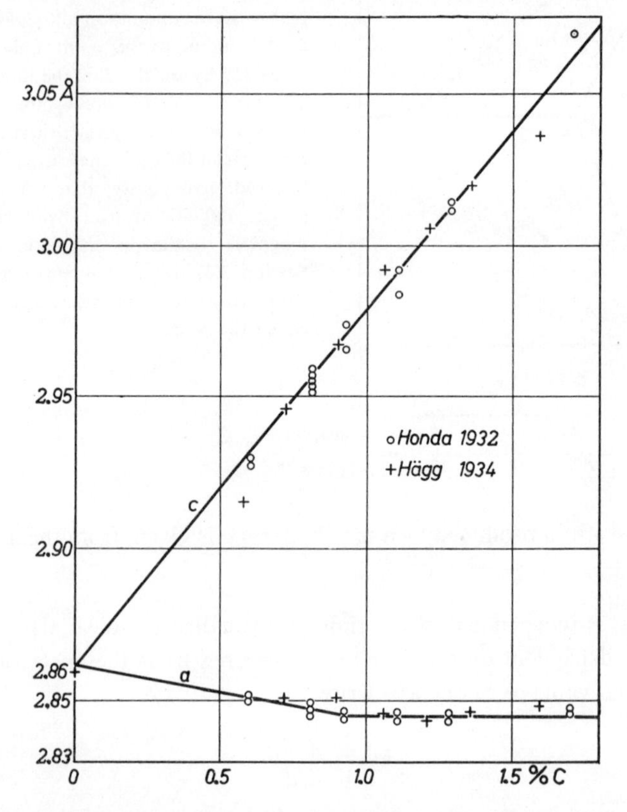

Fig. 30. Lattice constants (a and c) of martensite as a function of the carbon content in weight percent.

shows, this causes all the carbon to go into solution. Next, the metal is cooled very rapidly (quenched), so that the separation, prescribed in the diagram, into α crystals with little dissolved carbon and iron carbide Fe_3C does not take place. Round 200 °C, however, something else now happens. The lattice springs rapidly into the α state and the mechanism of this is such that all the carbon atoms are afterwards found to be accommodated in similar interstices, say z sites. The lattice constant of the α iron is therefore considerably increased in the z direction while, on the other hand, it is decreased in the x and y directions (Fig. 30). This hard, tetragonally deformed α iron is called martensite.

The carbon super-saturation in martensite is substantially greater than in the previously-discussed disordered solutions in which after-effects are usually studied. The latter are obtained by rapid cooling from the α region and thus, according to the phase diagram, contain 0.02 % by weight of carbon at the most. However, while the disordered solutions decompose at room temperature at a perceptible rate, separating out carbide, the ordered martensite solutions are comparatively stable. The explanation is to be found in the asymmetric deformation of the α iron lattice which takes place when a carbon atom is introduced. The effect of this is that the introduction of a second atom in certain neighbouring interstices of the same group (say z) requires a smaller deformation energy than the introduction of the first one. In fact, the first atom has already, to some extent, strained the space where the second is to be introduced. If the super-saturation in a completely ordered solution is so great that the carbon atoms are not widely separated, a mutual interaction thus occurs which has a stabilizing effect. This collective stabilization binds the carbon atoms in martensite to their positions; they almost never jump from one interstice to another. Thus martensite does not exhibit the after-effect discussed above, while there is almost no gradual precipitation of Fe_3C such as occurs in the disordered solutions.

The free enthalpy of elastically strained α iron containing carbon

Let us suppose that a crystal of α iron, in which carbon is present in a disordered solution, is stretched a little in a direction parallel to one of the cube axes. The tensile strain or, alternatively, the tensile stress may be kept constant in this stretching. In the first case it is the free energy, $F = U - TS$, which tends to a minimum value, while in the second case, with which we are concerned in this section, it is the free enthalpy. For 1 cm³ of iron this free enthalpy takes the form (cf. Section 3.3)

$$G = U - TS - \sigma\varepsilon \tag{3.6.4}$$

Immediately after a tensile stress has been applied along the z axis of the carbon-containing iron crystal in question, the concentrations of the carbon atoms in x, y and z sites will still be equal: $C_x = C_y = C_z = C_0$. Here, C_x is defined as the number of occupied x sites, divided by the total number of x sites available. Redistribution changes the concentration C_z to $(C_0 + c)$ and consequently C_x and C_y to $(C_0 - c/2)$. We still have

$$C_x + C_y + C_z = 3\ C_0 \tag{3.6.5}$$

We suppose for a moment that c and σ can be varied independently of each other and consider first the case where c has a constant value. Differentiating (3.6.4) we obtain:

$$\mathrm{d}G = \mathrm{d}U - T\mathrm{d}S - S\mathrm{d}T - \sigma\mathrm{d}\varepsilon - \varepsilon\mathrm{d}\sigma. \tag{3.6.6}$$

If the process is carried out reversibly, then

$$\mathrm{d}U = T\mathrm{d}S + \sigma\mathrm{d}\varepsilon$$

Substituting this expression in (3.6.6), we obtain

$$\mathrm{d}G = -S\mathrm{d}T - \varepsilon\mathrm{d}\sigma \tag{3.6.7}$$

It follows from (3.6.7) that the strain is

$$\varepsilon = -\left(\frac{\partial G}{\partial \sigma}\right)_{T,c} \tag{3.6.8}$$

This strain consists of an elastic part $\varepsilon_1 = \sigma/E$, wheie E represents the (unrelaxed) modulus of elasticity in a $\langle 100 \rangle$ direction, and an anelastic part ε_2, due to the excess of carbon atoms in the z sites. As a result of the linear relationship between the carbon content and the lattice constants of martensite (Fig. 30), we can write the anelastic strain in the form

$$\varepsilon_2 = \beta c. \tag{3.6.9}$$

where β is a constant. For (3.6.8), we can thus write

$$-\left(\frac{\partial G}{\partial \sigma}\right)_{T,c} = \frac{\sigma}{E} + \beta c.$$

or, after integration:

$$G = G_0 - \frac{\sigma^2}{2E} - \beta\sigma c. \tag{3.6.10}$$

The term $-\beta\sigma c$ in this expression is analogous to the term $-\mu_B \mathrm{H} n$ in the preceding section.

For our purpose it is desirable to split the integration constant G_0 into a part that is dependent and a part that is independent of c. The part that is dependent on c consists for by far the greater part of the term $-T\Delta S_m$, which accounts for the change ΔS_m which occurs in the entropy of mixing S_m as a result of the displacement of the carbon atoms. No term, dependent on c, need be introduced to take into account the interaction between carbon atoms, since this interaction can be neglected at the minute concentrations of carbon with which we are dealing in connection with the after-effect (max. about 0.1 atom per cent). (3.6.10) can thus be written:

$$G = G_0' - \frac{\sigma^2}{2E} - \beta\sigma c - T\Delta S_m. \tag{3.6.11}$$

Entropy of mixing for α iron containing carbon

The entropy of mixing for a solution of carbon in α iron is calculated as follows. The number of available x sites is N_0 per mole of iron, N_0 being Avogadro's number. Of these positions C_xN_0 are occupied and $(1 - C_x)N_0$ are unoccupied. The number of ways in which C_xN_0 carbon atoms can be distributed among the N_0 available positions is

$$\frac{N_0!}{(C_xN_0)!\,\{(1 - C_x)N_0\}!}. \tag{3.6.12}$$

Corresponding expressions are valid for the possibilities of distribution among the y and z sites. The total number of ways in which the dissolved carbon atoms can be distributed amongst the x, y and z sites is given, for any arbitrary set of values C_x, C_y and C_z by

$$m = \frac{N_0!}{(C_xN_0)!\,\{(1 - C_x)N_0\}!} \cdot \frac{N_0!}{(C_yN_0)!\,\{(1 - C_y)N_0\}!} \cdot \frac{N_0!}{(C_zN_0)!\,\{(1 - C_z)N_0\}!}. \tag{3.6.13}$$

By applying the equations $S = k \ln m$ and $\ln N! = N \ln N - N$, we can derive the entropy of mixing directly from the above equation, giving per cm³ of iron:

$$S_m = -\frac{kN_0}{V}\{C_x \ln C_x + C_y \ln C_y + C_z \ln C_z + (1 - C_x)\ln(1 - C_x) + \\ + (1 - C_y)\ln(1 - C_y) + (1 - C_z)\ln(1 - C_z)\}. \tag{3.6.14}$$

where V is the molar volume of the iron. When no stress is applied to the iron $C_x = C_y = C_z = C_0$ and thus

$$S_m^0 = -\frac{3R}{V} \{C_0 \ln C_0 + (1 - C_0) \ln (1 - C_0)\} \tag{3.6.15}$$

where the relationship $kN_0 = R$ (2.7.4) has been employed. When the carbon atoms are rearranged under the influence of a tensile stress, we have $C_z = C_0 + c$ and $C_x = C_y = C_0 - c/2$. The entropy of mixing is then

$$S_m = -\frac{R}{V} \Big\{ 2(C_0 - \frac{c}{2}) \ln (C_0 - \frac{c}{2}) + (C_0 + c) \ln (C_0 + c) +$$

$$+2 (1 - C_0 + \frac{c}{2}) \ln (1 - C_0 + \frac{c}{2}) + (1 - C_0 - c) \ln (1 - C_0 - c) \Big\} \tag{3.6.16}$$

Since c is always small with respect to C_0 (see below) we can write

$$\ln (C_0 - \frac{c}{2}) \cong \ln C_0 - \frac{c}{2C_0} - \frac{c^2}{8C_0^2}$$

$$\ln (C_0 + c) \cong \ln C_0 + \frac{c}{C_0} - \frac{c^2}{2C_0^2}$$

$$(1 - C_0 + \frac{c}{2}) \cong (1 - C_0) \cong (1 - C_0 - c)$$

Substituting, (3.6.16) becomes:

$$S_m = S_m^0 - \frac{3Rc^2}{4VC_0} \tag{3.6.17}$$

in which S_m^0 is given by (3.6.15).

The change in the entropy of mixing resulting from the redistribution is therefore given by

$$\Delta S_m = S_m - S_m^0 = -\frac{3Rc^2}{4VC_0} \tag{3.6.18}$$

Magnitude of the anelastic strain and of the relaxation strength

According to (3.6.9), the final value (equilibrium value) of the anelastic strain is given by

$$\varepsilon_{2e} = \beta c_e \tag{3.6.19}$$

and is thus known as soon as β and c_e are known.

The value of the quantity β can be found directly from Fig. 30. As appears from the relationship $\beta = \varepsilon_2/c_z$, β can be deduced from the strain parallel to the z axis which occurs when the carbon concentration in the z sites increases by c_z, at the expense of that in the x and y sites. We find β by dividing this strain by c_z ($c \ll 1$). Now we can imagine that the quantity of carbon corresponding to c_z is transferred from x and y sites to z sites in two operations. It is first removed from the x (y) sites and taken outside the crystal. According to Fig. 30, this causes the strain in the z direction to increase. The contribution of this first step to β is

$$\beta_1 = -\frac{\varepsilon_x}{2c_z} - \frac{\varepsilon_y}{2c_z} = -\frac{\varepsilon_x}{c_z}$$

where ε_x represents the (negative) strain in the x direction of an α iron crystal, due to the presence in z sites of c_z gram-atoms of carbon per gram-atom of iron. The second step consists of returning the carbon atoms to the iron and distributing them among z sites. From Fig. 30, this causes a further increase in the strain in the z direction. The contribution to β of this second step is

$$\beta_2 = \frac{\varepsilon_z}{c_z}$$

where ε_z is the strain in the z direction in an α iron crystal due to the presence in z sites of c_z gram-atoms of carbon per gram-atom of iron. The total value of β is thus

$$\beta = \frac{\varepsilon_z - \varepsilon_x}{c_z} \tag{3.6.20}$$

In this expression, the quantity $\varepsilon_z - \varepsilon_x$ for a particular concentration is given by the vertical distance between lines a and c in Fig. 30. For 0.6% C, $c_z = 0.028$ and $\varepsilon_z - \varepsilon_x = 0.080/2.86 = 0.028$. Thus $\beta = 1.00$.

To calculate ε_{2e} from (3.6.19), we must also know c_e. This can be found if we remember that for any given value of σ carbon atoms will continue to be displaced until the free enthalpy is a minimum. For the equilibrium state, therefore, we have

$$\left(\frac{\partial G}{\partial c}\right)_{\sigma,T} = 0$$

and thus, applying (3.6.11) and (3.6.18):

$$c_e = \frac{2\beta V\sigma C_0}{3RT} \tag{3.6.21}$$

It was stated earlier that c_e/C_0 is much smaller than unity. This statement can now be justified by substituting in Equation (3.6.21) the numerical values of the various quantities, taking for σ the maximum tensile stress which can be employed in the after-effect experiments without causing plastic deformation.

The final value of the anelastic strain is given, from (3.6.19) and (3.6.21), as

$$\varepsilon_{2e} = \frac{2\beta^2 V \sigma C_0}{3RT}. \tag{3.6.22}$$

while the relaxation strength, according to (3.6.3), is given by

$$\Delta_r = \frac{2\beta^2 V C_0 E}{3RT}. \tag{3.6.23}$$

Substituting (3.6.20) this becomes

$$\Delta_r = \frac{2VC_0E}{3RT}\left(\frac{\varepsilon_z - \varepsilon_x}{c_z}\right)^2. \tag{3.6.24}$$

The experimentally determined value of the relaxation strength agrees most satisfactorily with the value calculated from (3.6.24). According to the equation, Δ_r is directly proportional to the carbon concentration and inversely proportional to the absolute temperature. This, too, is confirmed by experiment.

If the tensile stress is applied, not in a ⟨100⟩ direction, but in some arbitrary direction, then (3.6.24) becomes [1]:

$$\Delta_r = \frac{2VC_tE}{9RT}\left(\frac{\varepsilon_z - \varepsilon_x}{c_z}\right)^2 \left\{1 - 3(\alpha_1^2\alpha_2^2 + \alpha_2^2\alpha_3^2 + \alpha_3^2\alpha_1^2)\right\}. \tag{3.6.25}$$

Here α_1, α_2, α_3 are the direction cosines of the applied stress with respect to the cubic axes of the crystal. In the equation, C_0 is replaced by $C_t/3$, where C_t is the total concentration of the dissolved carbon. E is the modulus of elasticity in the direction α_1, α_2, α_3.

Equation (3.6.25) predicts that the Snoek effect must be absent if the tensile stress is applied in a ⟨111⟩ direction, which can also be predicted, without the help of the equation, from the fact that a ⟨111⟩ direction is symmetrical with respect to the three cubic axes. The disappearance of the effect in a ⟨111⟩ direction has been confirmed experimentally by Dijkstra.

---

[1]) D. Polder, Philips Res. Rep. **1**, 5 (1945); J. Smit and H. G. van Bueren, Philips Res. Rep. **9**, 460 (1954).

3.7. SUBSTITUTIONAL ALLOYS

Energy of mixing and entropy of mixing in a disordered alloy

Metal alloys play an important part both in technology and in our daily lives. The previously discussed interstitial alloys of iron and carbon are of primary importance. Besides these, innumerable substitutional alloys (see Section 3.6) are also employed. An enormous number of these alloys can be produced by combining the many metallic elements in various proportions to form binary, ternary, etc. alloys.

The central problem in the metallurgy of alloys is that of the miscibility of the various metals, in particular the miscibility in the solid state. If we restrict ourselves to alloys of two metals (binary alloys), we find that most pairs of metals form a homogeneous mixture in any proportion in the *liquid* state, while in the solid state the miscibility is usually limited. Miscibility in all proportions in the *solid* state is only possible if both metals have the same crystal structure and if their atoms show only small differences in size and electron configuration.

Homogeneous mixing of two elements is accompanied by an increase in entropy, i.e. it is encouraged by the entropy effect. On the other hand, the energy may either increase or decrease during mixing, i.e. the energy-effect may either oppose or promote mixing.

In order to obtain a mathematical expression for the energy of mixing of a binary alloy, we start from two assumptions. In the first place, we assume that the distribution of the A and B atoms in the lattice positions is entirely random. Strictly speaking this is only true when the energy of mixing is zero. If the energy of mixing is relatively small (see below), the distribution will differ but little from complete disorder. In the second place we assume that the internal energy at absolute zero temperature can be written as the sum of the binding energies of nearest neighbours. The terms of this sum will thus contain only the three interaction energies E_{AA}, E_{BB} and E_{AB} between neighbouring pairs A—A, B—B and A—B.

We would ultimately arrive at the same Equation (3.7.3) — still to be derived — if we also took into account the interaction between *more remote* pairs. The energies E_{AA}, E_{BB} and E_{AB} would then represent sums of interaction energies between pairs A-A, B-B or A-B. The formula would only change fundamentally if we were to consider the interaction between e.g. groups of three or four atoms.

In order to find the sum of the interaction energies of nearest neighbours we must first know the numbers of the three types of neighbour bonds. If the molecular fraction of B in the alloy is denoted by the symbol x, the first assumption means that the chance of finding a B atom in an arbitrary lattice point is given by the value of x. The chance of finding an A atom there is given, consequently, by $(1-x)$. The probability of finding the combination $A—A$ in two adjacent positions is given by $(1-x)^2$, the probability for $B—B$ by x^2 and the probability for either $A—B$ or $B—A$ by $2x(1-x)$. The total number of neighbour bonds between the N atoms in the alloy is $Nz/2$ if z is the number of nearest neighbours for each atom ($z = 8$ for the body-centred cubic structure and $z = 12$ for the close-packed cubic and hexagonal structures). The $Nz/2$ bonds in the case under consideration are distributed as follows:

$$\frac{Nz}{2}(1-x)^2 \quad A—A \text{ bonds}$$

$$\frac{Nz}{2}x^2 \quad B—B \text{ bonds}$$

$$Nzx(1-x) \quad A—B \text{ bonds}$$

The energy of the alloy at 0 °K is therefore given by

$$U_{AB} = \frac{NzE_{AA}}{2}(1-x)^2 + \frac{NzE_{BB}}{2}x^2 + NzE_{AB}x(1-x) \qquad (3.7.1)$$

For pure A ($x = 0$) this expression gives:

$$U_{AA} = \frac{NzE_{AA}}{2}$$

For pure B ($x = 1$):

$$U_{BB} = \frac{NzE_{BB}}{2}$$

For a *heterogeneous* mixture of the pure metals, the internal energy is therefore given by

$$\frac{NzE_{AA}}{2}(1-x) + \frac{NzE_{BB}}{2}x \qquad (3.7.2)$$

The energy of mixing (or integral heat of solution) at 0 °K is given by the difference between (3.7.1) and (3.7.2):

$$\Delta U = x(1-x)Nz\left\{E_{AB} - \frac{E_{AA}+E_{BB}}{2}\right\} \qquad (3.7.3)$$

In general, the vibrational energy will not change much during mixing, so that (3.7.3) can also be used at higher temperatures.

Since $x(1-x)Nz$ is always positive, the sign of ΔU is determined by that of $E_{AB} - \frac{1}{2}(E_{AA} + E_{BB})$.

The entropy of mixing ΔS of the alloys under consideration follows directly from the number of ways m in which Nx atoms of B and $N(1-x)$ atoms of A can be distributed over the N available lattice sites:

$$m = \frac{N!}{(Nx)!\,\{N(1-x)\}!} \tag{3.7.4}$$

$$\Delta S = -Nk\,\{x \ln x + (1-x)\ln(1-x)\} \tag{3.7.5}$$

Strictly speaking, this expression only gives the entropy of mixing at 0 °K. If we assume that the vibrational entropy (see Chapter 2) alters but little or not at all on mixing, then the entropy of mixing will be given by (3.7.5) at other temperatures as well.

Mixing, phase separation and precipitation

The binding energies occurring in (3.7.3) are all negative: Heat is liberated on isothermal condensation from the gas phase. If E_{AB} is equal to the mean of E_{AA} and E_{BB} (which is never exactly the case), then $\Delta U = 0$. As far as energy is concerned there will be no preference for any of the three bonds and the entropy effect will result in a homogeneous mixture in the equilibrium state. The copper-nickel system, both in the liquid and the solid state, corresponds closely to this condition; the phase diagram has the simple form shown in Fig. 31.

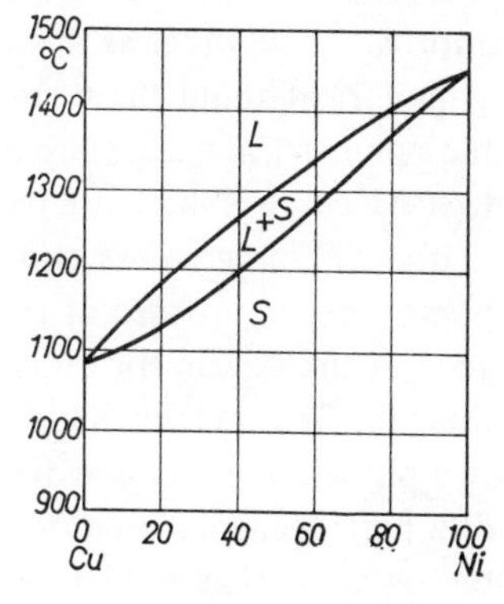

Fig. 31. Phase diagram for the system copper-nickel. These metals form an uninterrupted series of solutions both in the liquid and the solid state. The two-phase region ($L + S$) would be narrower if the energies of mixing in the two phases were exactly equal to zero. The concentrations on the horizontal axis are given in atom percent.

If E_{AB} is greater (less strongly negative) than the mean of E_{AA} and E_{BB}, then ΔU is positive. Identical atoms then attract each other more strongly than do dissimilar atoms. Isothermal mixing causes a rise in internal energy, i.e. heat is absorbed from the surroundings. The entropy effect promotes mixing, the energy effect favours separation. At low temperatures the latter predominates and the system consists of two phases: almost pure A and almost pure B. The entropy effect, however, results in increasing mutual solubility at rising temperatures. For a system which satisfies (3.7.3) and (3.7.5), the mutual solubilities can be calculated as a function of the temperature.

From (3.7.3) and (3.7.5) it follows that the increase in the free energy due to homogeneous mixing is given by

$$\Delta F = NCx(1-x) + NkT\{x \ln x + (1-x)\ln(1-x)\} \tag{3.7.6}$$

where

$$C = z\left\{E_{AB} - \frac{E_{AA}+E_{BB}}{2}\right\}. \tag{3.7.7}$$

It will be obvious that (3.7.3), (3.7.5) and (3.7.6) are only justified for small values of C/kT. In fact, if C differs from zero a certain degree of order will be established, causing the entropy of mixing to become smaller than is calculated for a random distribution, while the energy of mixing also changes. Furthermore, strictly speaking (3.7.6) is only valid for a constant value of the volume. If the pressure is kept constant, ΔG tends to a minimum, so that a further term $p\Delta V$ must be taken into account. This, however, can generally be neglected.

Fig. 32 shows ΔF according to Equation (3.7.6) as a function of x for six values of the ratio kT/C. The curve for $T = 0$ ($kT/C = 0$) is identical with the ΔU curve.

Over a wide range of temperatures the ΔF curves exhibit two minima, which approach each other as the temperature rises. The ΔF curves are completely symmetrical about the vertical axis $x = 0.5$, which is obviously caused by the symmetrical occurrence of x and $(1-x)$ in (3.7.6). The double tangent PQ to a ΔF curve is therefore parallel to the x axis.

If a ΔF curve shows two minima (or rather, two points of inflection) a homogeneous mixture of the two components will not be the most stable state of the system for all concentrations. For a whole range of concentrations the free energy will decrease if the homogeneous mixture splits up into two phases. This is immediately clear if one remembers that the free energy of a heterogeneous mixture of two phases is the sum of the free energies of the separate phases. If, in Fig. 33, a mixture of $(1-x)$ moles of A and x moles of B is split into two homogeneous phases with B concentrations x_1 and x_2 and if the free energies of these phases are f_1 and f_2 per mole, the free energy (per mole) f of the heterogeneous mixture will be given by a point on the straight line f_1f_2. The quantities (in mole) of the two phases are given by the well-known lever rule

$$m_1 : m_2 = (x_2 - x) : (x - x_1)$$

In this particular case

$$m_1 + m_2 = 1$$

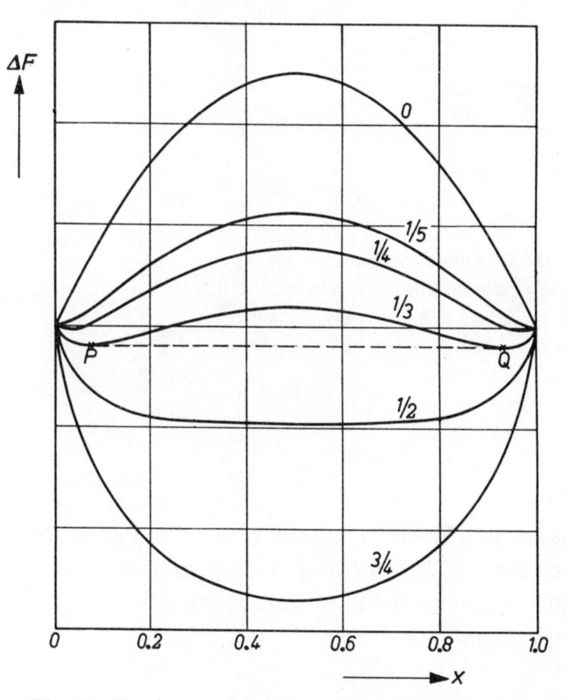

Fig. 32. Free energy of mixing, ΔF, according to Equation (3.7.6), as a function of the mixing ratio x for six values of kT/C. Phase separation occurs at low temperatures since this is energetically more favourable. At any temperature the compositions of the two coexistent phases are given by the points P and Q at which the double tangent (broken line) touches the ΔF curve.

If we consider an arbitrary ΔF curve with two points of inflection (Fig. 34), we see that the free energy of a homogeneous mixture of composition x (indicated by point T) can be reduced to V if the mixture splits into two phases whose compositions are given by the points R and S. The greatest possible reduction of the free energy (to the point W) is obtained if the homogeneous mixture separates into two phases of compositions x_1 and x_2. These compositions do *not* correspond to the two minima of the ΔF curve, which have no physical significance, but to the points P and Q at which the double tangent touches the curve. Only in the case of the symmetrical curve in Fig. 32 do the points P and Q coincide with the minima.

For all the compositions lying between P and Q, a heterogeneous mixture of two phases with compositions given by P and Q is thus more stable than

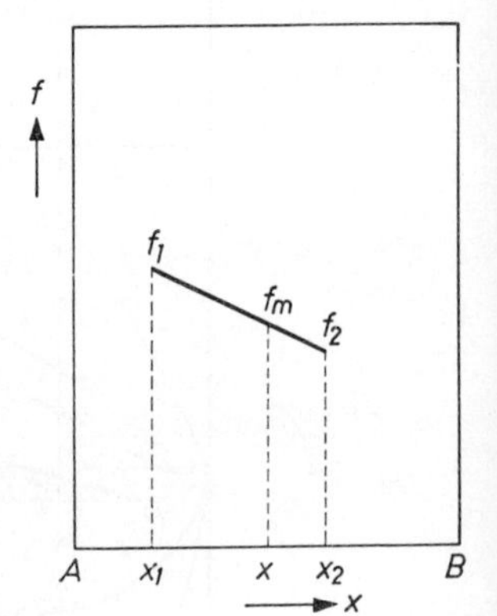

Fig. 33. If 1 mole of a solution of composition x separates into two solutions of compositions x_1 and x_2, the free energies of which are f_1 and f_2 respectively per mole, the free energy of the whole will be given by the point f_m on the straight line f_1f_2.

a homogeneous mixture and also more stable than any other heterogeneous mixture. For points to the left of P and to the right of Q, on the other hand, the homogeneous mixture is the most stable state. The points P and Q thus give the required solubilities.

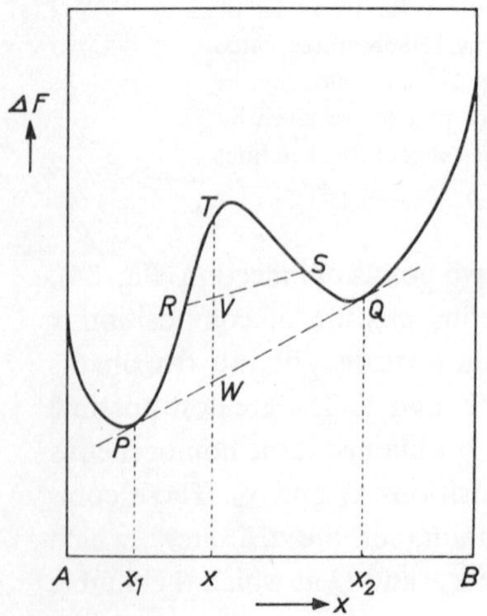

Fig. 34. Arbitrary ΔF-curve with two points of inflection.

For the symmetrical case in Fig. 32, we can find the positions of these points by equating the first derivative of (3.7.6) to zero:

$$\frac{\mathrm{d}(\Delta F)}{\mathrm{d}x} = NC(1-2x) + NkT\{\ln x - \ln(1-x)\} = 0 \qquad (3.7.8)$$

From this it follows that

$$\frac{x}{1-x} = e^{-C(1-2x)/kT} \qquad (3.7.9)$$

In order to construct the curve giving solubility as a function of temperature, it is more convenient to write the relationship between x and T in the form

$$T = \frac{C}{k}\,\frac{(1-2x)}{\ln\{(1-x)/x\}} \qquad (3.7.10)$$

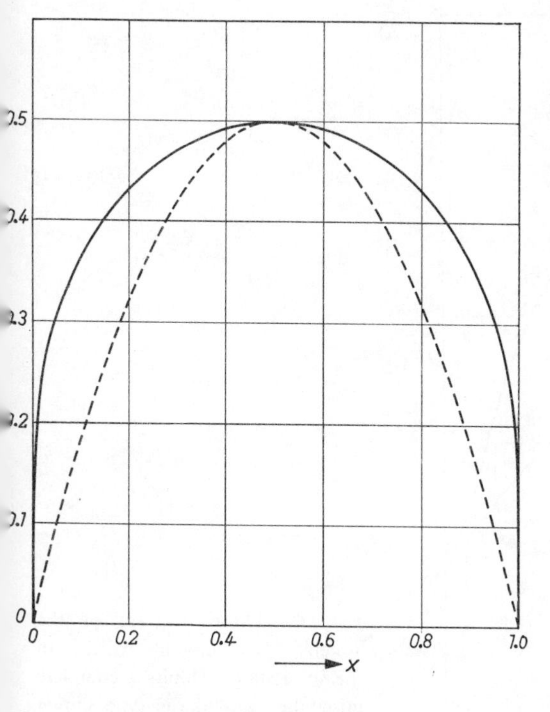

Fig. 35. The full line is the solubility curve $\mathrm{d}(\Delta F)/\mathrm{d}x = 0$ according to (3.7.6) and (3.7.7). It gives the relationship between the compositions (P and Q in Fig. 32) of the co-existent phases α_1, α_2 and kT/C. The broken line is the spinodal curve $\mathrm{d}^2(\Delta F)/\mathrm{d}x^2 = 0$.

Fig. 35 gives the solubility curve corresponding to (3.7.10).

It will be seen that above a certain temperature complete miscibility occurs. This "critical" temperature is found as follows. The ΔF curve with two minima will have two points of inflection. At these points

$$\frac{d^2(\Delta F)}{dx^2} = -2NC + NkT\left(\frac{1}{x} + \frac{1}{1-x}\right) = 0$$

or

$$x(1-x) = \frac{kT}{2C}. \tag{3.7.11}$$

The dotted curve in Fig. 35 represents this function graphically. Curves of this sort, for which $d^2(\Delta F)/dx^2 = 0$, are often referred to as spinodal curves. It can be seen that the points of inflection approach one another as the temperature rises, finally coinciding at $x = \frac{1}{2}$ when the critical temperature, T_C, is reached. From (3.7.11) it follows that

$$T_C = \frac{C}{2k} \tag{3.7.12}$$

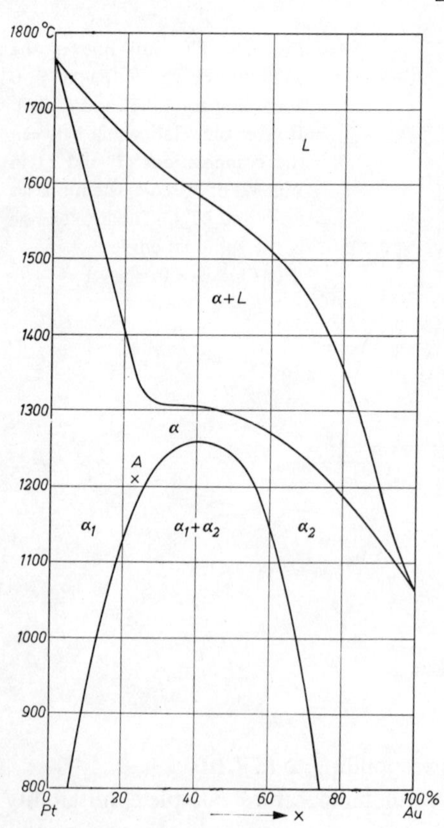

Fig. 36. Phase diagram of the system platinum-gold which, in the solid state, exhibits a complete miscibility gap of the type shown in Fig. 35; α, α_1 and α_2 indicate the solid homogeneous phases; L refers to the liquid phase.

Actual solubility curves never have the symmetrical form of Fig. 35. The main reason for this is that it is never strictly true that the energy may be taken as the sum of interactions between pairs only. The energy of mixing is therefore never given accurately by (3.7.3). In fact the *entropy* of mixing may also deviate considerably from the Gibbs entropy of mixing given by (3.7.5). Some liquid binary alloys (e.g. some K—Pb alloys) even show negative values of ΔS, although (3.7.5) only permits positive values.

A complete phase-separation curve, in the sense described above, occurs for example in the system platinum-gold (see Fig. 36). In most cases of binary systems with a positive value of ΔU, however, melting phenomena appear long before the critical mixing temperature is reached. An example of this is the system silver-copper (Fig. 37). Not every other eutectic system of the type in Fig. 37 possesses a critical mixing temperature. In order to be able to speak of this temperature both components must have the same crystal structure.

The decreasing solubility at decreasing temperature, which occurs in systems of the type discussed here, is of commercial importance in the har-

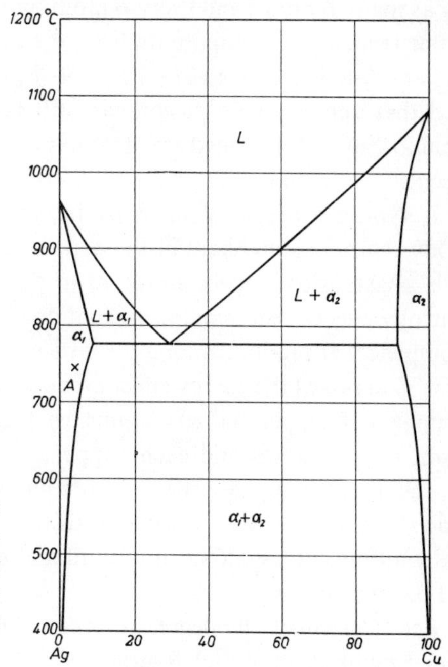

Fig. 37. Phase diagram of the system silver-copper. Melting phenomena occur here long before the critical mixing temperature is reached. The concentrations on the horizontal axis are given in weight percent.

dening of alloys by "precipitation". This is done by rapidly cooling an alloy from the region in which it is homogeneous (e.g. from point A in Fig. 36 or

37) to a temperature at which it is heterogeneous in the equilibrium state and at which virtually no diffusion takes place. The rapid cooling prevents the separation into two phases which is thermodynamically required and a super-saturated homogeneous solution is obtained. By subsequently heating the alloy to a suitable, intermediate temperature, incipient precipitation is induced, accompanied by a considerable increase in hardness.

Ordering

If E_{AB} in (3.7.3) is less (more negative) than the mean of E_{AA} and E_{BB}, ΔU will be negative. In that case atoms will be more strongly attracted by unlike than by like atoms. Isothermal mixing of A and B liberates heat. Both the tendency of entropy to a maximum and the tendency of energy to a minimum promote mixing. When the mixing has taken place, however, the "struggle" between entropy and energy at once manifests itself in the mixture formed. Maximum entropy corresponds to a completely random distribution of the atoms of A and B among the lattice sites, minimum energy, on the other hand, to a distribution in which every atom of A is surrounded by as many B atoms and every B atom by as many A atoms as possible. This latter tendency will impose itself on the system at temperatures so low that the absolute value of $E_{AB} - \frac{1}{2}(E_{AA} + E_{BB})$ is large compared to kT, provided that appreciable diffusion can still occur at these temperatures. If the composition of the mixture is suitably chosen, long-range ordering will occur.

A typical example of an alloy in which ordering occurs is β brass, a body-centred cubic phase in the copper-zinc system. The stability range of this phase extends from about 40 to 50% zinc. At high temperatures the entropy effect predominates, so that copper and zinc atoms are distributed completely at random among the lattice sites. At lower temperatures (below 460 °C approx.) the energy effect predominates. In the simple case of equal numbers of copper and zinc atoms, each zinc atom is finally surrounded by eight copper atoms and each copper atom by eight zinc atoms (Fig. 38). In the disordered phase there is only one kind of site. In the ordered phase known as β' brass it is possible to distinguish between a sites where practically all the copper atoms are found and b sites accommodating practically all the zinc atoms.

Let us calculate the degree of order as a function of the temperature for equal numbers of A and B atoms, situated in the a and b sites of a body-centred cubic lattice. The fraction of the a sites occupied by A atoms will be indicated by f. When $f = 1$, the alloy is completely ordered, when $f = \frac{1}{2}$ it is completely disordered. When $f = 0$ the alloy is again fully ordered, but the

A and B atoms have changed places with respect to the distribution characterized by $f = 1$. Of course, both distributions are identical.

A parameter which is often chosen to indicate the degree of order is

$$\omega = 2f - 1$$

a quantity which takes the value zero for complete disorder and 1 or -1 for complete order. In our case it is more convenient to indicate the degree of order directly by f. Like ω it only gives the degree of long-range order: it gives information about the distribution of the atoms among the two sub-lattices (the a and b sites) without saying anything at all about the immediate neighbourhood of an atom.

If there are N atoms in all, thus $N/2$ of each sort, and N lattice points, also $N/2$ of each sort, then the number of A atoms in a sites is $fN/2$ and in b sites $(1-f)N/2$. The number of B atoms in a sites is therefore $(1-f)N/2$ and in b sites $fN/2$.

In order to be able to write the energy as the sum of binding energies of nearest neighbours we must know how many neighbour pairs A-A, B-B and

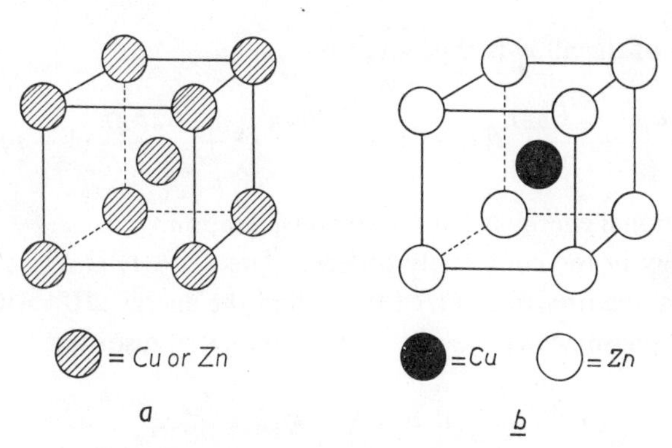

Fig. 38. The unit cell of brass with the composition 50 atomic % Cu + 50 atomic % Zn.
(a) At high temperatures the entropy effect predominates and the atoms are distributed at random on the lattice sites (β brass).
(b) At low temperatures the energy effect prevails and the atoms are ordered (β' brass).

A-B there are. All the eight nearest neighbours of an atom in the sub-lattice a are situated in sub-lattice b. The number of A-A bonds is therefore given by eight times the number of A atoms in a sites, $8fN/2$, multiplied by the fraction of the b sites which are occupied by A atoms, i.e. by $(1-f)$:

$$N_{AA} = \frac{Nz}{2} f(1-f) \tag{3.7.13}$$

where the co-ordination number 8 has been replaced by the more general symbol z. Equation (3.7.13) involves the assumption that the distribution of atoms in one sub-lattice is completely independent of that in the other sub-lattice. This assumption would mean that the nearest neighbours of an A atom in the a sub-lattice do not include a greater percentage of B than is present in the whole b sub-lattice. Here, as in the above discussion on separation, no account is taken of short-range order. This is characteristic of the theory of Bragg and Williams [1]), which is given here in a slightly modified form.

In the same way as (3.7.13) was obtained, we get

$$N_{BB} = N_{AA} = \frac{Nz}{2} f(1-f)$$

$$N_{AB} = \frac{Nz}{2} f^2 + \frac{Nz}{2} (1-f)^2.$$

The energy of the alloy is thus given by

$$U = \frac{Nz(E_{AA}+E_{BB})}{2} f(1-f) + \frac{NzE_{AB}}{2} f^2 + \frac{NzE_{AB}}{2} (1-f)^2 \tag{3.7.14}$$

if the vibrational energy is left out of consideration.

The energy in the completely ordered state ($f = 1$) is $NzE_{AB}/2$. By subtracting this quantity from (3.7.14), we find the energy ΔU_w with respect to the ordered phase, which we may call the energy of disorder:

$$\Delta U_w = -f(1-f)\, Nz \left[E_{AB} - \frac{E_{AA}+E_{BB}}{2} \right] \tag{3.7.15}$$

If the vibrational energy depends on the degree of order, (3.7.15) will only give part of the energy change caused by the disruption of the order.

The number of ways of distributing the A and B atoms over the two sub-lattices is given by

$$m = \left[\frac{(N/2)!}{(Nf/2)!\,\{(1-f)N/2\}!} \right]^2 \tag{3.7.16}$$

The entropy ΔS_w with respect to the ordered state is thus

[1]) W. L. Bragg and E. J. Williams, Proc. Roy. Soc. A **145**, 699 (1934).

$$\Delta S_w = k \ln m = -Nk \{f \ln f + (1-f) \ln (1-f)\} \tag{3.7.17}$$

The advantage of the choice of f as order parameter becomes quite clear if we compare Equations (3.7.3) and (3.7.5) with (3.7.15) and (3.7.17). By replacing x by f in the first two, we obtain (with the exception of the sign) the last two. The negative sign in (3.7.15) ensures that ΔU_w, like ΔU, takes positive values. The expression between braces in (3.7.3) is positive, while that in (3.7.15) is negative.

The free energy ΔF_w as a function of f is given, according to the above, by (3.7.6) if x is replaced by f. C in this formula is given in this case by

$$C = -z \left\{ E_{AB} - \frac{E_{AA} + E_{BB}}{2} \right\} \tag{3.7.18}$$

Thus Fig. 32 is also immediately applicable to this particular case of ordering if x is replaced as abscissa by f. In this case either the left- or right-hand side of the figure is sufficient, since points on the curve which lie symmetrically with respect to the vertical axis $f = \frac{1}{2}$, actually correspond to the same state. The double tangents have no physical significance this time. Only the minima are important; they give directly the degree of order at the temperature of the curve. As the temperature rises, the minima move inwards in the direction of $f = \frac{1}{2}$. All order in the alloy has disappeared above the temperature given by (3.7.12). This critical temperature corresponds to the Curie temperature in ferromagnetism (see Section 3.8) and is often referred to, in this particular case too, as the Curie temperature T_C:

$$T_C = \frac{C}{2k} = -\frac{4}{k} \left\{ E_{AB} - \frac{E_{AA} + E_{BB}}{2} \right\} \tag{3.7.19}$$

The left- or right-hand half of the full-drawn curve in Fig. 35 gives the values of the order parameter f as a function of kT/C. For $kT/C > 0.5$ we always have $f = 0.5$ (complete disorder), as can be seen in Fig. 32. For the transition from order to disorder, half of the curve shown by the full line in Fig. 35 is usually given in a somewhat different way, viz. with T/T_C as abscissa and $\omega = 2f - 1$ as ordinate (Fig. 39). Up to $T/T_C = 0.5$ there is little loss of order, after which it decreases more and more rapidly as the temperature approaches the Curie point. The shape of the curve demonstrates the co-operative nature of the transition from order to disorder. As long as the alloy is well ordered, it requires a great deal of energy to put a few atoms into the "wrong" sites, since all the neighbours of the atoms in question resist this movement. However, the greater the deviation from the ordered state, the easier it becomes to create still greater disorder, provided that maximum disorder has not yet been reached.

Due to the disorder created, the internal energy of the alloy is greater than the vibrational energy. This extra energy, according to (3.7.15) and (3.7.18), is given for complete disorder by

$$\Delta U_w = \tfrac{1}{4} NC \tag{3.7.20}$$

The extra energy per mole ($N = N_0$ = Avogadro's number) is thus, from (3.7.19):

$$\Delta U_w = \tfrac{1}{2} RT_C \tag{3.7.21}$$

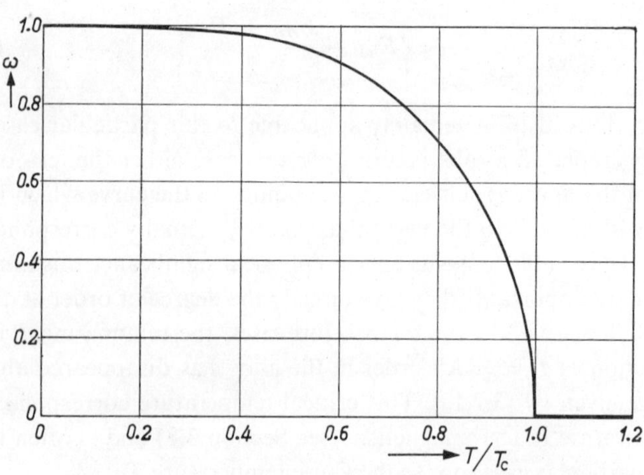

Fig. 39. Value of the order parameter ω in an alloy of the type AB as a function of T/T_C, where T_C is the Curie temperature. The curve has been calculated on the basis of the theory of Bragg and Williams.

The experimentally-found value of T_C is 733 °K for β brass. The theoretical value of ΔU_w is therefore 730 cal per mole. A value of 630 cal per mole is found experimentally [1]. The difference must be attributed to the assumption that the alloy is in complete disorder as soon as the long-range order has disappeared. In fact, even above the Curie temperature there is still short-range order present which is not negligible, so that the internal energy of the alloy is smaller than would be expected from the theory of Bragg and Williams.

The disorder energy makes an extra contribution to the specific heat. This is clearly shown in Fig. 40, which shows the specific heat of the alloy

[1] C. Sykes and H. Wilkinson, J. Inst. Metals **61,** 223 (1937).

50% Zn + 50% Cu as a function of the temperature. According to these measurements the transition from order to disorder takes place over a smaller range of temperatures than would be expected from the theory of Bragg and Williams. In contradiction to this theory, furthermore, the extra specific heat has not disappeared completely above the Curie temperature. This is again due to the short-range order which was neglected in the theory.

Bethe developed a theory of the transition from order to disorder, based on the short-range order. A detailed exposition of this theory and of more recent work is beyond the scope of this discussion [1]).

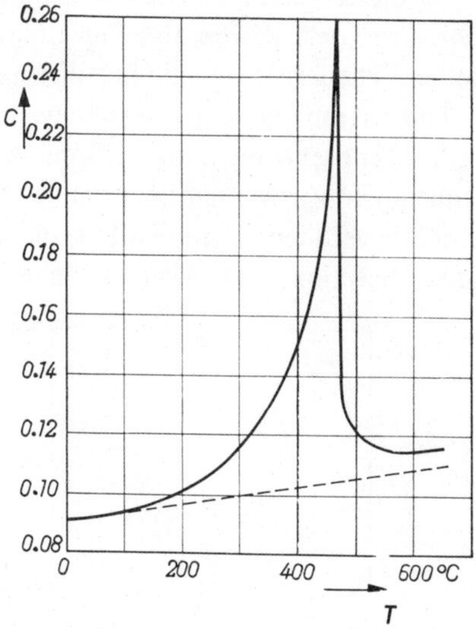

Fig. 40. Specific heat (in cal per gram per degree C) of brass containing approx. 50 atomic % Zn, as a function of the temperature. The broken line relates to a heterogeneous mixture of Cu and Zn.

3.8. FERROMAGNETISM

In Section 3.6 we discussed alloys in which a certain degree of order can be produced by means of external forces. Section 3.7, on the other hand, dealt with alloys in which order occurs under the influence of internal forces. In the same way, besides the paramagnetic substances discussed in Section 3.5, in which spin-order can be produced by the influence of an external magnetic field, there also exist so-called ferromagnetic substances in which this order is produced by internal forces. They may be regarded as paramagnetic materials in which certain electrons exert forces of a quantum-mechanical nature upon each other, which align their spins. As a result of this

[1]) See e.g. F. C. Nix and W. Shockley, Revs. Mod. Phys. **10,** 1 (1938); E. A. Guggenheim, Mixtures, Oxford, 1952; G. S. Rushbrooke, Changements de Phases, Paris, 1952 (p. 177); A. Münster, Statistische Thermodynamik, Berlin, 1956.

"spontaneous magnetization", a ferromagnetic at low temperatures is magnetized to saturation at every point even without an applied field.

The latter is also true for the demagnetized state of a ferromagnetic. In this state the spontaneous magnetization has different directions in different regions of the material, such that the resultant overall magnetization is zero. In stress-free, demagnetized iron the magnetic vector in each separate domain (Weiss domain) points in one of the six directions of the cube edges of the unit cell. In nickel the preferential directions coincide with the eight directions of the body diagonals of the cubic unit cell.

As the temperature rises, the magnitude of the spontaneous magnetization decreases. Above the Curie temperature T_C it has disappeared altogether. The ferromagnetic has then become paramagnetic.

The ferromagnetic-paramagnetic transition is strongly analogous to the order-disorder transition just discussed for binary alloys of the AB type. The tendency towards a minimum value of the energy encourages order, so that at absolute zero temperature all the elementary magnetic moments point in the same direction (Fig. 41). At higher temperatures the striving towards a

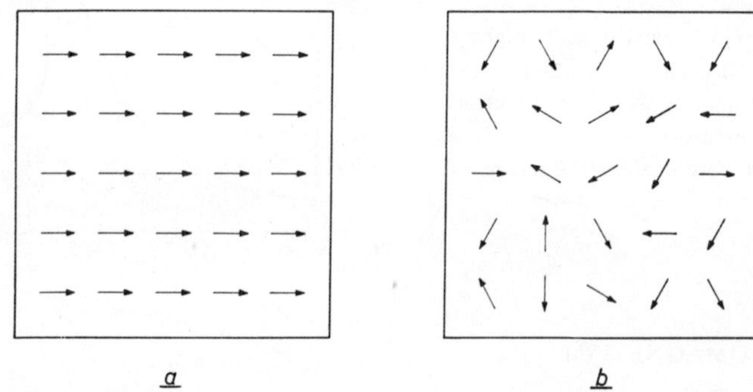

Fig. 41. (*a*) Within a single Weiss domain in a ferromagnetic material the elementary magnetic moments are similarly orientated at low temperatures. (*b*) When the Curie temperature is exceeded, the magnetic order is destroyed; the material becomes paramagnetic.

maximum value of the entropy begins to exert its influence. In the simplest theory, corresponding to the theory of long-range order in the preceding section, it is supposed that as the temperature rises an increasing number of electrons reverse their spin direction. This reduces the spontaneous magnetization. We are once more dealing with a co-operative phenomenon. At low temperatures, when all the spins lie parallel, it requires a great deal of energy to reverse the direction of one spin since all the neighbours resist this action. However, the greater the number of electrons which have already reversed their spin, the easier it becomes for other electrons to do the same. As the temperature rises the spontaneous magnetization thus decreases very slowly at first, then more rapidly until ultimately it approaches zero at the Curie point. This point lies at 780 °C for iron, at 365 °C for nickel and at 1075 °C for cobalt.

In the simple case of N electrons, of which the fraction f have the "right" and thus $(1 - f)$ have the "wrong" spin direction, the extra entropy with respect to complete spin order (the disorder entropy) is given at once by (3.7.17). The extra energy resulting from the spin-disorder is also given by a formula corresponding to that for atomic disorder in the previous section. In fact, we are again dealing with a three-dimensional lattice, in which each site can be occupied in the manner A or the manner B ($\uparrow$ or $\downarrow$). The three interaction energies E_{AA}, E_{BB} and E_{AB} now relate to the pairs $\uparrow\uparrow$, $\downarrow\downarrow$ and $\uparrow\downarrow$, so that $E_{AA} = E_{BB}$. This analogy means that the curve in Fig. 39 gives not only the atomic order in β brass, but also the spin-order in a ferromagnetic in a qualitative way as a function of T/T_C. The ordinate in the latter case is I/I_0, the spontaneous magnetization divided by its maximum value at absolute zero temperature.

The energy necessary to disturb the spin-order, like that needed to disturb the atomic order, makes an extra contribution to the specific heat. This anomaly in the specific heat is clearly shown for nickel in Fig. 42. Contrary to the simple theory, the extra specific heat has not disappeared completely at the Curie point. Here again, the explanation is to be found in the persistence of a certain short-range order after the long-range order has disappeared.

It might be argued that the paramagnetic-ferromagnetic transition, which occurs on a lowering of the temperature, shows a closer resemblance to phase separation than to ordering, for the following reasons. The occurrence of separation or ordering was seen to be determined by the sign of the expression

$$2\,E_{AB} - (E_{AA} + E_{BB}), \tag{3.8.1}$$

which gives the increase in energy for the change

$$AA + BB \rightarrow 2\,AB \tag{3.8.2}$$

in the solid under consideration. The magnetic analogy to this is the change

$$\uparrow\uparrow + \downarrow\downarrow \rightarrow 2\uparrow\downarrow \tag{3.8.3}$$

Phase separation at low temperatures occurs if (3.8.1) is positive, since this means that $AA + BB$ is more stable, energetically, than $2AB$. Also in the ferromagnetic case, (3.8.1) is positive, since similar pairs ($\uparrow\uparrow$ or $\downarrow\downarrow$) are

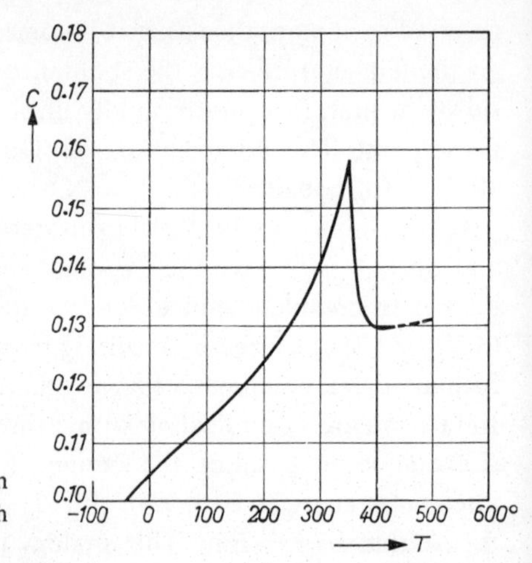

Fig. 42. Specific heat of nickel as a function of the temperature (compare this with Fig. 40).

energetically more stable than dissimilar pairs (↑↓). From this point of view, the paramagnetic-ferromagnetic transition thus corresponds to the separation into two phases as described in Section 3.7. The important difference, however, is that the "two phases" are now identical.

Ordering takes place with falling temperature if (3.8.1) is negative, i.e. if unlike pairs (*AB* or ↑↓) are energetically more stable than like pairs (*AA*, *BB* or ↑↑, ↓↓). The magnetic analogy for β brass would thus be a material in which the interaction between electrons was such that the elementary magnetic moments were antiparallel at low temperatures. Indeed, materials of this sort do exist (Fig. 43). They are called antiferromagnetics.

Just as in the case of ferromagnetics, the number of spins with the "wrong" orientation in an antiferromagnetic increases at rising temperatures. Above a certain temperature (here often called the Néel temperature) the entropy

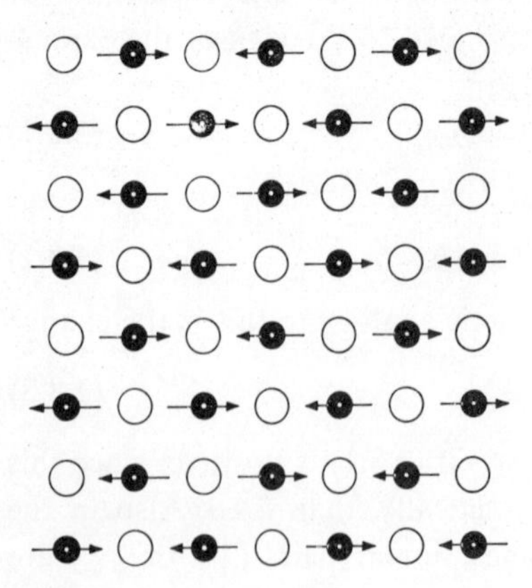

Fig. 43. In MnO the magnetic moments of the Mn ions are anti-parallel. The arrows in the figure indicate their orientations in a (100) plane.

predominates and the material behaves paramagnetically. Below this temperature each spin direction has its own sub-lattice. The transition paramagnetic-antiferromagnetic thus corresponds exactly to the disorder-order transition in Section 3.7.

3.9. MORE ABOUT THE EINSTEIN SOLID

In the previous chapter (see Sections 2.9 to 2.14) the specific heat of an Einstein solid was calculated. The calculation was based on the most probable distribution of the atoms over the available energy levels. The equation already found (2.14.4) can be obtained in a simpler way by proceeding from the *total* number of micro-states and from the tendency of the free energy to a minimum value. This will be shown below.

It will be remembered that Chapter 2 first dealt with a simpler model of a solid than that of Einstein. The identical atoms were supposed to behave as linear harmonic oscillators which, virtually independent of each other, perform their oscillations about fixed centres and can absorb equal energy quanta or phonons of magnitude $h\nu$. The energy absorbed can be divided among the oscillators in a great number of ways. The total number of distributions (micro-states) m is given, according to Equation (2.13.3), by

$$m = \frac{(q + N - 1)!}{q!(N - 1)!}$$

where N is the number of atoms (oscillators) and q the number of quanta absorbed.

In an Einstein solid, as in an actual solid, each atom possesses not one but three degrees of vibrational freedom. This type of solid can therefore be regarded as a system of $3N$ linear oscillators. Since the solids in which we are interested contain at least 10^{19} atoms, unity may safely be neglected with respect to N. For an Einstein solid the above equation thus becomes:

$$m = \frac{(q + 3N)!}{q!\,(3N)!} \tag{3.9.1}$$

The energy of the solid, after absorbing q vibrational quanta, is raised above the zero-point energy by an amount

$$U = qh\nu \tag{3.9.2}$$

while the entropy, $S = k \ln m$, according to (3.9.1) and Stirling's formula in the approximation (2.4.4), is given by

$$S = k\{(q + 3N)\ln(q + 3N) - q\ln q - 3N\ln 3N\} \tag{3.9.3}$$

The Helmholtz free energy, $F = U - TS$, can therefore be written as

$$F(q) = qh\nu - kT\{(q + 3N)\ln(q + 3N) - q\ln q - 3N\ln 3N\} \tag{3.9.4}$$

In the foregoing we have assumed that the number of quanta absorbed was known. In fact, usually it is not the number of quanta which is primarily known, but the temperature to which the crystal is heated by thermal contact with surroundings at some constant temperature. What we wish to know, thus, is the number of quanta q present in an Einstein solid at a given absolute temperature T. If q is known as a function of T it is possible to calculate the specific heat at every temperature.

If the solid were able to submit completely to its "striving" towards a minimum value of the energy, it would absorb no quanta at all. If, on the other hand, it were governed completely by the striving towards maximum entropy, the number of quanta absorbed would continue to increase. The compromise (the equilibrium state) lies, for constant values of the volume and temperature, at the point where the Helmholtz free energy is a minimum, thus where

$$\frac{\mathrm{d}F(q)}{\mathrm{d}q} = 0$$

From this follows, in combination with (3.9.4):

$$h\nu - kT\ln\frac{q + 3N}{q} = 0$$

or, rearranging:

$$q = \frac{3N}{e^{h\nu/kT} - 1} \tag{3.9.5}$$

The heat capacity at constant volume is given by

$$C_v = \frac{\mathrm{d}Q}{\mathrm{d}T} = \frac{\mathrm{d}U}{\mathrm{d}T} = \frac{\mathrm{d}(qh\nu)}{\mathrm{d}T}.$$

Substituting for q from (3.9.5) gives the required relationship:

$$C_v = 3kN\left(\frac{h\nu}{kT}\right)^2 \frac{e^{h\nu/kT}}{(e^{h\nu/kT} - 1)^2}. \tag{3.9.6}$$

which is identical with (2.14.4). This also demonstrates once more and by a different method to that in Section 2.16, that the "statistical" temperature and the absolute temperature are identical. In fact, in (2.14.4) T was the statistical temperature, defined by (2.11.1). In (3.9.6) on the other hand, T originates in the free energy, $F = U - TS$, and in this relationship, according to Section 3.2, T is the classical absolute temperature from Chapter 1.

The entropy of an Einstein solid as a function of N, ν and T is found by substituting (3.9.5) in (3.9.3). If kT is very much greater than $h\nu$, then

$$e^{h\nu/kT} \simeq 1 + h\nu/kT$$

Thus for the entropy we have

$$S \simeq 3\,Nk \ln\frac{kT}{h\nu}\,, \text{ if } kT \gg h\nu \tag{3.9.7}$$

3.10. VACANCIES AND DIFFUSION IN SOLIDS

In many processes which can occur in solids, atomic diffusion plays an important part. In particular, diffusion in metals and alloys has been extensively studied in the last few decades. It has been discovered that the presence of defects in the crystal lattice, especially vacancies or interstitial atoms, is a necessary condition for the occurrence of diffusion.

As we have seen in the discussion of the Einstein solid, the vibrations of the atoms about their equilibrium positions become more violent as the temperature rises. Even before the melting point is reached, a fraction of the atoms will possess sufficient energy to leave their lattice sites altogether. This creation of *vacancies* may start at the surface, some atoms leaving their original positions and occupying new positions on top of the surface layer (Fig. 44). Atoms from deeper layers can subsequently jump into the vacancies created, and so on. We may equally well say that the vacancies originating at the surface diffuse into the interior. In this process new lattice planes are formed on the outside of the crystal.

Even in the state of equilibrium, lattice defects are bound to arise at high temperatures due to the increase of entropy with which this is accompanied. Disregarding the extremely small macroscopic volume changes which occur through the formation of the vacancies, it can be said that at any high temperature T, vacancies will be formed until the Helmholtz free energy, $F = U - TS$, is a minimum. In other words, although the introduction of vacancies causes the internal energy U to rise by an amount ΔU, they will continue to be formed as long as $T\Delta S > \Delta U$.

If we know the energy, ε, required to produce a vacancy we can calculate the equilibrium percentage of vacancies at any temperature. This is done as follows.

If an ideal crystal of N atoms is transformed into a disturbed state with n vacancies, then the accompanying increase in the entropy, $\Delta S = k \ln m$, is given by the number of different ways, m, in which N similar atoms can be placed in a lattice with $(N + n)$ available sites. This number is

$$m = \frac{(N+n)!}{N!n!} \tag{3.10.1}$$

Using Stirling's formula in the approximation (2.4.4), we find

$$\Delta S = k\{(N+n)\ln(N+n) - N\ln N - n\ln n\}$$

As long as the concentration of vacancies is so small that interaction between them can be neglected, the increase in energy is given by $\Delta U = n\varepsilon$ and thus the change in the free energy by

$$\Delta F = \Delta U - T\Delta S = n\varepsilon - kT\{(N+n)\ln(N+n) - N\ln N - n\ln n\} \tag{3.10.2}$$

In the equilibrium state the free energy is a minimum: thus $\partial F/\partial n = 0$, or, from (3.10.2):

$$\varepsilon - kT\{\ln(N+n) - \ln n\} = 0$$

$$\frac{n}{N+n} = e^{-\varepsilon/kT}$$

Since n is negligible with respect to N, this becomes

$$\frac{n}{N} = e^{-\varepsilon/kT} = e^{-E/RT} \tag{3.10.3}$$

where E is the energy required to produce $6 \cdot 10^{23}$ (Avogadro's number) vacancies. According to calculations by Huntington and Seitz [1]) ε has a value of approximately 1 electron volt for copper, i.e. $E \cong 23\,000$ cal per "gram-atom of vacancies". Thus, for 1000 °K, we find:

$$\frac{n}{N} = e^{-23000/2000} = 10^{-5}$$

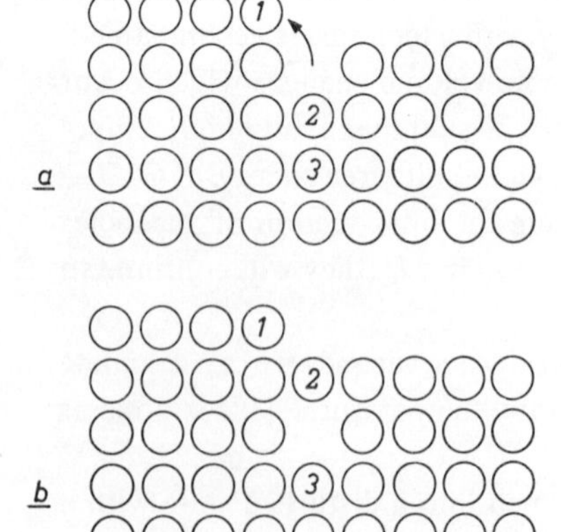

Fig. 44. Vacancies in a solid can originate at the outer surface or at internal surfaces. In (*a*) a vacancy is formed by the displacement of atom 1. In (*b*) the vacancy diffuses inwards by the displacement of other atoms.

---

[1]) H. B. Huntington and F. Seitz, Phys. Rev. **61**, 315 (1942).

In this case, therefore, one in every hundred thousand lattice sites would be vacant. This corresponds to an average distance between neighbouring vacancies of some tens of interatomic distances.

The extra contribution to the internal energy due to the formation of vacancies is accompanied by an extra contribution to the specific heat, given by

$$\frac{\mathrm{d}\Delta U}{\mathrm{d}T} = \frac{\mathrm{d}n\varepsilon}{\mathrm{d}T} = \frac{\mathrm{d}(N\varepsilon e^{-\varepsilon/kT})}{\mathrm{d}T} = R\left(\frac{\varepsilon}{kT}\right)^2 e^{-\varepsilon/kT} \text{ per gram-atom}$$

In the above it is assumed that the increase in entropy caused by the transition from the ideal to the disturbed arrangement depends exclusively on the number of possible distributions m, given by Equation (3.10.1). In reality, the vibration frequencies of the atoms surrounding the vacancies also change, making a further contribution to the entropy. The reason is that an atom adjacent to a vacancy is more weakly bound than an atom which is completely surrounded by similar atoms. Consequently, it will have a lower vibration frequency, at least in the direction of the vacancy. From Equation (3.9.7), the introduction of vacancies not only causes the increase in entropy calculated from (3.10.1), but furthermore an increase in the vibrational entropy. The concentration of the vacancies is therefore greater than that calculated above.

At high temperatures the vacancies move at random through the lattice. A vacancy is displaced by one interatomic distance if an adjacent atom jumps into it. For such a jump an amount of energy is required, at least equal to the activation energy q, which in general will be large with respect to the average energy of the atoms. The fraction of the atoms which possess at least this activation energy q is given by

$$f = e^{-q/kT} = e^{-Q/RT}$$

where $Q = N_0 q$ is the "activation energy per gram-atom". The diffusion coefficient D of the *vacancies* will be approximately proportional to this factor:

$$D = D_0 e^{-Q/RT} \tag{3.10.4}$$

where D_0 is a constant which depends little or not at all on the temperature.

Primarily, however, we are not so much interested in the diffusion of the vacancies as in the diffusion of the atoms of the pure metal, the self-diffusion, or of foreign atoms occupying lattice sites (substitutional atoms).

Self-diffusion can be studied by means of certain radioactive isotopes of the atoms of the pure metal. The coefficient of self-diffusion is smaller than that of the vacancies, since the atoms can only jump at a moment when there happens to be a vacancy beside them. The number of times per second that this occurs is proportional to the diffusion coefficient of the vacancies and to their relative number, thus to (3.10.4) and (3.10.3).

The self-diffusion coefficient, according to this reasoning, will be given by an expression of the form

$$D = D_0' e^{-Q/RT} e^{-E/RT}$$

or

$$D = D_0' e^{-W/RT}, \tag{3.10.5}$$

where

$$W = Q + E \tag{3.10.6}$$

and D_0' is a constant depending little or not at all on the temperature.

The foregoing is also valid for substitutional atoms provided that their properties and size differ so little from the atoms of the solvent that the vacancies show no preference for either type of atom and that the activation energy required for a jump is the same for both.

3.11. ELASTICITY OF RUBBER

The molecules of the substances known as high-polymers take the form of long chains with many hundreds of links, each of which consists of a group of relatively few atoms (monomeric groups). A simple example of this kind of substance is polyethylene, of which the molecular structure is shown in

```
  H  H  H  H  H  H  H          H  H
  |  |  |  |  |  |  |          |  |
—C—C—C—C—C—C—C—          C=C
  |  |  |  |  |  |  |          |  |
  H  H  H  H  H  H  H          H  H
           a                    b
```

Fig. 45. The simplest polymer is polyethylene (a), which is produced by the joining together of ethylene molecules (b).

Fig. 45. The monomeric groups are bound together by real chemical bonds. Nevertheless, the atoms in these molecules often have a certain mobility with respect to each other in the sense that one bond can rotate about another at a constant angle. This is shown schematically for a simple case in

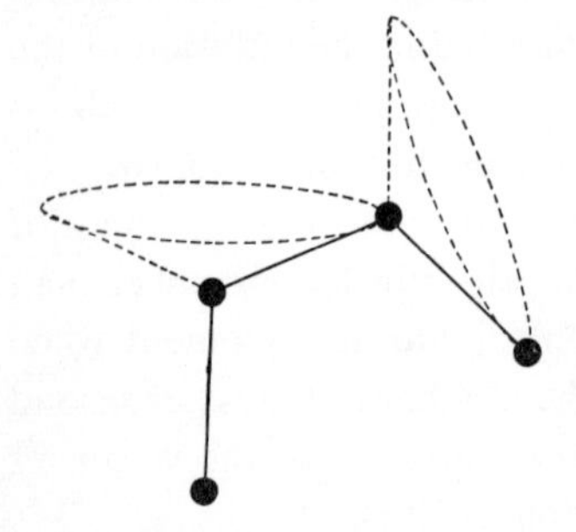

Fig. 46. Polymer molecules often possess considerable flexibility due to the fact that the chemical "bonds" with their constant valency angle, can still take number of different positions with respect to each other.

Fig. 46, in which the black dots represent atoms of the chain. This freedom of rotation enables the polymer molecules to assume an enormous number of different forms. Against the one possibility of the molecule taking the form of a stretched chain, there are innumerable possibilities of the molecule occurring in a crooked or tangled shape (see Fig. 47).

The entropy of a high-polymer, according to the relationship $S = k \ln m$

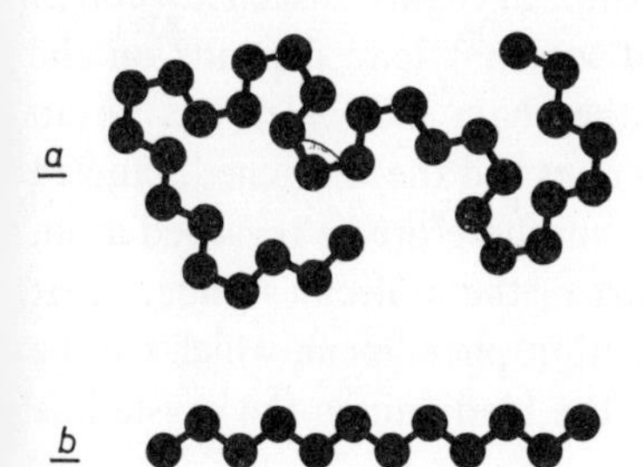

Fig. 47. Part of a molecular chain: (*a*) twisted at random, (*b*) stretched.

(m = number of micro-states), will be greater when the molecules take an irregular form than when they are stretched and lie parallel to each other. In some high-polymers, e.g. cellulose, the latter state is so much more favourable from the point of view of energy than the former, that the tendency towards maximum entropy is overwhelmed by the tendency to a minimum value of the energy and the ordered (stretched) state occurs.

In rubber, the difference in energy between the two states is only very slight, so that the entropy effect prevails and the molecules have a tangled

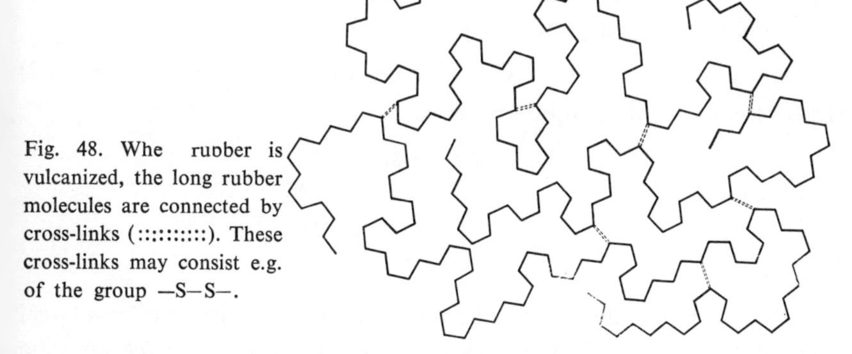

Fig. 48. Whe ruober is vulcanized, the long rubber molecules are connected by cross-links (:::::::::). These cross-links may consist e.g. of the group —S—S—.

form. By applying a tensile stress to a rubber rod, we can stretch the chains more or less, so that a considerable elongation of the whole rod occurs. When the stress is released, the thermal agitation of the chain elements return the chains to the coiled, thus shortened, form. The fact that the rod as a whole also returns to its original shape is due to the presence of cross-links between the polymeric chains. These are produced by the vulcanization of the rubber, which forms a wide-meshed network (Fig. 48). In normal cases this process establishes cross-links between about 1% of the monomeric groups.

Some of the unusual thermal properties of rubber are explained by the above, e.g. the negative coefficient of expansion of highly stretched rubber. The contraction with rising temperature and constant load depends on the increasing thermal motion of the links of the chain molecules, i.e. on an increase in the tendency of the molecules to revert to the shortened, tangled forms. The negative coefficient of expansion can therefore be regarded as an entropy effect. The parallel alignment of part of the molecules under great stress produces, in natural rubber, crystallization phenomena which can be observed by means of X-rays. Removal of the load causes the crystalline parts to "melt" again.

The thermodynamic treatment starts from the first law

$$\mathrm{d}U = \mathrm{d}Q + \mathrm{d}W \tag{3.11.1}$$

where $\mathrm{d}Q$ is the heat supplied to the system and $\mathrm{d}W$ the work performed upon it. If the system consists of a rubber rod of length l, which is stretched to a length $l + \mathrm{d}l$, then (neglecting the very small amount of work required to bring about the small change in volume) $\mathrm{d}W$ is given by

$$\mathrm{d}W = K\mathrm{d}l$$

where K is the tensile force. If the stretching process is carried out reversibly, $\mathrm{d}Q = T\mathrm{d}S$ (S being the entropy of the rubber rod); in that case K is the equilibrium value of the force. Equation (3.11.1) can now be written:

$$\mathrm{d}U = T\mathrm{d}S + K\mathrm{d}l \tag{3.11.2}$$

Thus at constant temperature we have

$$\mathrm{d}(U - TS) = \mathrm{d}F = K\mathrm{d}l$$

and therefore

$$K = \frac{\partial F}{\partial l} = \frac{\partial U}{\partial l} - T\frac{\partial S}{\partial l} \tag{3.11.3}$$

For many sorts of rubber it has been shown experimentally that for a large, constant elongation Δl, K is approximately proportional to the absolute temperature T. This confirms what has been implied above, viz. that $\partial U/\partial l$ (the increase in the internal energy due to the elongation, thus due to the unrolling of the molecular chains) is very small for this class of materials. In other words: the elastic force depends, in this concept, almost entirely on the decrease in entropy on elongation and hardly at all on the accompanying changes in energy. In most solids, e.g. metals, the reverse is true. The latter cool slightly when stretched adiabatically and elastically, while rubber becomes warmer during this process. The heating effect can be explained partly by the fact that (since $\mathrm{d}S = \mathrm{d}Q_{\mathrm{rev}}/T$) the total entropy does not change in this reversible-adiabatic process. The decrease in entropy, corresponding to the order produced by stretching, is compensated by an increase in that portion of the entropy which is caused by the thermal agitation of the molecules (the vibrational entropy); this implies a rise in temperature.

The behaviour of a "perfect" rubber, for which $\partial U/\partial l = 0$ and whose elasticity may be regarded as a pure entropy effect, is completely analogous to that of a perfect gas. For a reversible volume change of any gas

$$\mathrm{d}U = T\mathrm{d}S - p\mathrm{d}V$$

and thus at constant temperature

$$\mathrm{d}(U - TS) = \mathrm{d}F = -p\mathrm{d}V$$

$$p = -\frac{\partial F}{\partial V} = -\frac{\partial U}{\partial V} + T\frac{\partial S}{\partial V}$$

At constant volume the pressure of a gas is found to be approximately proportional to the absolute temperature, i.e. the energy U of a gas is virtually independent of the volume. We speak of a perfect gas when $\partial U/\partial V = 0$. The pressure of such a gas can thus also be described as an entropy effect.

3.12. SOLUTIONS OF POLYMERS

We have already seen that the entropy of mixing of a homogeneous solid alloy, insofar as this depends on the configuration of the atoms, is given by Equation (3.7.5). Strictly speaking, this equation is only valid when the energy of mixing $\Delta U = 0$. If $\Delta U \neq 0$, the formula can still be used as an approximation at those temperatures for which kT is much larger than $E_{AB} - (E_{AA} + E_{BB})/2$.

No reference was made above to possible differences in size between the atoms or molecules to be mixed. This was unnecessary in the case of the *solid* alloys in question, since two metals with greatly different atomic radii show no appreciable mutual solubility in the solid state. The same is true for non-metals.

Liquid homogeneous mixtures, however, often consist of molecules (or atoms) of widely different size. Solutions of polymers are extreme examples of such mixtures. An approximate expression for the entropy of mixing of these solutions will by derived below.

Let us first consider a liquid mixture of molecules A and B, which have the same shape and size and, furthermore, show no preference energetically for similar or dissimilar neighbouring molecules ($\Delta U = 0$). In this type of mixture a molecule A can be replaced by a molecule B without affecting the surroundings of A. It is known from X-ray research on liquids that these surroundings differ but little from those in a crystalline mixture. The arrangement is only slightly less orderly, so that the order does not extend over such a long range as in the crystal. Since we are principally interested in the bonds between neighbours, liquids of the type under consideration can be represented by a quasi-crystalline model (Fig. 49). Equation (3.7.5) can then be

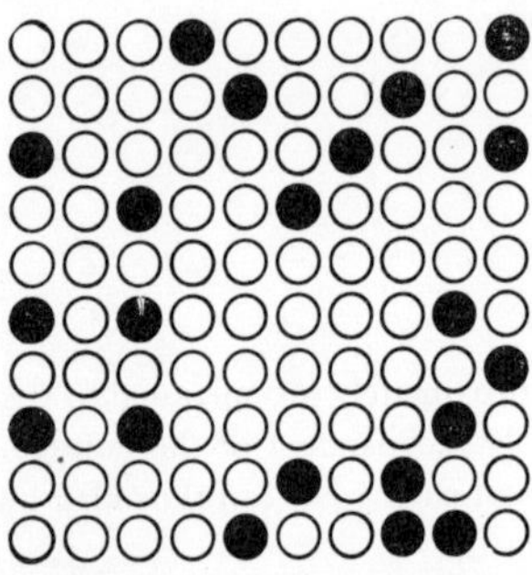

Fig. 49. Schematic representation of a solution of molecules B (black) in a liquid of molecules A (white). Molecules A and B are about the same size and distributed at random over the quasi-lattice.

applied unmodified to this model. The equation depends on the interchangeability of molecules A and B and on the absence of any preference for like or unlike molecules. If the molecules differ considerably in size or shape, the entropy of mixing will no longer be given by (3.7.5), even when the energy of mixing $\Delta U = 0$. For example, direct interchangeability of A and B molecules is out of the question if A molecules are spherical and occupy one space in the diagram in Fig. 49, while B molecules are dumb-bell shaped and occupy two spaces (Fig. 50).

The extreme case, already mentioned, of polymer molecules (macro-mole-

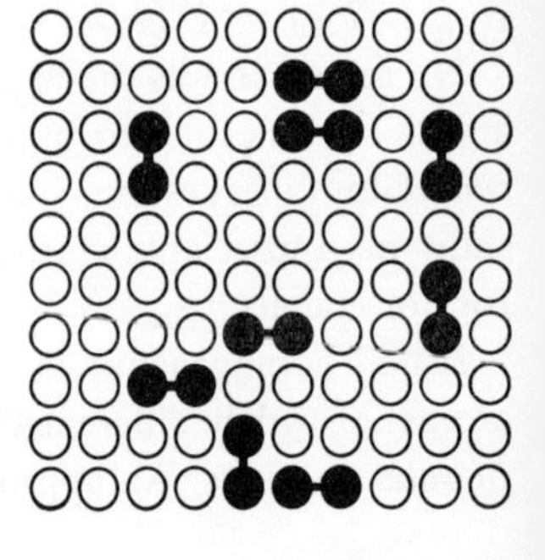

Fig. 50. Schematic representation of a random distribution of dumb-bell shaped molecules B (black) in a liquid of spherical molecules A (white). The dumb-bell molecules occupy two adjacent points in the quasi-lattice.

cules) dissolved in a micro-molecular solvent, can be represented as follows [1]. It is assumed that a polymer behaves *as if* each of its molecules were divided into a large number of mobile segments, each of the same size as a molecule of the solvent. It is further assumed that each segment occupies one space in the quasi-lattice, and that adjacent segments of a chain occupy adjacent spaces (Fig. 51). If the solution contains N_1 molecules of the solvent and N_2 molecules of the polymer, each consisting of p segments, there will be $(N_1 + pN_2)$ spaces in the sense of Figs. 49 to 51. The fractions of the volume occupied by the two components will be given by

$$\phi_1 = \frac{N_1}{N_1 + pN_2}, \quad \phi_2 = \frac{pN_2}{N_1 + pN_2} \tag{3.12.1}$$

An expression for the entropy of mixing, in which ϕ_1 and ϕ_2 will be found to play an important part, can be derived as follows. We start with the state in which all of the $(N_1 + pN_2)$ spaces are empty. The first segment of the first

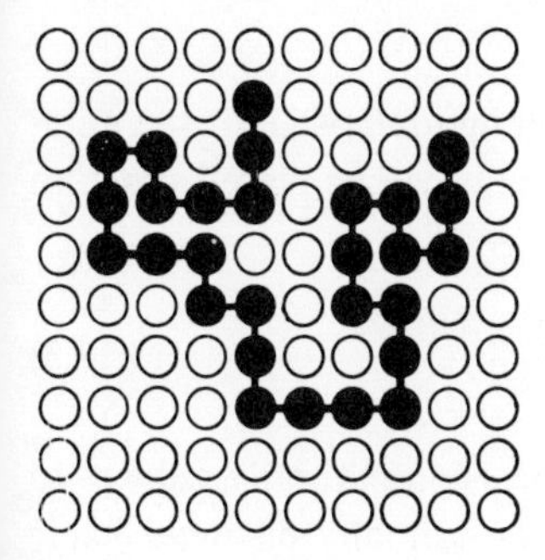

Fig. 51. Schematic representation of a polymer molecule (black) twisted at random in the liquid described in the captions to Figs. 49 and 50.

polymer molecule can thus be placed in $(N_1 + pN_2)$ different ways. Once a certain position has been chosen for this segment, the second segment of the same molecule can be placed in z different ways, if z is the number of spaces adjacent to any given space (z = co-ordination number). For each of the following segments (the 3rd, 4th, . . . , pth) of the same molecule there are $(z - 1)$ ways in which they can be placed, since one of the z adjacent places is already occupied by the preceding segment. The first polymer molecule can thus be accommodated in

$$(N_1 + pN_2)z(z - 1)^{p-2} \tag{3.12.2}$$

different ways.

---

[1] Compare e.g. P. J. Flory, Principles of Polymer Chemistry, Cornell University Press, New York, 1953.

We have here neglected certain factors, to which we shall return at the end of this section.

Suppose that $(k-1)$ polymer molecules have already been placed in the quasi-lattice. How many possibilities remain for the placing of the kth molecule? The first segment of this molecule can be accommodated in

$$N_1 + pN_2 - (k-1)p = N_1 + p(N_2 - k + 1)$$

ways. The second segment of the kth molecule of the polymer cannot be placed in z different ways (as was the case for the first polymer molecule), since there is a chance that one of the spaces surrounding the first segment is already occupied by a segment of one of the $(k-1)$ molecules now in position. This can be taken into account to a reasonable approximation by multiplying z by the fraction of the number of spaces which are still available, thus by

$$\frac{N_1 + p(N_2 - k + 1)}{N_1 + pN_2}$$

The same reduction factor can be applied to each factor $(z-1)$ for the number of ways in which the 3rd, 4th, . . . , pth segments can be placed. The kth polymer molecule can thus be placed in

$$\frac{\{N_1 + p(N_2 - k + 1)\}^p \, z(z-1)^{p-2}}{(N_1 + pN_2)^{p-1}} \tag{3.12.3}$$

different ways in the lattice. For $k = 1$, this expression naturally reduces to (3.12.2). The total number of ways in which N_2 mutually distinguishable polymer molecules can be accommodated in the lattice is given by the product of the N_2 terms obtained by substituting the values 1, 2, . . . , N_2 for k in Equation (3.12.3). Since the polymer molecules in the particular case under consideration are identical, no new situation is created when two molecules are interchanged. We must therefore divide by $N_2!$. Consequently, the total number of different arrangements of N_2 identical polymer molecules in $(N_1 + pN_2)$ spaces is given by

$$m = \frac{\{z(z-1)^{p-2}\}^{N_2}}{N_2!} \times$$
$$\times \frac{(N_1 + pN_2)^p [N_1 + p(N_2 - 1)]^p \ldots (N_1 + p)^p}{(N_1 + pN_2)^{(p-1)N_2}} \tag{3.12.4}$$

In each of these arrangements the remaining spaces (N_1) are unoccupied. Since these N_1 spaces can be filled with solvent molecules in only one way for each possible arrangement of the polymer molecules, the insertion of the solvent molecules makes no contribution to m. In other words: expression (3.12.4) also gives the number of possible configurations m_{12} of the mixture of N_1 solvent molecules and N_2 polymer molecules. Rearranging, we obtain the expression

$$m_{12} = \frac{\{z(z-1)^{p-2}\}^{N_2} p^{N_2}}{N_2!\left(\dfrac{N_1}{p}+N_2\right)^{(p-1)N_2}} \frac{\left\{\left(\dfrac{N_1}{p}+N_2\right)!\right\}^p}{\left\{\left(\dfrac{N_1}{p}\right)!\right\}^p} \tag{3.12.5}$$

The required entropy of mixing is given by

$$\Delta S = k \ln m_{12} - k \ln m_1 - k \ln m_2 \tag{3.12.6}$$

where m_1 refers to the number of possible arrangements of the N_1 molecules in the pure solvent (N_1 sites) and m_2 to the number of possible arrangements of the N_2 polymer molecules in the pure polymeric substance (pN_2 sites). It is clear that $m_1 = 1$ and thus that $k \ln m_1 = 0$. This does not apply to m_2, due to the many possible tangled arrangements. If one assumes that the polymer molecules in the undiluted state behave in a manner analogous to that in solution, m_2 can be found by putting $N_1 = 0$ in (3.12.5):

$$m_2 = \frac{\{z(z-1)^{p-2}\}^{N_2} p^{N_2} (N_2!)^p}{N_2! N_2^{(p-1)N_2}} \tag{3.12.7}$$

It follows from the last three equations that

$$\Delta S = k \ln \left[\frac{\left\{\left(\dfrac{N_1}{p}+N_2\right)!\right\}^p}{\left\{\left(\dfrac{N_1}{p}\right)!\right\}^p \left(N_2!\right)^p} \frac{N_2^{(p-1)N_2}}{\left(\dfrac{N_1}{p}+N_2\right)^{(p-1)N_2}} \right]$$

Applying Stirling's formula in the approximation (2.4.4), we find

$$\Delta S = -kN_1 \ln \frac{N_1}{N_1+pN_2} - kN_2 \ln \frac{pN_2}{N_1+pN_2} \tag{3.12.8}$$

If, in the above derivation, we had erroneously regarded the polymer molecules as being equivalent to the molecules of the solvent, then the earlier Equation (3.7.5) would have been found which, in the same notation as (3.12.8), has the form:

$$\Delta S = -kN_1 \ln \frac{N_1}{N_1 + N_2} - kN_2 \ln \frac{N_2}{N_1 + N_2} \tag{3.12.9}$$

The same formula is obtained by substituting $p = 1$ in (3.12.8).

On the other hand, if we had considered the pN_2 segments of the polymer molecules as separate molecules, we should have obtained the relationship

$$\Delta S = -kN_1 \ln \frac{N_1}{N_1 + pN_2} - kpN_2 \ln \frac{pN_2}{N_1 + pN_2} \tag{3.12.10}$$

It will be clear that Equation (3.12.8) gives results which lie between those for (3.12.9) and (3.12.10).

Using Equation (3.12.1) we may also write (3.12.8) in the form

$$\Delta S = -kN_1 \ln \phi_1 - kN_2 \ln \phi_2. \tag{3.12.11}$$

where $\phi_2 = 1 - \phi_1$.

This formula is similar in form to the earlier Equation (3.7.5) for the entropy of mixing of an alloy; instead of the mole fractions x_1 and $x_2 = 1 - x_1$, the volume fractions ϕ_1 and $\phi_2 = 1 - \phi_1$ now occur.

From the derivation of (3.12.8) it follows that a possible symmetry factor may be neglected in (3.12.2), since this factor would appear in the expressions for both m_{12} and m_2 and would thus cancel out in the expression for ΔS (cf. Equation (3.12.6)). For the same reason we may disregard also the fact that there are less than $(z - 1)$ possible positions in an empty lattice for the 3rd, 4th,. . . , pth segments of a polymer molecule, since in general a polymer molecule is not completely flexible and moreover, part of the $(z - 1)$ spaces may already be occupied by previous segments.

Equation (3.12.8) helps us to understand many thermodynamic properties of solutions of polymers, such as osmotic pressure and phase-separation phenomena, and also to see why solutions of this kind differ from normal, micro-molecular solutions.

3.13. RADIATION OF HEAT AND LIGHT

All bodies emit electromagnetic radiation. From photoelectric and other phenomena it is known that this radiation is always emitted and absorbed in the form of energy quanta of finite magnitude $h\nu$ (ν being the frequency). These quanta are called photons.

The amount of energy radiated in unit time from unit surface area increases rapidly with rising temperature and depends on the nature of the radiating body. The "blacker" a body is, i.e. the more radiation it absorbs, the more it emits. Maximum emissive power occurs in the "perfect black body", a body which absorbs all the radiation falling upon it. Such a body does not exist, but a close approximation to it is formed by a small opening in the wall of a hollow body. Although every beam entering through the aperture will be partially reflected by the point of incidence on the inner wall, this reflected light will ultimately be absorbed to a very great extent owing to multiple reflection and absorption by other parts of the same wall (see Fig. 52).

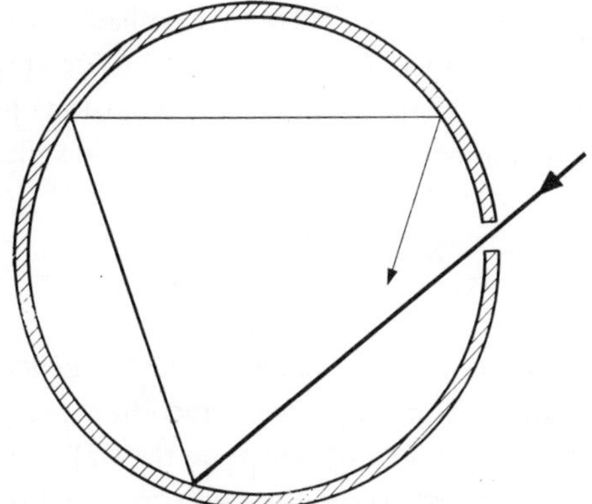

Fig. 52. A small aperture in a hollow body radiates and absorbs as a perfect black body.

If the hollow body is at a temperature T, the aperture will thus radiate as a perfect black body at this temperature. What is emitted is the same radiation as that present inside the cavity. Further study of the latter will thus give an answer to the question of the radiation of a black body. This question is of great importance for all problems of the radiation of light and heat. We shall not be concerned here with the distribution of the radiation among the various wavelengths, but merely use thermodynamic considerations to derive the Stefan-Boltzmann law for the total radiation of a black body.

Consider a cavity, cylindrical in shape and fitted with a weightless, frictionless movable piston. The bottom, or some other portion of the vessel is a black body, the other walls and also the inner surface of the piston are perfectly reflecting. The black portion of the cylinder continuously emits and absorbs radiation. Since the radiation has a finite speed of propagation, a quantity of radiant energy U is always present in the cavity. The space filled with radiation is in many ways analogous to a space filled with gas. An important difference, however, is that the density $u = U/V$ of the radiation depends entirely on the temperature T, while the density of the gas can only be changed by changing the volume. The total number of photons in a cavity changes with the temperature, the total number of gas molecules remains constant. The pressure of the radiation (photons) on the wall can easily be calculated.

We shall consider a cubic vessel with perfectly reflecting walls. The edge of the cube is a cm long. The space contains a single photon of frequency ν and energy $h\nu$. According to the theory of relativity, the mass of the photon is $h\nu/c^2$, where c is the speed of propagation of light. The momentum of the photon is thus $h\nu/c$. We suppose that the photon moves perpendicular to two opposing faces and thus continuously flies to and fro along the same path. At each collision the direction of the velocity of the photon is reversed and the change in its momentum is $2h\nu/c$. Between two successive collisions with the same face, the photon travels a distance $2a$. The time which elapses between these two collisions is $2a/c$, so that the number of collisions per second will be $c/2a$. The change of momentum which the photon suffers per second at one face is therefore

$$\frac{c}{2a}\frac{2h\nu}{c} = \frac{h\nu}{a}$$

If the cube considered contains N photons of frequency ν, then in general these will move in all possible directions. Since no single direction is preferred above any other, the same pressure would be obtained on the wall if one third moved parallel to the x axis, one third parallel to the y axis and one third parallel to the z axis. The change of momentum suffered by the photons per second at one face is thus

$$\frac{N}{3}\frac{h\nu}{a}$$

From mechanics we know that the change of momentum per second is equal to the force exerted, i.e.

$$K_\nu = \frac{N}{3}\frac{h\nu}{a}$$

The force per cm^2 is the pressure, i.e.

$$p_\nu = \frac{N}{3}\frac{h\nu}{a^3} = \frac{nh\nu}{3} \tag{3.13.1}$$

where n is the number of photons per cm^3.

Since $nh\nu$ is the energy density (energy per cm^3) u_ν of the photons of frequency ν, (3.13.1) can also be written in the form

$$p_\nu = \tfrac{1}{3}u_\nu \tag{3.13.2}$$

This relationship is valid for every value of ν. The total pressure in a space filled with photons of many different frequencies is therefore given by

$$p = \Sigma p_\nu = \tfrac{1}{3} \Sigma u_\nu$$

or

$$p = \tfrac{1}{3} u \tag{3.13.3}$$

where u is the total energy per cm³. If part of the wall is black, so that the incident photons are absorbed, the result remains the same provided that radiation equilibrium has been established. In order to maintain equilibrium, the number of photons emitted must be the same as that absorbed and each photon emitted imparts an impulse to the emitting wall.

Work can be done on the photon "gas" in the cylinder by moving the piston. Heat can be supplied to it by placing the outside of the black wall in contact with a heat reservoir at a higher temperature. Using (3.13.3), the first law takes the form

$$\mathrm{d}U = \mathrm{d}Q - \tfrac{1}{3}\, u \mathrm{d}V$$

The radiation also possesses a certain entropy. This statement can be justified statistically by considering the radiation as a gas composed of photons, as was done above. On the basis of classic thermodynamics it can be directly justified in an even simpler way as follows. If the heat reservoir supplies a quantity of heat $\mathrm{d}Q$ reversibly to the cavity, the entropy of the reservoir drops. Since the total entropy remains constant in a reversible change, the radiation (the photon gas) must have an entropy S which increases during the transfer. Combining the first and second laws we obtain the relationship

$$\mathrm{d}U = T\mathrm{d}S - \tfrac{1}{3}\, u \mathrm{d}V \tag{3.13.4}$$

where T is the temperature of the black-body radiation, i.e. the temperature of the black wall with which the radiation is in equilibrium. Employing the relationship $U = Vu$ or $\mathrm{d}U = V\mathrm{d}u + u\mathrm{d}V$, we obtain

$$\mathrm{d}S = \frac{V}{T}\frac{\mathrm{d}u}{\mathrm{d}T}\mathrm{d}T + \frac{4}{3}\frac{u}{T}\mathrm{d}V \tag{3.13.5}$$

$\mathrm{d}S$ is a real differential in the sense of Chapter 1 (a total differential), so that, from (1.4.7):

$$\frac{\partial}{\partial V}\left(\frac{V}{T}\frac{\mathrm{d}u}{\mathrm{d}T}\right) = \frac{\partial}{\partial T}\left(\frac{4}{3}\frac{u}{T}\right)$$

or

$$\frac{1}{T}\frac{\mathrm{d}u}{\mathrm{d}T} = \frac{4}{3}\frac{1}{T}\frac{\mathrm{d}u}{\mathrm{d}T} - \frac{4}{3}\frac{u}{T^2},$$

$$\frac{\mathrm{d}u}{u} = 4\frac{\mathrm{d}T}{T},$$

$$u = a\,T^4, \qquad (3.13.6)$$

where a is an integration constant. It will be seen that (3.13.6) is the Stefan-Boltzmann law. Since the intensity of radiation, the quantity of radiation falling per second upon 1 cm^2, is proportional to the quantity of energy per cm^3, we can also formulate this law as follows: the total radiation of a black body is proportional to T^4.

Substituting (3.13.6) in (3.13.5), we obtain

$$\mathrm{d}S = 4aVT^2\mathrm{d}T + \frac{4}{3}aT^3\mathrm{d}V.$$

$$\left(\frac{\partial S}{\partial V}\right)_T = \frac{4}{3}aT^3 \qquad (3.13.7)$$

The energy per unit volume of the radiation is thus proportional to the fourth power of the absolute temperature, the entropy per unit volume to its cube.

3.14. FUEL CELLS AND HEAT PUMPS

The consumption of energy in the world is increasing so rapidly and the stores of fossil fuels (with their accumulated solar energy) are decreasing so rapidly, that the generation of sufficient energy threatens to be one of mankind's greatest problems [1]). Apart from the development of new sources of energy, such as nuclear reactors, it is possible, in principle, to make much more economical use of the available reserves of coal, oil and natural gas than is at present the case. Great economies could be achieved if the fuel cell and heat pump can be brought to a satisfactory technical development.

Before discussing these possibilities we may remind the reader of what was said about heat engines in Chapter 1. We saw that only part of the heat of combustion of a fuel can be converted into work. The remainder flows to a lower temperature level. Or, expressed in terms of entropy: the drop in entropy, corresponding to the conversion of a quantity of heat into work,

[1]) See e.g. P.C. Putnam, Energy in the Future, Van Nostrand Co., New York, 1953; A. M. Weinberg, Phys. Today **12**, 18 (Nov. 1959).

must be compensated by an increase in entropy and this occurs when another quantity of heat flows to a lower temperature. In the Carnot cycles discussed, the working substance (e.g. steam) absorbed a quantity of heat Q_1 at a high temperature T_1, and released a quantity Q_2 at the lower temperature T_2. The difference $|Q_1| - |Q_2|$ was gained as work. In accordance with the second law of thermodynamics in the form $\oint dQ/T \leqslant 0$ the efficiency of a heat engine was found to be given by the equation

$$\frac{|Q_1| - |Q_2|}{|Q_1|} \leqslant \frac{T_1 - T_2}{T_1}. \tag{3.14.1}$$

The fuel cell

The limitation, expressed in Equation (3.14.1), of the possibility of transforming heat into work is inevitable since this transformation amounts to the production of order (e.g. in the form of the similarly-directed movements of the atoms in a piston) from disorder, viz. from the thermal agitation of atoms or molecules. The limitation of the Carnot cycle can, however, be avoided by leaving out the intermediate stage of disorder. Use is then made of the fact that a fuel is a source of free energy and that this free energy (at least in principle) can be converted directly into electrical energy without the intermediary of heat. It should be remembered in this connection that according to Section 3.2, the change in free energy gives the maximum of work which can be obtained during the isothermal and *reversible* course of a reaction. In the second place it should be remembered that, according to Sections 1.5, 1.18 and 3.4, the reversible progress of a chemical reaction can be approximated by allowing the reaction to proceed under suitable conditions in a galvanic cell. The work is then obtained directly in the form of electrical energy. If the isothermal-reversible reaction proceeds at constant pressure, this maximum electrical energy will be given, according to Section 3.4, by the free enthalpy of the combustion reaction:

$$-zFE = \Delta G = \Delta H - T\Delta S. \tag{3.14.2}$$

Here E represents the electromotive force of the cell in which the reaction takes place, F the charge on one gram-ion of univalent positive ions (i.e. 1 faraday = 96500 coulomb) and z the number of faradays transported in the fuel cell during the combustion reaction.

It is further recalled that the quantity ΔH in (3.14.2) is the normal heat of combustion, i.e. the heat absorbed by the system under consideration when the combustion occurs *irreversibly* at constant pressure. The combustion reactions in question are exothermal, so that ΔH is negative. The heat of combustion of coal and oil is employed in central power stations to generate electrical energy. To do this, the heat is first converted into mechanical energy in steam engines or turbines with the efficiency given by (3.14.1):

$$\frac{W}{\Delta H} \leqslant \frac{T_1 - T_2}{T_1} \tag{3.14.3}$$

The conversion of the mechanical energy into electrical energy is achieved with the help of generators (dynamos) and is accompanied by much smaller losses. The overall efficiency, i.e. the ratio of the quantity of electrical energy produced to the heat of combustion ΔH, rarely exceeds 30%.

On the other hand, the theoretical yield of electrical energy when the reaction takes place *reversibly* is given, according to (3.14.2), by

$$\frac{\Delta G}{\Delta H} = 1 - T\frac{\Delta S}{\Delta H} \tag{3.14.4}$$

Since ΔH is negative, the theoretical efficiency is greater than 1, i.e. greater than 100%, if ΔS is positive. In Section 3.4 it was seen that this is the case, for example, in the reaction

$$2\,C + O_2 \rightarrow 2\,CO \tag{3.14.5}$$

where the number of gas molecules increases. Equation (3.14.4) shows, furthermore, that in this case the efficiency increases with increasing temperature. It is 125% at 300 °K, 150% at 600 °K. The reverse is true when the number of gas molecules decreases during the reaction. The efficiency is then smaller than 100% and decreases as the temperature rises. For the reaction

$$C + O_2 \rightarrow CO_2 \tag{3.14.6}$$

where the number of gas molecules neither increases nor decreases, the theoretical efficiency is virtually 100% and almost independent of the temperature.

For a positive value of ΔS, not only the chemical energy ΔH is converted into electrical energy but also the heat $Q = T\Delta S$, which is absorbed from the surroundings during the isothermal-reversible course of the reaction. This explains the fact that the energy yield can be greater than 100%.

Combining Equations (3.14.2) and (3.4.33), we obtain

$$\Delta S = zF\left(\frac{\partial E}{\partial T}\right)_p \tag{3.14.7}$$

If the entropy change, ΔS, which occurs during combustion is positive, then not only the efficiency, but also the electromotive force of the cell, increases with rising temperature.

The above discussion shows how extravagant it is to employ intermediate stages in converting the free energy of fuel into electrical energy. If the energy available in waterfalls was treated in the same way, the potential energy of the water would first, by free fall, be converted into heat which would then be used to supply a heat engine, instead of using the water power directly. However, decades of research have shown that it is easier to propound the above critical reasoning than to develop technically usable fuel cells. Nevertheless, the importance of these considerations, dating back to the last century, should not be underestimated on that account. It should be realized that the great strength and fundamental significance of thermodynamics is most clearly demonstrated in its power, quite independent of the state of technology, to draw conclusions of such great importance as those discussed. These conclusions are still a stimulant for the research on fuel cells which is being carried out in several countries.

The fuel cell does not differ fundamentally from the galvanic cell which has already been mentioned in Section 1.5. Instead of starting, as in the case discussed, with an expensive metal "fuel" (zinc), it would be desirable to use a cheap conventional fuel (preferably coal) and to supply this continuously to the anode. Also the required oxygen should be supplied continuously to the cell at the cathode. In the ideal case, the cell would supply electricity as long as it is fed with fuel and oxygen (or air).

The simplest conceivable carbon-oxygen cell will contain an oxide as electrolyte, in which electricity will be conducted by the transport of oxygen ions. The reaction

$$C + 2O^{=} \rightarrow CO_2 + 4e^- \tag{3.14.8}$$

will occur at the anode. The electrons formed will flow through the external circuit to the cathode. On their way from anode to cathode they can be made to perform the desired work whilst avoiding the randomization of the chemical energy. At the cathode the electrons must react with oxygen according to the reaction

$$O_2 + 4e^- \rightarrow 2O^{=} \tag{3.14.9}$$

The circuit is completed by the oxygen ions flowing through the electrolyte to the anode (see Fig. 53).

This all sounds very simple, but the difficulties involved in the realization are so great that up to now it has not been possible to develop the carbon-oxygen cell as a practical generator of electrical energy. A more advanced stage of development has been reached with cells fed by gaseous fuels (H_2,

CO or hydrocarbons). It is hardly to be expected, however, that they will be found in service as fuel cells in power stations within a predictable period. But it may very well be that they will find applications in special fields (space vehicles, automobiles, locomotives, etc.) where convenience rather than cost is the primary factor. They are fairly light in weight and supply energy without producing noise or polluting the atmosphere.

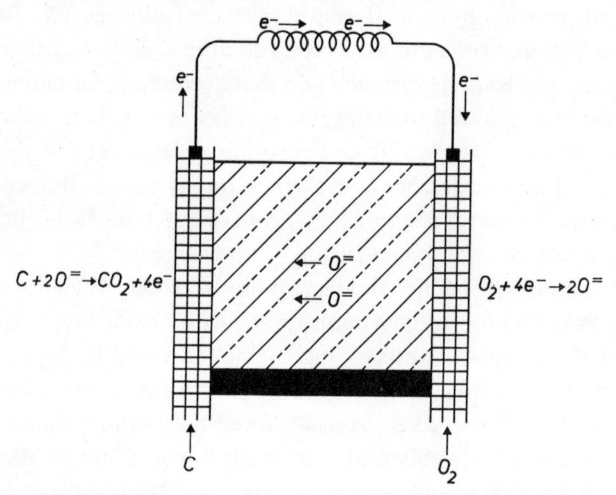

Fig. 53. Schematic representation of the carbon-oxygen fuel cell.

Hydrogen-oxygen fuel cells have already been used in a number of space vehicles because of their high efficiency, high power-to-weight ratio and usable reaction product (potable water). The most highly developed hydrogen-oxygen systems are those with an aqueous solution as electrolyte. In the Bacon cell hydrogen and oxygen, under pressures between 30 and 50 atm, are introduced into porous nickel electrodes immersed in a concentrated aqueous KOH solution at 200 °C. At the hydrogen electrode (anode) H_2 molecules diffuse through the porous structure and are chemisorbed at the surface in the form of atoms. The adsorbed atoms react with hydroxyl ions in the electrolyte according to the reaction

$$2H + 2OH^- \rightarrow 2H_2O + 2e^-$$

At the cathode, oxygen diffuses through the electrode and reacts with water from the KOH solution and with electrons to form hydroxyl ions:

$$\tfrac{1}{2}O_2 + H_2O + 2e^- \rightarrow 2OH^-$$

The latter then migrate through the electrolyte to the anode. The porous electrodes have a matrix of coarse pores, coated on the side facing the electrolyte with a layer of nickel having finer pores. In order to obtain a large area of electrode-electrolyte interface, at which electron transfer occurs, the reactant gases are under sufficient pressure to displace the electrolyte from the coarse pores, but not from the fine pores.

The large-scale generation of cheap electrical energy with the aid of fuel cells is still a distant dream. Nevertheless, the possibility, in principle, of generating electricity with a considerably greater efficiency than is reached in existing power stations, will undoubtedly remain a strong incentive to continue with research on fuel cells. [1])

The heat pump

After considerable trouble has been taken to produce electrical energy in the customary wasteful fashion (via the heat of combustion of coal), this energy is, in many cases, eventually degraded to low-temperature heat for heating purposes with the aid of electrical resistance heaters. In the latter process the electrical energy is *wholly* converted into heat and a layman in the field of thermodynamics might therefore think that it could not be done more economically. In fact, this is also extremely wasteful. Several hundred percent of the input energy can be gained by making use of a heat pump.

In order to be able to understand the heat pump, we return for the moment to the idealized heat engine which was recalled at the beginning of this section. The processes occurring there can also be made to take place in reverse. Instead of allowing heat to flow from a reservoir at a high temperature to one at a low temperature and converting as much of it as possible into work "on the way", we now perform work (via a compressor) to transfer heat from the low temperature reservoir to that at the high temperature. If the working substance (gas or vapour) in the cycle of the heat engine absorbed more heat at the high temperature than it delivered at the low temperature, it will now deliver more heat at the high temperature than it absorbs at the low one. The difference is the external work W which is at least necessary to drive the engine.

The machine just described is called a *refrigerator* when its task is to extract heat from the low-temperature reservoir, and a *heat pump* when it is

[1]) Of the recent literature on fuel cells we mention only the following: W. Mitchell (ed.), *Fuel Cells*, Academic Press, New York and London, 1963; W. Vielstich, *Brennstoffelemente*, Verlag Chemie, Weinheim, 1965; K. H. Spring (ed.), *Direct Generation of Electricity*, Academic Press, New York and London, 1965; K. Williams (ed.), *Introduction to Fuel Cells*, Elsevier, Amsterdam, 1966; A. B. Hart and G. J. Womack *Fuel Cells: Theory and Application*, Chapman and Hall, London, 1967.

required to supply heat to a high temperature reservoir. In the first case we are interested in Q_2 and the efficiency is

$$\frac{|Q_2|}{|W|} \leqslant \frac{T_2}{T_1 - T_2} \tag{3.14.10}$$

In the case of the heat pump Q_1 is the important quantity and the efficiency is

$$\frac{|Q_1|}{|W|} \leqslant \frac{T_1}{T_1 - T_2} \tag{3.14.11}$$

Let us consider, for example, the case where $T_1 = 300$ °K and $T_2 = 270$ °K. If the refrigerator is for household use, we thus require to keep its internal temperature at 270 °K = −3 °C, while it is placed in a room at 300 °K = 27 °C. In the case of the heat pump, we are extracting heat from the surroundings at −3 °C and using it to keep the inside of a building at 27 °C. The upper limit of the efficiency is 270/30 = 9 for the refrigerator and 300/30 = 10 for the heat pump. The actual efficiency, of course, will not be as high as this. This is expressed by means of the *figure of merit*, which gives the ratio between the real and the maximum efficiency and which is, therefore, always less than unity.

The principle of the heat pump is still only used in a few places in the world. Its practical form is as follows: a liquid with a high heat of evaporation (e.g. CF_2Cl_2) is allowed to evaporate at a low temperature, thus extracting the required heat from the surroundings. Condensation takes place at a higher temperature thus giving off heat of condensation to the inside of a building [1]). The heat pump is particularly suitable for heating buildings because its efficiency, according to Equation (3.14.11), is very high for small values of $(T_1 - T_2)$. If properly designed, the heat pump can be used to heat a building in the winter and to cool it in the summer.

One can certainly forsee that the interest in fuel cells and heat pumps will increase as the energy sources which are now being consumed become still further depleted.

---

[1]) See e.g. the books: P. Sporn, E. R. Ambrose and T. Baumeister, *Heat Pumps*, John Wiley and Sons, New York, 1947; E. N. Kemler and S. Oglesby, *Heat pump applications*, McGraw Hill Book Co., New York, 1950; See also a series of articles on *Trends in Heat Pump Systems*, ASHRAE Journal, September 1967.

CHAPTER 4

QUANTUM MECHANICS AND STATISTICS

4.1. INTRODUCTION

In the previous chapters it was stated that the entropy of a system can be determined by statistical means by counting the number of micro-states in which the system may occur under the given conditions. By failing to enlarge upon this point, we have falsely given the impression that the way in which the count should be made was definitely established. Even in the simple case of the various configurations of 50 white and 50 red billiard balls (Fig. 11), counting is more ambiguous than was suggested in Chapter 2. If two white balls are interchanged, a state is reached which, to all appearances, is indistinguishable from the original state and which was thus not counted separately. However, a more thorough investigation with the help of scales and measuring instruments would certainly make it possible to distinguish two similarly coloured balls from one another. The number of visually-distinguishable configurations is thus much smaller than the number — 100! — which can be distinguished on closer investigation.

In classical physics the counting of the number of micro-states was carried out as if atoms were a sort of billiard balls on a very reduced scale. Atoms were thought of as individual objects whose identity was permanently preserved. They were treated during the counting as though they were numbered. In this manner a given micro-state could be converted into a new micro-state by interchanging two similar atoms. Even before the birth of quantum mechanics, doubts could have been — and, indeed, were — expressed as to the validity of this method of counting. Finally, quantum mechanics made it certain that this method was incorrect. As early as Chapter 2, in counting the micro-states of an idealized solid, we had anticipated some results obtained by quantum mechanics, not only by allowing energy to be absorbed and given up by the oscillators only in quanta, but also by dispensing with the individuality of the oscillators. States which could be derived from one another by interchanging similar atoms were not considered different if these atoms were in the same energetic state, i.e. were at the same energy level (cf. Fig. 14 and 15). Anticipating what will be discussed below, we should like to remark that the decisive factor is not "being at the same energy level" but "being in the same quantum state". In the particular case of an Einstein solid each energy level corresponds to one quantum state, so that no ambiguity is introduced by speaking of levels rather than "states". There are, however, many cases (see e.g. Section 4.7) in which different states have the same energy. These are often referred to as "multiple energy levels". To prevent confusion we shall avoid this expression as far as possible for the present. Wherever possible, we prefer to speak of separate (though possibly coinciding) levels.

In the solid model under consideration a new state is brought about by interchanging particles which are at different energy levels. In Fig. 14 the situation changes when the particle with 5 energy quanta is to be found in position A_1, for example, instead of E_5. It should be noted that, in making this distinction, it is not the particles which are numbered but the lattice sites where they perform their oscillations. This way of distinction is absent in a gas. In contrast to the situation in a solid, the particles in a gas do not each have their own volume, but the whole volume is the common "living space" for all of the particles. This collective occupation of the volume means that the number of micro-states must be counted in a way which is completely different from that used for a solid. In the case of the solid, the interchange of two particles at different levels produced a new state, in a gas this is not the case. In gases it is as though each particle occupies the whole of the available space at any moment; according to modern views (see below) one can only speak of the probability of finding it in the various parts of the available space. "Interchanging two particles" thus produces no new state, in fact, the expression has no physical meaning. If Fig. 15 related to a gas, it would represent only one micro-state instead of about 10^{12} in the case of a solid.

What has been said above should not be taken to indicate that there is only a limited number of micro-states for a gas. In the sections which follow we shall see that the separation of the energy levels becomes smaller as the volume in which a particle can move becomes larger. While in an Einstein solid the separation of the levels is so large that at all but very high temperatures not more than a few tens of them will be noticeably occupied by particles (cf. Fig. 16), in a gas the separations are so extremely small, except at extremely high pressures, that the number of energetically accessible levels is very much greater than the number of particles (see Section 4.7). The number of distribution possibilities is therefore very large, despite the fact that each distribution can only be achieved in one way.

The manner in which the number of micro-states of a system of non-localized particles (i.e. a gas) must be determined is not wholly specified in the above. It will presently be seen that there are systems of such a nature that it is fundamentally impossible for any quantum state to contain more than one particle. In this case we speak of a system to which Fermi-Dirac statistics are applicable. There are also systems of such a nature that an arbitrary number of particles may be found in every possible quantum state. These are systems to which Bose-Einstein statistics are applicable.

A further distinction is made for the classic Maxwell-Boltzmann statistics. This may be applied to a gas model in which the particles are modelled after the visible world, i.e. particles whose properties are likened to those of macroscopic bodies. In this model the particles can be individualized. There are no real gases which fit this classic model and thus no systems for which the Maxwell-Boltzmann statistics are exactly valid. Although fundamentally untenable, the M.B. statistics can be applied to ordinary gases to give results which do not differ appreciably from the results obtained by the use of the correct statistics (see later). In other cases, however, e.g. when applied to a photon gas in a hollow space or an electron gas in a metal, M.B. statistics lead to wholly incorrect results.

Strictly speaking, it is incorrect to talk of different "statistics". The statistical methods remain unchanged, but they are applied to systems of different types.

Before going on to the calculation of the energy distribution in Fermi-Dirac, Bose-Einstein and Maxwell-Boltzmann gases, we shall first devote the following sections to a very elementary treatment of the basic principles of quantum mechanics.

4.2. HEISENBERG'S UNCERTAINTY PRINCIPLE

No completely satisfactory pictorial representation can be made of either matter or electromagnetic radiation, although some aspects of both physical realities show analogy with a corpuscular model and others with a wave model. For example, not only beams of electromagnetic radiation, but also electron beams and even atomic and molecular beams can produce interference phenomena, which immediately demonstrates that a gas does not satisfy the model to which Maxwell-Boltzmann statistics are applicable.

From this it is clear that the corpuscular model of matter and of radiation may only be regarded as an analogy, which is restricted to certain aspects of the phenomena. such as the conservation of energy and momentum.

To take the corpuscular model too literally, i.e. to regard photons, electrons, etc. as normal bodies on a greatly reduced scale, is to make the same mistake that would be made by concluding from the well-known analogy between electrical currents and hydraulic currents that electricity is a normal fluid.

In order to avoid lengthy descriptions it is often convenient to continue to make use of the phraseology of the corpuscular model and, up to a point, of the model itself. Thus we continue to use the word "particles" to indicate electrons, protons, atoms, etc., but must bear in mind that they do not possess all the properties which are associated with the normal concept of a material particle or corpuscle. Since the corpuscular model is not an accurate representation of reality, its use necessitates the application of a correction for the error introduced. This correction is comprised in Heisenberg's uncertainty (or indeterminacy) relation:

$$\Delta x.\Delta p_x \simeq h \tag{4.2.1}$$

where Δx and Δp_x are the uncertainties in the co-ordinate of position and the associated momentum and h is Planck's constant. (Similarly, we have $\Delta y.\Delta p_y \simeq h$ and $\Delta z.\Delta p_z \simeq h$).

The relationship (4.2.1) expresses the fact that the fundamental property of a particle in point mechanics, viz. its having a well-defined path, is not valid for an elementary "particle". If, for instance, we consider the one electron in a hydrogen atom, we cannot specify its position with respect to the nucleus at any moment. In other words, we may not speak of the path of the electron. It is, however, possible to give the probability of finding the electron in a particular volume element. It should be noted that the concept of probability enters the question here not, as in classical theory, because of the practical (*not* fundamental) impossibility of knowing all the factors necessary for a calculation of the paths of the particles, but because of the nature of the phenomenon itself, because of the experimental fact that the "particles" are not really particles in the visual (macroscopic) sense of the word.

The uncertainty principle is thus based on the inapplicability of the corpuscular model and indicates the limits within which it is valid. Strictly speaking, it applies not only to particles of atomic or subatomic dimensions, but also to ordinary bodies. In the latter case, however, this has no practical consequences, since Planck's constant is so small that the uncertainties Δx and Δp_x in position and momentum are negligible with respect to the errors of measurement.

4.3. THE PHYSICAL SIGNIFICANCE OF THE UNCERTAINTY PRINCIPLE

Heisenberg's uncertainty principle by no means indicates that no accurate measurements can be made on atomic objects. Nevertheless, if the position of an atom is being accurately investigated, the action of the measuring instrument on the atom produces an unverifiable and undefined change in the momentum of this atom. If, for example, one tries to determine the co-ordinate x with an accuracy Δx, one is obliged to impart to the atom an impulse Δp_x of the order of magnitude of $h/\Delta x$, but the exact value of which it is fundamentally impossible to know. The more accurately x is measured, the less we know about p_x. If, subsequently, the component p_x of the momentum of the particle is to be measured with an accuracy of Δp_x, one is obliged to allow a displacement along the x-axis by a fundamentally undetermined amount of the order of magnitude of $h/\Delta p_x$. Our knowledge of x, gained from the first experiment, is thus partially lost as a result of the second experiment.

The methods used to measure a co-ordinate and those for the corresponding momentum are thus mutually exclusive. In each experiment only one of the two can be measured accurately, while the other is changed by the measurement in a manner which cannot be verified.

That a particle, strictly speaking, is not a particle at all, reveals itself in the fact that the two concepts of position and velocity, each of which separately has an accurate physical significance, cannot simultaneously be attributed in a precise manner to the same particle. This relationship between two concepts of classical physics was termed "*complementarity*" by Bohr [1]). The finite magnitude of Planck's constant forces us to use a form of description which we call *complementary* in the sense that each application of concepts from the visual models of classical physics, excludes the simultaneous use of other classical concepts. Experiments which lay emphasis on the "position of a particle" leave the concept "velocity of a particle" undetermined and vice versa. For a detailed discussion of these experiments, the reader is referred to a well-known book by Heisenberg [2]).

From the foregoing it will be clear that one may *not* say "It is true that we cannot accurately *determine* both the position and the velocity of a particle at a given moment, but naturally the particle does *have* a definite position

---

[1]) N. Bohr, Atomtheorie und Naturbeschreibung, Berlin, 1931.

[2]) W. Heisenberg, Die physikalischen Prinzipien der Quantentheorie, Leipzig, 1930.

and a definite velocity". According to this (incorrect) assertion, the uncertainty (indeterminacy) in the Heisenberg relation would be lack of knowledge. In reality, however, there is a profound difference in meaning between the assertion that the elements of the motion of a particle (position, velocity, etc.) are "uncertain" or "indeterminate" and the assertion that these elements are "unknown". We are dealing with indeterminacy in the sense that in principle there is nothing in nature to which a definite position *and* a definite velocity can be simultaneously attributed, because no particles, in the classical meaning of the word, exist.

The uncertainty relation does not limit our ability to acquire knowledge about the real properties of the elementary "particles", it only sets limits to the applicability of classical mechanics to these "particles".

4.4. SCHRÖDINGER'S EQUATION

In Section 4.2 it was remarked that the pictorial representation of the continuous movement of a particle along a fixed path must be abandoned and replaced by the consideration of the probability of finding the particle in the various volume elements. In order to elucidate this point, we first direct our attention to the particles already known to us by the name photons or light quanta (see Section 3.13).

If, with the aid of some optical arrangement, we project an image, e.g. an interference pattern, onto a screen, we can employ the wave theory of light (using the classical wave equation) to calculate the intensity of illumination I at all points on the screen. If the light is monochromatic, with a frequency ν, then the intensity I divided by $h\nu$ gives the number of photons arriving at the screen per unit of time and area. Thus we know that the number of photons striking each element $d\sigma$ of the surface in a time dt will be given by $I d\sigma dt/h\nu$.

In most optical experiments the photons are very numerous and in that case there is excellent agreement between the measured and the calculated intensity. Wave optics, however, fails as soon as the number of photons is small. For example, if we suppose that a light source (placed behind a screen with two slits) which will produce an interference pattern on another screen, should emit only one photon, the wave theory predicts an intensity distribution producing the same interference pattern (though of extremely low intensity), while in fact the entire effect of the photon is felt at one single point on the screen. The complete interference pattern does not appear until the light source emits many photons; in that case it appears even when the source is so weak that the photons pass through the apparatus one by one.

From this it can be seen that the intensity I must be interpreted statistically. In principle it is not a measure of the *number* of photons reaching the screen per unit time and area, but of the *probability* that a photon will arrive at the surface element under consideration. When large numbers of photons are involved the "law of large numbers" blurs the difference between the two interpretations.

It is well-known that with electron and atomic beams the same interference phenomena can be obtained as observed with electromagnetic radiation. Matter thus exhibits the same duality as light (properties of both corpuscles and waves) and also in the mechanics of material particles, the calculation of the path of a particle must therefore be replaced by a calculation of the probability that it will be found in particular volume elements at the moment of observation. The strong analogy between light and matter led to the (successful) attempt to calculate the required probability density with the aid of an equation of the same type used in wave optics to calculate the intensity I. This equation is associated with the names of Schrödinger, who introduced it, and De Broglie, who was the first to recognize the wave character of matter. Schrödinger's equation forms the basis of quantum mechanics. Just as geometrical optics is only a first approximation to wave optics, so classical mechanics is only a first approximation to "wave mechanics" (quantum mechanics), an approximation which fails when dealing with microphysical problems.

For one particle of mass m, Schrödinger's equation has the form

$$\frac{h^2}{8\pi^2 m}\left(\frac{\partial^2\psi}{\partial x^2}+\frac{\partial^2\psi}{\partial y^2}+\frac{\partial^2\psi}{\partial z^2}\right)+(\varepsilon-V)\psi=0 \qquad (4.4.1)$$

or, more concisely written (with the help of the Laplace operator):

$$\frac{h^2}{8\pi^2 m}\nabla^2\psi+(\varepsilon-V)\psi=0 \qquad (4.4.1a)$$

Here V is the potential energy and ε the total energy of the particle; ψ is called the *amplitude of the probability*, in analogy to the amplitude of light waves. It has no direct physical significance, but provides us with the required probabilities in terms of the quantity $|\psi|^2$. (The function ψ itself cannot give a measure of the probability of finding the particle somewhere, since it can also take negative values).

This book will only employ real wave functions, so that instead of $|\psi|^2$, we could simply write ψ^2. In quantum mathematics, however, one is often dealing with complex wave functions ψ and the more general expression for the probability density is given by $|\psi|^2$, which for ψ complex, is given by ψ multiplied by its complex-conjugate ψ^*.

The probability of finding the particular particle in the volume element $d\omega$, is given by $|\psi|^2 d\omega$. Since there is certainty that the particle will be found somewhere, it must be true that

$$\int |\psi|^2 d\omega = 1 \tag{4.4.2}$$

if the integration is carried out throughout all space. Equation (4.4.2) represents the operation of "normalisation".

For a system consisting of more than one particle, Schrödinger's equation becomes:

$$\frac{h^2}{8\pi^2} \sum \frac{1}{m_i} \nabla_i^2 \psi + (U - V)\psi = 0 \tag{4.4.3}$$

Here, m_i is the mass of the ith particle, ∇_i^2 the Laplacian operator for the ith particle (thus $\partial^2/\partial x_i^2 + \partial^2/\partial y_i^2 + \partial^2/\partial z_i^2$), U the total energy of the whole system and ψ the wave function for the whole system.

Like the other basic equations of physics (e.g. Newton's equations in classical mechanics, Maxwell's equations for the electromagnetic field and the laws of thermodynamics) Schrödinger's equation is not derived but postulated. It is justified by the agreement between its results and the experimental data. We shall discuss this agreement in the coming sections with reference to the Schrödinger equation for one particle.

4.5. STATIONARY STATES

The ψ function must everywhere be finite. Furthermore, it must everywhere be continuous and unique, since the probability of finding the particle at a given point cannot have more than *one* value. From the theory of differential equations it is known that solutions which satisfy the requirements mentioned only exist for certain values of the parameter ε in (4.4.1). These values of ε are called the *eigenvalues* of the differential equation, the corresponding solutions the *eigenfunctions*. The eigenvalues of Schrödinger's equation in a particular case thus fix certain, discrete values of the energy. In other words, Schrödinger's equation produces, in a completely natural way, the discrete values of the energy (the energy levels) required by the experiments, so that the physical problem of the quantization of energy has been reduced to a purely mathematical problem. The eigenfunctions corresponding to the eigenvalues describe the *stationary states* (quantum states).

In order to appreciate this fine result of quantum mechanics, we recall the unsatisfactory manner in which the energy levels were found in the older theory of Bohr. The theory of Bohr and Sommerfeld starts with the classical laws of mechanics and then (to reach agreement with experiment) selects a limited number of the movements possible according to these laws, using the quantum condition

$$\oint p_x \mathrm{d}x = nh \tag{4.5.1}$$

in which x is the co-ordinate under consideration and p_x the corresponding momentum (h = Planck's constant, $n = 1, 2, 3, \ldots$). The integration extends over one period of the supposedly periodic motion. In each special case, there will be as many conditions of the type (4.5.1) as the system has degrees of freedom.

The quantum condition (4.5.1) from Bohr and Sommerfeld's theory can be derived as an approximate result from Schrödinger's theory. In many (but not all) cases it leads to correct results. The energy levels in a hydrogen atom, for example, can be found with the help of (4.5.1) with equal accuracy as with Schrödinger's equation. The circular and elliptic orbits associated with these levels in the theory of Bohr and Sommerfeld, however, lose their pictorial significance in quantum mechanics. The solutions of the Schrödinger equation for the electron in the hydrogen atom give a description of the stationary states in the sense that $|\psi|^2$ indicates the probability of finding the electron at any place around the nucleus. The "charge cloud" given by $|\psi|^2$ has a particular form for each stationary state. The most probable distances of the electron from the nucleus correspond roughly to the radii of the circular orbits of Bohr (see Chapter 5).

4.6. PARTICLE IN A BOX

As an example of the calculation of eigenvalues and eigenfunctions we shall choose a very simple case which is of importance to us within the scope of this book. We shall consider a particle, moving freely in a cubic box with walls impenetrable for the particle and edges of length a. This situation corresponds to a potential energy $V = 0$ inside the box and $V = \infty$ outside. In order to simplify the problem still further, we consider first the case of a particle which can only move along the x-axis between the two fixed points A and B. In other words, we shall first consider the problem of a particle in a "one-dimensional box" AB (length $AB = a$). In this "box" also we assume freedom of movement for the particle and complete impenetrability of the "walls" A and B, corresponding to $V=0$ in the box and $V=\infty$ outside. Since $V=0$, the Schrödinger Equation (4.4.1) for the particle in the box reads:

$$\frac{h^2}{8\pi^2 m}\frac{d^2\psi}{dx^2}+\varepsilon\psi=0 \tag{4.6.1}$$

This is satisfied by

$$\psi = A \sin 2\pi(2m\varepsilon)^{\frac{1}{2}}\frac{x}{h} + B \cos 2\pi\,(2m\varepsilon)^{\frac{1}{2}}\frac{x}{h} \tag{4.6.2}$$

as can be easily verified by differentiating twice.

We began with the assumption that the walls are completely impenetrable for the particle, which corresponds to an amplitude $\psi = 0$ outside the box. The requirement of continuity which, according to the beginning of this section, must be satisfied by the ψ function, means that inside the box ψ must approach zero as the walls A and B are approached. In other words, it must satisfy the boundary conditions

$$\psi = 0 \quad \text{at} \quad x = 0 \quad \text{and}$$
$$\psi = 0 \quad \text{at} \quad x = a$$

The first of these conditions means that B in (4.6.2) must be attributed the value zero, so that we obtain

$$\psi = A \sin 2\pi(2m\varepsilon)^{\frac{1}{2}}\frac{x}{h} \tag{4.6.3}$$

From the second boundary condition follows that

$$2\pi(2m\varepsilon)^{\frac{1}{2}}\frac{a}{h} = n\pi \tag{4.6.4}$$

or

$$\varepsilon_n = \frac{h^2}{8ma^2}n^2 \tag{4.6.4a}$$

where $n = 1, 2, 3, \ldots$

Substitution of (4.6.4) in (4.6.3) gives:

$$\psi_n = A \sin\frac{n\pi x}{a} \tag{4.6.5}$$

Each of these wave functions describes a state in which (from the classical viewpoint) the particle moves to and fro between A and B with a constant velocity. Quantum mechanics can only give the "expectation value" of a quantity. For the velocity or momentum, this is zero, since there is just as much chance of finding the particle moving to the left as to the right. The absolute value of the momentum $|p_x|$ can, however, be deduced from the energy because in this case the energy is purely kinetic ($\varepsilon = p_x^2/2m$). From (4.6.4a) follows:

$$|p_x| = \frac{h}{2a} n \tag{4.6.6}$$

It is not difficult to see that the same permissible values of the momentum are found with the help of the quantization rule (4.5.1) from the "old" quantum theory.

According to the eigenvalues given by Equation (4.6.4a), the particle can possess only very special values of the energy, which are in the proportion $1:4:9:\ldots$, i.e. it can only absorb or give up energy in the form of quanta, the size of which is given by (4.6.4a). Even for very light particles, however, this quantization only becomes important when the "box" is of atomic dimensions. If an electron (mass $m = 9.1 \times 10^{-28}$ g) occupies a box with a length $a = 1$ cm, then from (4.6.4a) the energy levels are given by

$$\varepsilon_n = \frac{(6.62 \times 10^{-27})^2}{8 \times 9.1 \times 10^{-28}} n^2 = 6.0 \times 10^{-27} n^2 \text{ erg} = 3.7 \times 10^{-15} n^2 \text{ eV}$$

As long as n is not extremely large, the energy levels are seen from this result to lie so close together, that the energy may be regarded as almost continuously variable. If, on the other hand, the box is of atomic dimensions, say $a = 10^{-7}$ cm, then

$$\varepsilon_n = 6.0 \times 10^{-13} \; n^2 \text{ erg} = 0.37 \; n^2 \text{ eV}$$

Even the separation of the first and second levels is, in this case, as much as 1 eV. It is important to note here that the normalization condition (4.4.2) must always be satisfied and that therefore n can never be equal to zero, even at absolute zero temperature. The energy corresponding to $n = 1$ is therefore called the zero-point energy of the particle in the box.

Each energy level (each eigenvalue) is associated with an eigenfunction, given by (4.6.5). The constant A which appears there can immediately be calculated by means of the normalization condition. It is easy to find that $A = (2/a)^{\frac{1}{2}}$. In these discussions, however, the value of this constant is of little consequence. Fig. 54 gives a graphical representation of the eigenfunctions for $n = 1$, 2 and 3. The figure demonstrates very clearly that quantization by means of (4.5.1) in the "old" quantum theory is replaced, in quantum mechanics, by the search for what (employing familiar concepts from vibration and wave theory) might be called standing waves. The left hand side of the figure could equally well relate to the eigenvibrations of a string, fastened at the points $x = 0$ and $x = a$. When $n = 1$, it gives the fundamental tone, $n = 2$ gives the first overtone, etc. For each characteristic mode, the length of the string is equal to a whole number of half wavelengths

$$a = \frac{n\lambda}{2}$$

This correspondence stems from the fact that Schrödinger's equation according to Section 4.4 is of the same type as the wave equations used in mathematical physics to calculate the standing waves in strings, membranes etc.

On one hand this formal agreement simplifies the mathematical problem, on the other hand it involves the danger that the analogy will be carried too far. In fact, as already discussed, the ψ waves have no direct physical significance but are purely symbolic. In the problem, illustrated by Fig. 54, of a particle in a box there is no question of some kind of undulating medium, but only of a graphical representation of ψ as a function of x.

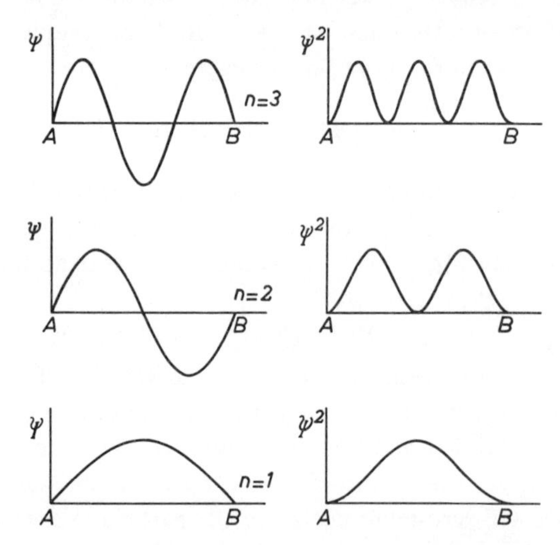

Fig. 54. Wave functions and their squares for the three lowest energy levels of a particle in a one-dimensional box.

In contrast to this, as already mentioned, a physical significance can be attributed to $|\psi|^2$. This quantity gives, for each point on the x axis, the probability that the particle will be encountered there. The variation of this quantity when $n = 1$, 2 and 3 is shown in the right-hand half of Fig. 54.

According to classical mechanics, the chance of finding the particle is equal for all points in the box. As the figure shows, this is not the case according to quantum mechanics. The greater n becomes, however, the closer the maxima and minima of the $|\psi|^2$ curve approach one another and thus the more closely the behaviour of the particle resembles that of the non-existent classical particle which visits all points with equal frequency. This is in agreement with Bohr's *correspondence principle*, according to which the quantum theory, in the limiting case of high quantum numbers, must lead to results which are in agreement with those of classical mechanics.

Finally we come to the treatment of the problem already posed of a particle moving freely in a cubic box with edge a and impenetrable walls. The Schrödinger equation for the particle in the box reads

$$\frac{h^2}{8\pi^2 m}\left(\frac{\partial^2\psi}{\partial x^2}+\frac{\partial^2\psi}{\partial y^2}+\frac{\partial^2\psi}{\partial z^2}\right)+\varepsilon\psi=0. \tag{4.6.7}$$

By differentiating twice it can easily be verified that this equation is satisfied by the three-dimensional form of (4.6.3):

$$\psi = A\sin 2\pi\frac{|p_x|}{h}x.\sin 2\pi\frac{|p_y|}{h}y.\sin 2\pi\frac{|p_z|}{h}z \tag{4.6.8}$$

where

$$\frac{1}{2m}(p_x^2+p_y^2+p_z^2)=\varepsilon \tag{4.6.9}$$

The boundary conditions are:
$\psi = 0$ for $x = 0$, for $y = 0$ and for $z = 0$ and
$\psi = 0$ for $x = a$, for $y = a$ and for $z = a$.

The first of these two conditions has already been satisfied by omitting the cosine term in (4.6.8), instead of writing the more general solution analogous to (4.6.2). The second condition supplies the relationships:

$$\left.\begin{aligned} 2\pi\frac{|p_x|}{h}a &= n_x\pi \quad \text{or} \quad |p_x| = \frac{h}{2a}n_x \\ 2\pi\frac{|p_y|}{h}a &= n_y\pi \quad \text{or} \quad |p_y| = \frac{h}{2a}n_y. \\ 2\pi\frac{|p_z|}{h}a &= n_z\pi \quad \text{or} \quad |p_z| = \frac{h}{2a}n_z \end{aligned}\right\} \tag{4.6.10}$$

where $n_x, n_y, n_z = 1, 2, 3, \ldots$. It will be seen that the relationships (4.6.10) are the three-dimensional extension of (4.6.6).

Combining (4.6.9) and (4.6.10):

$$\varepsilon_{n_x, n_y, n_z} = \frac{h^2}{8ma^2}(n_x^2+n_y^2+n_z^2) \tag{4.6.11}$$

Substituting (4.6.10) in (4.6.8):

$$\psi_{n_x, n_y, n_z} = A \sin \frac{n_x \pi x}{a} \sin \frac{n_y \pi y}{a} \sin \frac{n_z \pi z}{a} \qquad (4.6.12)$$

The kinetic energy of the particle can thus only assume the values prescribed by Equation (4.6.11), where n_x, n_y and n_z are the three (integral, positive) quantum numbers. The corresponding ψ functions are given by Equation (4.6.12). The stationary states (quantum states) in which the particle may exist are characterized by these ψ functions and thus also by the three quantum numbers mentioned.

In connection with the above it should be remarked that both in the one-dimensional and the three-dimensional case the second boundary condition ($\psi = 0$ at $x = a$ or at $x = a$, $y = a$ and $z = a$) is satisfied also by allowing negative values of n or n_x, n_y and n_z in (4.6.4) or (4.6.10). This, however, produces no new solutions of (4.6.1) or (4.6.7). The result of negative values of n in the one-dimensional case is simply that the ψ curves in Fig. 54 appear as their mirror image in the AB axis while the $|\psi|^2$ curves (which are the important ones) remain unchanged. The effect is the same in the three-dimensional case. All the required solutions can thus be obtained by allowing only positive values of n_x, n_y and n_z and consequently (4.6.8) and (4.6.10) do not contain the actual components of the momentum but only the absolute values of these components. This is in complete agreement with what was said in other words at the introduction of Equation (4.6.6).

4.7. QUANTUM STATES AND CELLS IN PHASE SPACE

Different quantum states of the particle just discussed have the same energy if the sum of the squares of their quantum numbers ($n_x^2 + n_y^2 + n_z^2$) is equal. If we speak of an energy level in the sense of Section 4.1, then there are but few levels which do not coincide with two or more others. This is shown for the lower levels by Table 8. In contrast to what was found in the one-dimensional case, the levels do *not* lie further apart as the quantum numbers become larger. On the contrary, an energy interval $\delta\varepsilon_i$ contains the more quantum states the higher the energy.

The situation is sometimes clarified by representing the quantum states by cells in the six-dimensional phase space (see Section 2.8). We shall show that each quantum state (e.g. each of the 23 states in Table 8) corresponds to a cell of magnitude h^3 in phase space and, conversely, each cell h^3 to one quantum state. The starting point for this will be (4.6.10) which gives the allowed absolute values of the three components of the "momentum" of the particle under discussion.

TABLE 8

n_x, n_y, n_z	$n_x^2 + n_y^2 + n_z^2$	Number of coincident levels
1, 1, 1	3	1
2, 1, 1 1, 2, 1 1, 1, 2	6	3
2, 2, 1 2, 1, 2 1, 2, 2	9	3
3, 1, 1 1, 3, 1 1, 1, 3	11	3
2, 2, 2	12	1
3, 2, 1 3, 1, 2 2, 3, 1 2, 1, 3 1, 3, 2 1, 2, 3	14	6
3, 2, 2 2, 3, 2 2, 2, 3	17	3
4, 1, 1 1, 4, 1 1, 1, 4	18	3

According to (4.6.10) the stationary states of the particle in a cubic box can be geometrically represented in a simple manner in the momentum space with Cartesian co-ordinates p_x, p_y and p_z. To do this, we choose $h/2a$ as the unit along all three axes and consider the lattice in which the points have the (positive) co-ordinates n_x, n_y and n_z (Fig. 55). According to (4.6.10) each of these points corresponds to a stationary state of the particular particle. One might also say that this octant of the momentum space is divided by the lattice into cells of magnitude $h^3/8a^3$ and that to every quantum state corresponds one of these cells.

The whole volume a^3 in the cubic box is open to the particle, so that the

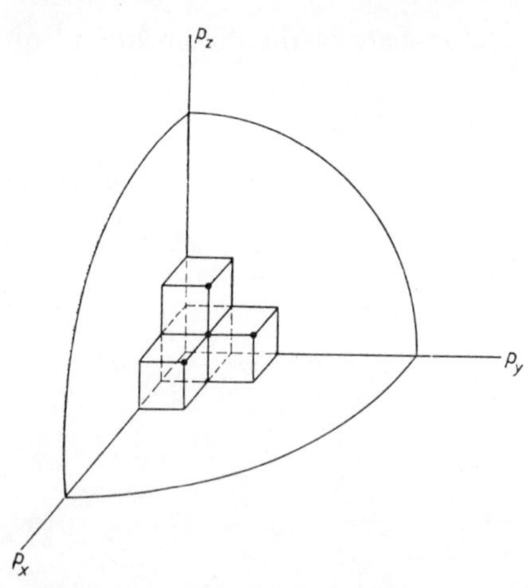

Fig. 55. The stationary states of a particle in a cubic box can be represented by the points of a simple cubic lattice in momentum space.

cells in the corresponding portion of the six-dimensional *phase* space have a volume $a^3 \times h^3/8a^3 = h^3/8$. The number of available quantum states is therefore found by dividing the accessible part of the octant of phase space by $h^3/8$.

The lattice of points, which was the starting point, can also be extended throughout the whole momentum space. A stationary state then has a representative point (or, if preferred, representative cell) in each octant of the momentum space. This does not affect the results since both the accessible momentum space or phase space and the cells (each now consisting of eight original cells) have been made eight times as large. In this treatment, the phase space is thus divided into units of magnitude h^3.

In the pictorial model of the "old" quantum theory all the cells (even those corresponding to negative values of p_x, p_y and p_z) have a direct physical meaning. In this model each collision of a particle with one of the walls reverses the sign of p_x or p_y or p_z. The particle therefore jumps continually in the momentum or phase space from one cell of magnitude $h^3/8a^3$ or $h^3/8$ to another. One might say that it is situated in the momentum space in a cell of magnitude h^3/a^3 and in the phase space in one of magnitude h^3 (each consisting of eight smaller units).

In order to obtain a rough impression of the number of quantum states (cells h^3 in the phase space) which are attainable by a gas molecule in a container of V litres volume, we shall count the number of cells for which the associated energy is less than (or equal to) the average energy $\varepsilon = 3kT/2$ of the molecules. This energy corresponds to a momentum $p = (2m\varepsilon)^{\frac{1}{2}} = (3mkT)^{\frac{1}{2}}$. The domain in the momentum space which contains these cells is a sphere of radius p and volume

$$\frac{4\pi p^3}{3} = \frac{4\pi(3mkT)^{3/2}}{3}$$

The corresponding volume in the six-dimensional phase space is V times as large, so that the required number of quantum states is given by

$$\frac{4\pi V(3mkT)^{3/2}}{3h^3}. \tag{4.7.1}$$

For a helium atom ($m = 6.7 \times 10^{-24}$ gram) in a box of 1000 cm^3 volume at a temperature of 300 °K, this formula supplies the number 10^{28}.

If the box contained not one helium molecule, but helium at 1 atm, then the volume of 1 litre would contain about 3×10^{22} molecules. The number of quantum states available to each particle in a gas at normal temperature and pressure is thus very much larger than the number of particles present. By far the greater part of the levels (cells in the phase space) will be unoccupied; only a minute fraction contains one particle. It is not practical to examine the populations of the separate levels as was done for a solid. It is more convenient to divide the levels (quantum states) into groups, each of which covers a small energy range (between ε_i and $\varepsilon_i + \delta\varepsilon_i$).

In the more pictorial model this amounts to a division of the momentum space into concentric shells of radius p_i and thickness δp_i. The volume of each shell is $4\pi p_i^2 \delta p_i$, while the corresponding volume in the six-dimensional phase space is $4\pi V p_i^2 \delta p_i$. The number of quantum states z_i in this range is again found by dividing this volume by h^3. For equal intervals δp_i, it increases as p_i increases.

If the energy range, i.e. $\delta\varepsilon_i$, is taken small enough, we may treat all the z_i levels (quantum states) belonging to one group as having the same energy ε_i. On an average they will all be occupied by a particle during the same fraction of the time. This fraction of the time (the population) is n_i/z_i, where n_i is the number of particles with energy ε_i. We shall see that — as in a solid — the levels become less densely populated as the energy ε_i becomes larger. In order to be able to derive the dependence of the populations on the energy, we must first study the problem of two or more particles in a container.

4.8. TWO IDENTICAL PARTICLES IN A BOX

We suppose that two identical particles move in a closed box and that the forces which these particles exert on each other may be neglected. The total energy U can then be written as

$$U = \varepsilon(1) + \varepsilon(2) \tag{4.8.1}$$

where $\varepsilon(1)$ and $\varepsilon(2)$ are the energies of the first and the second particle. This definition indicates that we are pretending to deal with particles in the classical sense of the word, which can be numbered. The artificial distinction thus introduced will later be undone.

The various energies which each separate particle may have, form, according to Section 4.6, a series of discrete values which we shall indicate by $\varepsilon_{k_1}, \varepsilon_{k_2}, \varepsilon_{k_3}, \ldots$, where each index k_r represents a particular combination of the three quantum numbers n_x, n_y, n_z. The solutions, pertaining to these energies, of the Schrödinger Equation (4.4.1) for *one* particle are the partial wave functions which describe the stationary states of the separate particles and which, correspondingly, are indicated by $\psi_{k_1}, \psi_{k_2}, \psi_{k_3}, \ldots$ If the first particle is in the r th quantum state and the second particle in the s th, then according to (4.8.1) we have for the total energy

$$U = \varepsilon_{k_r}(1) + \varepsilon_{k_s}(2). \tag{4.8.1a}$$

while the partial wave functions are $\psi_{k_r}(1)$ and $\psi_{k_s}(2)$.

The Schrödinger equation for the system of two similar particles is given, according to (4.4.3), by

$$\frac{h^2}{8\pi^2 m}(\nabla_1^2 + \nabla_2^2)\psi + U\psi = 0. \tag{4.8.2}$$

or, written in full:

$$\frac{h^2}{8\pi^2 m}\left(\frac{\partial^2\psi}{\partial x_1^2} + \frac{\partial^2\psi}{\partial y_1^2} + \frac{\partial^2\psi}{\partial z_1^2} + \frac{\partial^2\psi}{\partial x_2^2} + \frac{\partial^2\psi}{\partial y_2^2} + \frac{\partial^2\psi}{\partial z_2^2}\right) + U\psi = 0 \tag{4.8.2a}$$

A solution of this Schrödinger equation for the system of two particles is given by the product of the wave functions of the separate particles, thus by

$$\psi_A = \psi_{k_r}(1)\psi_{k_s}(2) \tag{4.8.3}$$

This is easily recognized by substituting (4.8.3) and (4.8.1a) in (4.8.2). The result of this substitution is:

$$\psi_{k_s}(2)\left[\frac{h^2}{8\pi^2 m}\nabla_1^2\psi_{k_r}(1) + \varepsilon_{k_r}(1)\psi_{k_r}(1)\right] +$$

$$+ \psi_{k_r}(1)\left[\frac{h^2}{8\pi^2 m}\nabla_2^2\psi_{k_s}(2) + \varepsilon_{k_s}(2)\psi_{k_s}(2)\right] = 0$$

Here the expressions between square brackets are equal to zero according to the Schrödinger equations for the separate particles, so that (4.8.3) is indeed a solution of (4.8.2).

For the same eigenvalue U, given by (4.8.1a), however, there is another solution of (4.8.2), viz.

$$\psi_B = \psi_{k_r}(2)\psi_{k_s}(1) \tag{4.8.4}$$

which is obtained from ψ_A by "interchanging the two particles".

Now it is a well-known, fundamental property of linear, homogeneous differential equations [1]), that every linear combination of two particular solutions is also a solution. The more general solution of (4.8.2) thus reads:

$$\psi = \alpha . \psi_{K_r}(1)\psi_{K_s}(2) + \beta . \psi_{K_r}(2)\psi_{K_s}(1). \tag{4.8.5}$$

4.9. THE INDISTINGUISHABILITY OF IDENTICAL PARTICLES

The general solution (4.8.5) found for Equation (4.8.2) would seem to indicate that, for the energy given by (4.8.1a), there are many stationary states of the system of two particles. This, however, is not the case due to the above-mentioned indistinguishability of identical particles.

In classical mechanics, identical particles do not lose their individuality. If, at a particular moment, the positions and velocities of a number of particles (supposed to be numbered) are known, then in principle it is possible to calculate their paths accurately and to say at any given moment which number belongs to each particle. According to the uncertainty relation (Section 4.2), the concept of the "path of a particle" has lost its meaning. In principle we can localize a number of elementary particles at a given moment, but it is quite impossible to identify them at a later time.

This indistinguishability means that, for the system in question, "interchange" of the two particles does not correspond to a physical reality. Consequently, since $|\psi|^2$, rather than ψ, has a direct physical significance, $|\psi|^2$ must not change as a result of this fictitious "interchange". This leaves two possibilities for ψ; either it changes sign when the two particles are "interchanged" or it remains unaltered. In the former case one speaks of an *antisymmetrical* wave function, in the latter case, of a *symmetrical* wave function. Experiments show that it depends on the nature of the particles whether they conform to one symmetry condition or the other.

For protons, neutrons, electrons and positrons, the eigenfunction is antisymmetric. For a system of two of these particles it therefore has the form

$$\psi_{\text{anti}} = \psi_{K_r}(1)\psi_{K_s}(2) - \psi_{K_r}(2)\psi_{K_s}(1) \tag{4.9.1}$$

This expression gives the wave function with the exception of a constant

---

[1]) Differential Equations (4.8.2) or (4.8.2a) are called linear because they contain no powers higher than the first of ψ or its derivatives and no products of these quantities; they are homogeneous because all terms contain either ψ or a derivative of ψ.

factor, which is determined by the normalization condition and does not interest us here.

Besides the change of sign when "interchanging" the particles, (4.9.1) shows us something else: When $k_r = k_s$ we obtain $\psi_{\text{anti}} = 0$. If the particles are subject to the antisymmetry condition, they can not be in the same quantum state. In the case of electrons, this is in complete agreement with the well-known Pauli exclusion principle from the "old" quantum theory. This, in fact, prohibits two electrons from ever being in the same quantum state, i.e. they may never have the same four quantum numbers (cf. Section 5.12). The antisymmetry condition is therefore a generalization of the Pauli principle.

For deuterons and other composite particles consisting of an even number of elementary particles, the eigenfunction is symmetrical for the particles. For a system made up of only two particles, this condition is fulfilled by the function

$$\psi_{\text{symm}} = \psi_{k_r}(1)\psi_{k_s}(2) + \psi_{k_r}(2)\psi_{k_s}(1). \tag{4.9.2}$$

Section 4.11 will deal in more detail with the symmetry requirements which apply to systems of identical complex particles. It will be shown that these requirements are a direct consequence of the antisymmetry requirement valid for the elementary particles.

The indistinguishability of the two identical particles is clearly expressed in the solutions (4.9.1) and (4.9.2). In contrast to the excluded solutions (4.8.3) and (4.8.4), they no longer give us any way of attributing a particular state to each separate particle.

a: $\psi_{k_r}(1) \cdot \psi_{k_s}(2)$

b: $\psi_{k_r}(2) \cdot \psi_{k_s}(1)$

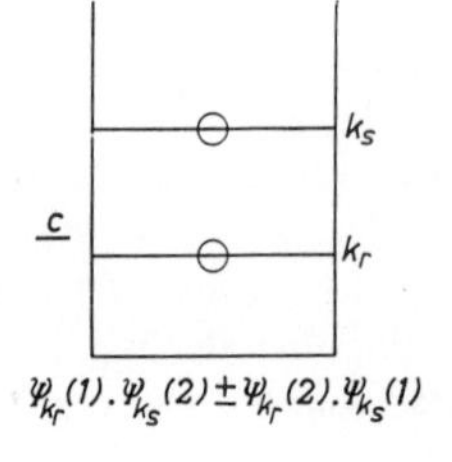

Fig. 56. The situation in which two identical particles are enclosed in a container, one in the state k_r and the other in the state k_s, can be realized in Maxwell-Boltzmann statistics in two ways. These are symbolically shown in *a* and *b*, but do not correspond to physical reality. In fact, the system can only exist in one state, shown by *c*, which is described by a symmetrical or antisymmetrical eigenfunction.

Since it is impossible to number the particles, the number of possible states in the new statistics, based on quantum mechanics, is smaller than in classical statistics. Particles which are described by antisymmetric functions are subject to the Fermi-Dirac statistics, while those described by symmetrical functions are subject to Bose-Einstein statistics. In our particular case of two identical particles, Fig. 56 gives a schematic representation of the states according to classical and new statistics in the case where one particle is at the level k_r and one at the level k_s. If the number of (single) energy levels is restricted to two, but the total energy no longer bound to remain constant, then the number of states of the two particles increases in the classical calculation of Maxwell-Boltzmann to four and in that of Bose-Einstein to three, because both particles may be found at the level k_r or both at the level k_s. In the Fermi-Dirac calculation the number of states remains unchanged at one; as a result of the Pauli exclusion principle the two particles can never be in the same state.

4.10. SYSTEMS OF MORE THAN TWO IDENTICAL PARTICLES

It is a simple matter to extend the foregoing discussions to systems of many similar particles. We shall consider a system of N identical particles moving in a closed container and we assume once more that the interaction between the particles can be neglected, so that

$$U = \sum \varepsilon(i)$$

A solution of the Schrödinger equation for the whole system is again given by the product of the wave functions of the separate particles:

$$\psi = \psi_{k_r}(1)\psi_{k_s}(2) \ldots \psi_{k_z}(N) \qquad (4.10.1)$$

Here, too, the "interchange" of two particles leads to a (generally different) eigenfunction with the same value of the energy U. In other words: all the eigenfunctions which are obtained from (4.10.1) by arbitrary permutation of the indices $k_r, k_s, \ldots$ are also solutions of the Schrödinger equation for the system at the same value U of the energy (i.e. at the same eigenvalue).

If all the N particles are in different quantum states $k_r, k_s, \ldots$ then permutation alone leads to $N!$ different eigenfunctions, all of which relate to the same energy U. If there are n_1 particles in the quantum state k_1, n_2 in k_2, and so on, the number of different eigenfunctions, obtainable by permutation is given by

$$P = \frac{N!}{n_1!\, n_2! \ldots} \tag{4.10.2}$$

However, the indistinguishability of the particles means that none of these P eigenfunctions describes a stationary state which can actually exist. When any two particles are "exchanged" the wave function must either remain unchanged or must change its sign. Consequently only one particular linear combination of *all* the P wave functions gives a description of the system which corresponds to the reality. The symmetrical combination, valid for instance for deuterons, is obtained by simply adding together all P functions $\psi(1)\psi(2) \ldots \psi(N)$:

$$\psi_{\text{symm}} = \sum_P \psi(1)\psi(2) \ldots \psi(N). \tag{4.10.3}$$

The letter P under the summation symbol indicates that the summation must be carried out over all the various products which can be obtained by permutation of the indices (not shown) $k_r, k_s, \ldots$

The antisymmetrical eigenfunction can be most easily written in the form of a determinant:

$$\psi_{\text{anti}} = \begin{vmatrix} \psi_{k_r}(1)\psi_{k_r}(2) \ldots \psi_{k_r}(N) \\ \psi_{k_s}(1)\psi_{k_s}(2) \ldots \psi_{k_s}(N) \\ \ldots\ldots\ldots\ldots\ldots\ldots\ldots\ldots \\ \ldots\ldots\ldots\ldots\ldots\ldots\ldots\ldots \\ \psi_{k_z}(1)\psi_{k_z}(2) \ldots \psi_{k_z}(N) \end{vmatrix} \tag{4.10.4}$$

(A property of a determinant is that it changes sign when either two rows or two columns are exchanged).

The determinant (4.10.4) represents the Pauli principle in its most general form. Indeed, every existing quantum state is represented by a row in the determinant, so that two rows of the determinant will be identical if two particles are in the same quantum state (e.g. $k_r = k_s$). It is known, however, that a determinant takes the value zero when two rows are the same. Thus, where the antisymmetry condition holds, no two particles can be in the same quantum state. Conversely, the behaviour of particles which obey the Pauli principle must be described by an antisymmetrical wave function.

The symmetrical wave function (4.10.3) does not disappear when $k_r = k_s$. In a system of particles which is described by a symmetrical wave function, it is therefore possible for an arbitrary fraction to be in the same state.

From the foregoing we see that a particular state of a system of many identical particles is described by a solution of the Schrödinger equation for the system, insofar as this solution is allowed by the symmetry requirements. A given wave function corresponds to both a particular energy and a particular "distribution" in the configuration space. In quantum statistics, thus, a particular wave function represents one particular micro-state of the system (one particular realization possibility in position and energy).

4.11. SYMMETRY RULES FOR SYSTEMS OF COMPLEX PARTICLES

It has already been remarked in Section 4.9 that the symmetry conditions applicable to systems of composite particles could be deduced from those applicable to elementary particles. We wish to demonstrate this now with the example of a gas consisting of hydrogen atoms and start by considering the simple case where only two hydrogen atoms, i.e. two protons and two electrons, are present in a container. If these elementary particles exerted no forces upon each other, then both the two electrons and the two protons would be governed by an eigenfunction of the type given by Equation (4.9.1). Each eigenfunction of the whole system of two protons and two electrons is thus a product

$$\psi = \psi_{\text{anti}} \cdot \psi'_{\text{anti}}, \tag{4.11.1}$$

where one antisymmetrical eigenfunction relates to the electrons and the other to the protons.

The function (4.11.1) changes sign when either the two electrons or the two protons are interchanged. It thus remains unchanged when both exchanges take place, which is the case when the two hydrogen atoms are interchanged. The question immediately arises whether this symmetry reasoning is also valid after the elementary particles have combined to form H atoms. This question must be answered in the affirmative, since the Coulomb forces give rise to an interaction which is symmetrical in the particles and therefore does not change the symmetry character of the wave function. And so we come to the important conclusion that the eigenfunction which describes a system of H atoms is symmetrical in the atoms.

Little is known, as yet, of the nature of the interaction between protons and neutrons in atomic nuclei. If this interaction, too, leaves the symmetry character of the wave functions unchanged, then one can understand the general rule, postulated by Ehrenfest and Oppenheimer [1]), that the eigenfunction describing a system of composite particles is symmetrical when the particles contain an even number of elementary particles and antisymmetrical when the number is odd.

The antisymmetry requirements (Fermi-Dirac statistics) are therefore

[1]) P. Ehrenfest and J. R. Oppenheimer, Phys. Rev. **37**, 433 (1931).

applicable to a gas consisting of D atoms (atoms of heavy hydrogen). In fact, exchanging two atoms means exchanging three elementary particles (proton plus neutron plus electron) with three others. The eigenfunction describing the system of D atoms, thus changes sign three times when two atoms change places and is therefore antisymmetrical in the atoms.

Application of the foregoing to ordinary gases leads us to expect that Bose-Einstein statistics will be valid for e.g. H_2, N_2, O_2 and He^4, and on the other hand that Fermi-Dirac statistics will be valid for e.g. HD, NH_3, NO and He^3. Up to now, all the experimental material, e.g. the spectra of the diatomic molecules, supports the rule of Ehrenfest and Oppenheimer (cf. Chapter 6). For quantum statistics, this rule is of great importance.

4.12. BOSE-EINSTEIN STATISTICS

The foregoing enables us to determine the number of micro-states in which a system of identical, independent, non-localized particles can exist. The simplest example of this type of system is an ideal gas of particles without internal degrees of freedom (see Chapter 5). In this section we shall discuss the calculation which must be employed when the particles are described by symmetrical wave functions.

In Section 4.7 we had already reached the conclusion that it is convenient to divide the quantum states, in which each separate particle may occur,

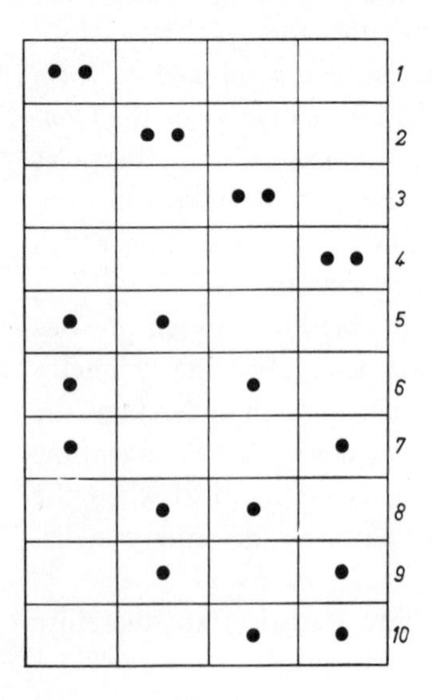

Fig. 57. According to Bose-Einstein statistics, two identical particles can arrange themselves in ten different ways over four states. Where Fermi-Dirac statistics are applicable the upper four arrangements are impossible.

into groups containing z_i states with approximately equal energy ε_i. Each of these states (or levels in the sense of Section 4.1) corresponds to a unit cell of magnitude h^3 in the six-dimensional phase space. The number of particles per group of z_i states was designated by n_i.

Of primary interest, now, is the number of ways in which the n_i particles in a group can be distributed over the appropriate z_i levels. In the case under consideration, each level can contain an unlimited number of particles. As a result it is easily seen that the required number of distributions can be found by the same reasoning as was used in deriving Equation (2.13.3). It is thus given by

$$\frac{(n_i + z_i - 1)!}{n_i!(z_i - 1)!} \tag{4.12.1}$$

In the case of two particles ($n_i = 2$) and four possible states ($z_i = 4$), this expression gives $5!/2!3! = 10$ different distributions. These are schematically represented in Fig. 57.

The groups of states are chosen such that z_i is very large with respect to 1. No appreciable error will be introduced by writing for (4.12.1)

$$\frac{(n_i + z_i)!}{n_i!z_i!} \tag{4.12.2}$$

The arrangement of the particles in a group is independent of that in the other groups. The number of micro-states relating to a particular distribution $n_0, n_1, n_2, \ldots$ over the various groups is thus given by the product of the corresponding expressions of the form (4.12.2):

$$g = \frac{(n_0 + z_0)!}{n_0!z_0!} \times \frac{(n_1 + z_1)!}{n_1!z_1!} \times \ldots \times \frac{(n_i + z_i)!}{n_i!z_i!} \times \ldots \tag{4.12.3}$$

or, more concisely:

$$g = \Pi \frac{(n_i + z_i)!}{n_i!z_i!} \tag{4.12.3a}$$

It is assumed that the system in question contains a constant number of non-localized particles and that the total energy is constant. Only those series $n_0, n_1, n_2, \ldots$ are then allowed which satisfy the restrictive conditions

$$\Sigma n_i = N \tag{4.12.4}$$

$$\Sigma n_i \varepsilon_i = U \tag{4.12.5}$$

(Compare the Equations (4.12.3a), (4.12.4) and (4.12.5) for a system of non-localized particles with (2.10.1), (2.10.2) and (2.10.3) for a system of localized particles).

Many distributions $n_0, n_1, n_2, \ldots$ are possible. The total number of microstates is given by the sum of all expressions of the form (4.12.3a), which satisfy the restrictive conditions (4.12.4) and (4.12.5).

In order to find the most probable distribution, one can use a method similar to that used in Section 2.10. Two particles are extracted from an arbitrary group of levels with energy ε_j and one of them is transferred to a group with higher energy ε_k, the other to a group at an equal interval lower with energy ε_i. In other words, a change is made which leaves the number of particles and the energy constant. Reasoning as in Section 2.10 and employing Equation (4.12.3) one finds that the condition for the occurrence of a maximum value of g is this time not that n_i, n_j and n_k shall form a geometrical progression, but that this shall apply to the three quantities

$$\frac{n_i + z_i}{n_i}, \frac{n_j + z_j}{n_j} \text{ and } \frac{n_k + z_k}{n_k}$$

For each energy ε_i may thus be written:

$$\frac{n_i + z_i}{n_i} = Ae^{\beta\varepsilon_i} \tag{4.12.6}$$

or

$$\frac{n_i}{z_i} = \frac{1}{Ae^{\beta\varepsilon_i} - 1} \tag{4.12.7}$$

in which A and β are constants.

Equation (4.12.7) can also be derived by the method of undetermined multipliers, described in Section 2.12. To do this, (4.12.3) or (4.12.3a) are written thus

$$\ln g = \ln \frac{(n_0 + z_0)!}{n_0!z_0!} + \ln \frac{(n_1 + z_1)!}{n_1!z_1!} + \ldots$$

Using Stirlings formula in approximation (2.4.4) we may write:

$$\ln g = \Sigma(n_i + z_i) \ln (n_i + z_i) - \Sigma n_i \ln n_i - \Sigma z_i \ln z_i \tag{4.12.8}$$

We are looking for the most probable distribution, i.e. the distribution for which g, and thus also $\ln g$, has a maximum value. When $\ln g$ is at a maximum, $\delta \ln g$ is zero for small variations, δn_i, of the occupation numbers n_i. From (4.12.8)

$$\delta \ln g = \sum \frac{\partial \ln g}{\partial n_i} \delta n_i = \Sigma \ln (n_i + z_i) . \delta n_i - \Sigma \ln n_i . \delta n_i$$

or:

$$\delta \ln g = \sum \ln \frac{n_i + z_i}{n_i} \cdot \delta n_i \tag{4.12.9}$$

The most probable distribution cannot yet be found by simply equating this expression to zero. Account must also be taken of the two restrictive conditions

$$\delta N = \sum \delta n_i = 0 \tag{4.12.10}$$

$$\delta U = \sum \varepsilon_i \delta n_i = 0 \tag{4.12.11}$$

These are taken into account by multiplying them by constants $-\alpha$ and $-\beta$, as yet unknown, and adding them to (4.12.9). For the most probable distribution this sum must have the value zero:

$$\sum \delta n_i \left(\ln \frac{n_i + z_i}{n_i} - \alpha - \beta \varepsilon_i \right) = 0 \tag{4.12.12}$$

Reasoning as in Section 2.12, and in Section 1 of the Appendix, it may be concluded from Equation (4.12.12) that

$$\ln \frac{n_i + z_i}{n_i} = \alpha + \beta \varepsilon_i$$

i.e.

$$\frac{n_i + z_i}{n_i} = e^{\alpha + \beta \varepsilon_i}$$

or

$$\frac{n_i}{z_i} = \frac{1}{e^{\alpha}\, e^{\beta \varepsilon_i} - 1} \tag{4.12.13}$$

This equation is in agreement with (4.12.7).

4.13. FERMI-DIRAC STATISTICS

The Fermi-Dirac method of counting must be used for a system of non-localized particles which are described by anti-symmetrical wave functions. It was stated earlier that two particles of such a system can never be in the same quantum state. Subject to this restrictive condition, the number of ways in which n_i particles can be divided among z_i quantum states is given, according to the derivation of Equation (2.5.1), by

$$\frac{z_i!}{n_i!(z_i - n_i)!} \tag{4.13.1}$$

For the two particles and four states in the preceding section this formula produces the result $4!/2!2! = 6$. This is in agreement with Fig. 57, in which the first four distributions are now forbidden.

Employing (4.13.1), Equation (4.12.3a) in the preceding section becomes, for a Fermi-Dirac system:

$$g = \Pi \frac{z_i!}{n_i!\,(z_i - n_i)!} \tag{4.13.2}$$

Proceeding as in the last section, this leads, for the occupation numbers of the most probable state of a Fermi-Dirac gas, to the expression

$$\frac{n_i}{z_i} = \frac{1}{Ae^{\beta\varepsilon_i} + 1} \tag{4.13.3}$$

In deriving this formula we have completely ignored the fact that the non-localized particles, to which Fermi-Dirac statistics are applicable, can possess an extra degree of freedom in the form of a "spin". This is true, for example, for electrons and means that the eigenfunction of an electron is not only dependent on the three co-ordinates x, y and z, but also on the "spin variable" s:

$$\psi = \psi(x, y, z, s).$$

In the pictorial model, an electron is an extremely small, electrically-charged sphere, which rotates about an axis through its centre of gravity. In this model the spin is the angular momentum associated with this rotation and can only orientate itself in two ways with respect to a magnetic field, viz. parallel or anti-parallel to the field. This pictorial representation may not be considered as more than an analogy (cf. Section 4.2). One should say that in many respects the electron behaves *as though* it were an electrically charged sphere, rotating about an axis. Quite apart from any pictorial representation, however, the fact remains that, besides x, y and z, another variable s is needed to describe the state of the electron.

Since the spin variable can only take two different values, it is often convenient to regard it more as an index than as a normal variable. Each wave function $\psi(x,y,z,s)$ is then replaced by two functions $\psi_1(x,y,z)$ and $\psi_2(x,y,z)$. The physical significance of this is that $|\psi_1|^2\mathrm{d}x\mathrm{d}y\mathrm{d}z$ indicates the probability that the electron will be found in the volume element $\mathrm{d}x\mathrm{d}y\mathrm{d}z$ with the first value of the spin variable, while $|\psi_2|^2\mathrm{d}x\mathrm{d}y\mathrm{d}z$ is the probability of its being found in the same volume element but with the second value of the spin variable.

As a result, the number of quantum states of an electron in a restricted space is twice as large as that found if the spin is neglected. If we wish, we may also say that each cell h^3 in the six-dimensional phase space (see Section 4.7) now corresponds to two, instead of one, eigenfunctions. Each cell can thus contain two electrons without breaking the Pauli rule. In this case z_i must be replaced by $2z_i$ in Equation (4.13.3).

4.14. MAXWELL-BOLTZMANN STATISTICS

The statistics to be treated in this section were developed by Maxwell and Boltzmann long before the birth of quantum mechanics. Therefore they relate to particles which are not bound by the rules of quantization and which, furthermore, can be individualized. In order to demonstrate the Maxwell-Boltzmann method of calculation, however, we shall apply it to particles, each of which can only exist in states allowed by the quantum theory. This means that the six-dimensional phase space is not divided into units of undefined magnitude, as was customary, but into units of magnitude h^3 (cf. Section 2.8). For a comparison of the three statistics there is absolutely no objection to this, since it does not affect the manner in which one proceeds to find the number of micro-states.

In the Maxwell-Boltzmann calculations an unlimited number of particles can be in the same state (in the same cell in the phase space). Interchange of two particles produces a new micro-state when they are in different quantum states (are situated in different cells in the phase space). Two particles, to which M.B. statistics may be applied, can thus be distributed in 16 different ways over the four states in Fig. 57 since distributions 5 to 10 must now be counted twice. In general, n_i M.B. particles can be distributed over z_i states in $z_i^{n_i}$ different ways.

The number of micro-states pertaining to a particular distribution n_0, n_1, n_2, . . . over the various groups of quantum states is *not* given simply by the product

$$\Pi\, z_i^{n_i}$$

Since the particles can be individualized, the exchange of two particles in different groups produces a new state. If the populations of the separate levels are prescribed (i.e. if the distribution within each group is fixed), then in both the B.E. and the F.D. statistics there is only one way in which it is possible to divide the N particles in groups n_0, n_1, n_2, . . . among these levels. This corresponds to the occurrence of only one eigenfunction, viz. the linear combination ψ_{symm} or ψ_{anti}. In M.B. statistics the distribution of the N particles in groups of n_0, n_1, n_2, . . . is possible in

$$\frac{N!}{n_0!n_1!\ldots}$$

ways. In each of these distributions, the n_i particles in a group can be distributed in $z_i^{n_i}$ different manners over the appropriate z_i levels. The total number of micro-states pertaining to a distribution $n_0, n_1, n_2, \ldots$ is thus

$$g = \frac{N!}{n_0!n_1!\ldots} z_0^{n_0} z_1^{n_1} \ldots \tag{4.14.1}$$

or, more concisely:

$$g = N! \Pi \frac{z_i^{n_i}}{n_i!} . \tag{4.14.1a}$$

By proceeding in the same manner as in Section 4.12 the occupation numbers of the most probable distribution of a Maxwell-Boltzmann gas can easily be deduced from this equation. We find:

$$\frac{n_i}{z_i} = \frac{1}{Ae^{\beta\varepsilon_i}} \tag{4.14.2}$$

4.15. COMPARISON OF THE THREE STATISTICS

The following formulae were found for the energy distribution in a system of identical, non-localized particles (e.g. a gas):

$$\frac{n_i}{z_i} = \frac{1}{Ae^{\beta\varepsilon_i} - 1} . \tag{B.E.}$$

$$\frac{n_i}{z_i} = \frac{1}{Ae^{\beta\varepsilon_i}} . \tag{M.B.}$$

$$\frac{n_i}{z_i} = \frac{1}{Ae^{\beta\varepsilon_i} + 1} \tag{F.D.}$$

A "statistical" temperature T can be introduced into the three formulae by means of the definition

$$\beta = \frac{1}{kT} \tag{4.15.1}$$

By the method followed in Chapter 2, it can be shown that this "statistical" temperature is identical with the absolute temperature from Chapter 1.

Even without looking at the above formulae, it is possible to make an important statement about the behaviour of F.D. gases at very low temperatures. In the neighbourhood of the absolute zero of temperature, the particles will have settled down into the lowest levels. According to the anti-symmetry condition, however, each level (each cell h^3) can only contain one particle. However low the temperature to which a F.D. gas is cooled, its energy can never fall below a paiticular value, which is called the *zero-point energy*. Also, the domain of the momentum space which is occupied by particles (or rather, points representing particles) can never be smaller than a particular minimum value.

Since the pressure of a gas is a result of the momentum of the particles, a F.D. gas, even at absolute zero temperature, must have a finite pressure, which is called the *zero-point pressure*. Validity of Fermi-Dirac statistics thus leads to a pressure greater than that given by the equation $pV = nRT$. Considered from the classical viewpoint, it seems as if repulsive forces are at work between the particles.

Conversely, validity of Bose-Einstein statistics leads to reduced pressure, so that it appears as if attractive forces are active between the paiticles.

If a gas is regarded as perfect only if it obeys the thermodynamic equation $pV = nRT$ (see Section 1.3), then F.D. and B.E. gases are never perfect, not even when the forces between the particles can be neglected. Only the application of M.B. statistics leads to the equation of state $pV = nRT$ (see Chapter 5). Deviations from the classically perfect behaviour is referred to as *degeneracy of a gas*. From the following it will appear that this degeneracy plays no appreciable role in normal gases.

Employing (4.15.1) we can write the distribution functions in the following form:

$$\frac{z_i}{n_i} + 1 = Ae^{\varepsilon_i/kT} \qquad \text{(B.E.)}$$

$$\frac{z_i}{n_i} = Ae^{\varepsilon_i/kT} \qquad \text{(M.B.)}$$

$$\frac{z_i}{n_i} - 1 = Ae^{\varepsilon_i/kT} \qquad \text{(F.D.)}$$

The three formulae are practically identical if $z_i/n_i \gg 1$. Now it has already been seen in Section 4.7 that this condition is certainly fulfilled for helium at 1 atm and 300 °K. Consideration of Equation (4.7.1) makes it clear that even a temperature drop to 10 °K, for example, is unable to produce appreciable signs of degeneracy in helium at 1 atm. The same is true for hydrogen and thus certainly for all heavier gases. According to (4.7.1) degeneracy phenomena cannot be expected to appear in a gas except at extremely low temperatures and/or extremely high pressures. Under such conditions, however, the deviations from classical behaviour as a result of the Van der Waals attractive forces are much greater than the deviations resulting from the modified statistics. Therefore, for calculating the energy distribution in a gas it is pointless to prefer one of the new types of statistics above the M.B. statistics, although the latter is, in principle, incorrect.

The relationships in the case of the "gas" of free electrons in a metal are quite another matter. Due to the tiny mass and high concentration of the particles, application of M.B. statistics in this case leads to completely incorrect results (see following section).

4.16. ELECTRONS IN SOLIDS

Energy bands

In order to explain the electrical conductivity of a metal, we must assume that it contains free or quasi-free electrons. Sodium can be taken as an example to demonstrate what is meant by this. One of the electrons in a free sodium atom, the valency electron, can be comparatively easily separated from the atom. This leads to the supposition that the lattice points in solid sodium are occupied by Na^+ ions, while the valency electrons can move freely through the lattice. The movements are disordered, but under the influence of an electric field a very small preference appears for the direction of the field.

In the old electron theory of metals which was developed by Drude and Lorentz at the beginning of this century, the disordered motion of the electrons in a metal was comparable in every way to the translational movements of the molecules in a gas. This assumption made it easy to understand why metals conduct both electricity and heat so well. The great difficulty immediately arose that the theory was apparently contradicted by measurements of specific heat. A metal with one free electron per atom should, according to the above model, possess an extra specific heat of $3R/2$ per gram-atom above the specific heat of $3R$ per gram-atom according to Dulong and Petit (see Section 2.14). In reality, metals exhibit specific heats which only deviate slightly from those of non-conducting elements. This apparently insurmountable difficulty could only be resolved with the help of quantum statistics.

Quantum-mechanical calculations and experimental results have led to the following conclusions with regard to the energy levels of electrons in solids. The innermost electrons in an atom are hardly affected when a large number of free atoms are brought together to form a crystal. Consequently these electrons remain localized and maintain their sharply defined energy levels. On the other hand, the electrons nearer to the outside of an atom are strongly influenced by the electrons in neighbouring atoms. This results in complete de-localization of these electrons: they move from atom to atom through the crystal. Instead of the equal energy levels of the electrons localized in the atoms, there occur new, no longer coincident, levels for the non-localized electrons. An energy level is thus transformed into an "energy band", in which the number of levels corresponds to the number of atoms in the crystal. The levels lie so close together that the energy within the band may be regarded as almost continuously variable. To each level in a band corresponds a wave function which is a linear combination of the wave functions relating to the separate atoms. The separation of the levels is greater and the bands thus broader when they lie at higher values in the energy spectrum (see Figs. 58 and 59).

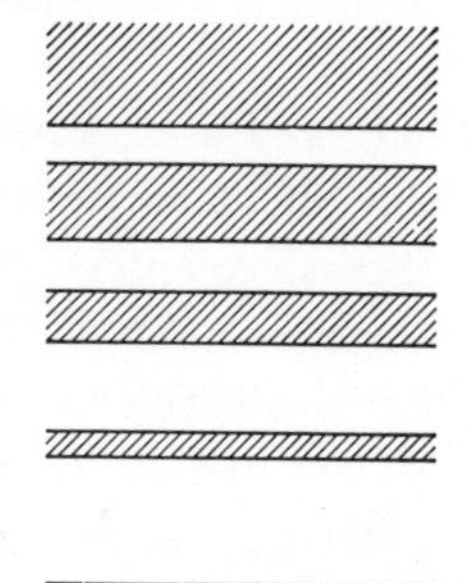

Fig. 58. Schematic representation of the energy bands in a crystal for constant separation of the atoms. In certain cases the bands may overlap (cf. following figure).

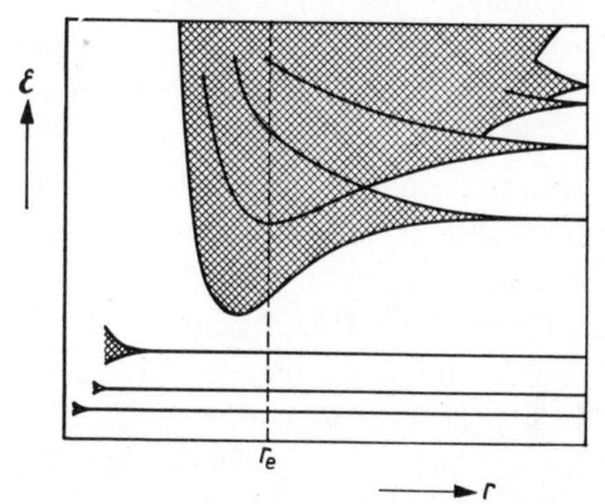

Fig. 59. When free atoms approach each other to form a crystal the energy levels in the atoms are transformed into energy bands. The broken line indicates the equilibrium separation (r_e) of the atoms in the crystal.

According to Section 4.13, not more than two electrons can be accommodated in each separate level. At absolute zero temperature the energy is minimum, so that up to a certain limit all the levels are occupied by two electrons with opposite spins, while above this limit every level is empty. This filling of the levels up to a certain limit may mean that in several bands all the levels are fully occupied, while the following band in the energy spectrum does not contain a single electron. (In this case, the highest full band is often referred to in short as "the" full band and the lowest empty band as "the" empty band). The result may also be that the highest band containing electrons is only partially filled.

In the solids under consideration, whether they be metals or insulators, the valency electrons move from atom to atom through the crystal. The distribution of these electrons over the permitted velocities is completely symmetrical: for every electron which moves with a particular velocity in a particular direction there is another which moves with the same velocity in the opposite direction. The occurrence of an electric current in a solid corresponds to an asymmetrical velocity distribution of the electrons. In other words: in the current bearing condition the average velocity of the electrons in the direction of the field is not zero. Consequently, it is only possible to generate an electric current carried by electrons in a material if the symmetrical velocity distribution can be changed into an asymmetrical one. If the upper band containing electrons is completely filled, this change cannot be brought about since to do this it is necessary to transfer electrons from occupied states to empty states. This transfer can take place only if the highest band containing electrons is only partly filled.

According to the above, materials with a partly filled band are metals. An example of this type of material is sodium containing one valency electron per atom. Its valency-electron band is only half filled since an energy band in a crystal of N atoms contains N levels and can accommodate $2N$ electrons. On these grounds it would seem surprising that bivalent elements such as magnesium, calcium and zinc are not insulators but metals. They are metals because there is a high degree of overlapping of their upper energy bands. This amounts to a fusion of several different bands to one broader band, which is only partially filled. Fig. 59 shows an example of overlapping energy bands.

In a material which only has (non-overlapping) full and empty bands, a weak electric field can only produce a current if the thermal motion has transferred an appreciable number of electrons from the full to the empty band. If the forbidden zone between full and empty bands is broad, extremely high temperatures would be needed to make the material electrically conducting in this way. Materials of this type are insulators. A weak electrical conductivity occurs at not too low temperatures if the forbidden zone is narrow or if atoms of an impurity are present to donate electrons to an empty band or withdraw them from a full one. In such cases one speaks of semi-conductors; in the first case of intrinsic semi-conductors and in the latter cases of extrinsic (or impurity) semi-conductors.

Electrons in metals

In a metal the conductivity electrons move in the periodic potential field caused by the metal ions. Various properties of metals, however, can be satisfactorily explained without considering this periodic field. In this "Sommerfeld approximation" the electrons are regarded as a simple Fermi-Dirac gas enclosed in an otherwise empty container and to which Equation (4.6.11) may be applied in order to find the allowed energies of the separate electrons.

The forces of attraction between the ions and the electrons and the mutual repulsion of the electrons are accounted for in a simplified manner in Sommerfeld's theory by supposing that the electrons in the interior of the metal have a (low) constant potential energy. Application of Equation (4.6.11) means that this constant energy is taken as zero point when considering the allowed kinetic energies of the electrons.

Sommerfeld's model makes it immediately clear why the classical statistics of Maxwell and Boltzmann are not applicable to the "gas" of free electrons in a metal. According to the previous section, the Fermi-Dirac statistics valid for the electrons may be replaced by Maxwell-Boltzmann statistics only when $z_i/n_i \gg 1$. Now, in sodium more than 70% of the metal volume is occupied by the electron gas (see Fig. 60). Therefore 6×10^{23} free electrons

Fig. 60. Arrangement of ions in the most closely packed lattice planes of sodium (left) and copper (right). The space between the ions is taken up by the "gas" of free electrons.

occupy a space of $0.7 \times 24 = 16.8$ cm^3. We will suppose for a moment that the number of quantum states available to the electrons in this space at 300 °K can be calculated in the manner of Section 4.7. Employing (4.7.1) with $V = 16.8$ cm^3, $m = 9.1 \times 10^{-28}$ g, $k = 1.38 \times 10^{-16}$ erg/degree and $h = 6.62 \times 10^{-27}$ erg.sec we find a number of approximately 3×10^{20}.

The number of levels calculated cannot be correct since it would only accommodate a very small fraction of the 6×10^{23} electrons. It is therefore quite out of the question that the electrons have an average translational energy of only $3kT/2$, as in the case of molecules in a gas and which was used as the starting point for the derivation of (4.7.1). Even at absolute zero temperature there are 3×10^{23} occupied levels in 1 mole sodium and the zero point energy based on this corresponds to a very high average energy of the electrons (see (4.16.8)). At room temperature, as will be seen below, there is still very little change in this situation. The condition for the validity of the M.B. statistics, viz. that in every energy domain only a very small fraction of the levels contains a particle, is therefore certainly not yet satisfied at room temperature. We are obliged to use the Fermi-Dirac distribution formula, which we shall now write in the form

$$\frac{n_i}{z_i} = \frac{1}{e^{(\varepsilon_i - \mu)/kT} + 1} \tag{4.16.1}$$

where n_i/z_i is the degree of occupation of the various levels ($n_i/z_i = 1$ indicates complete occupation, i.e. occupation by two electrons). The quantity μ in (4.16.1) has been introduced by replacing A in (4.13.3) by

$$A = e^{-\mu/kT}$$

The significance of μ becomes clear if we substitute the value zero for T in (4.16.1). The power of e in (4.16.1) in that case is infinitely great if $(\varepsilon_i - \mu)$ is positive, on the other hand it becomes zero if $(\varepsilon_i - \mu)$ is negative. Consequently n_i/z_i has the value zero when $\varepsilon_i > \mu$ and the value unity when $\varepsilon_i < \mu$. This means that all the levels of energies less than μ are completely occupied (by two electrons), while all levels with energies greater than μ are empty. In order to calculate n_i/z_i for $\varepsilon_i = \mu$, we start with a value of T very slightly above 0 °K and for $T \to 0$ we find $n_i/z_i = \frac{1}{2}$. The distribution at 0 °K is given in Fig. 61a. In this figure a continuous line has been drawn, although there are only discreet energy levels for the electrons. These levels, however, lie so close together that they cannot be shown separately in the figure.

This curve takes a completely different form if one plots the electron distribution $N(\varepsilon)$ as a function of the energy ε instead of the degree of occupation. This distribution is so defined that $N(\varepsilon)\mathrm{d}\varepsilon$ gives the number of electrons with energies between ε and $\varepsilon + \mathrm{d}\varepsilon$ present in the crystal in question. According to Section 4.7 the required distribution is given by the numbers of electrons present in concentric spherical shells of the momentum space. If all the cells of volume h^3/V in these shells are completely occupied (with two electrons), we have

$$N(p)\mathrm{d}p = \frac{8\pi V p^2 \mathrm{d}p}{h^3} \tag{4.16.2}$$

or, expressed in terms of energy (from $p = (2m\varepsilon)^{1/2}$):

$$N(\varepsilon)\mathrm{d}\varepsilon = \frac{4\pi V (2m)^{3/2} \varepsilon^{1/2} \mathrm{d}\varepsilon}{h^3} \tag{4.16.3}$$

In the energy range where electrons are to be found (i.e. below energy μ), the electron distribution at 0 °K is thus given by:

$$N(\varepsilon) = \frac{4\pi V(2m)^{3/2}\varepsilon^{1/2}}{h^3} \tag{4.16.4}$$

At energies greater than μ, $N(\varepsilon) = 0$. The parabolic relation given by (4.16.4) is reproduced in Fig. 61b.

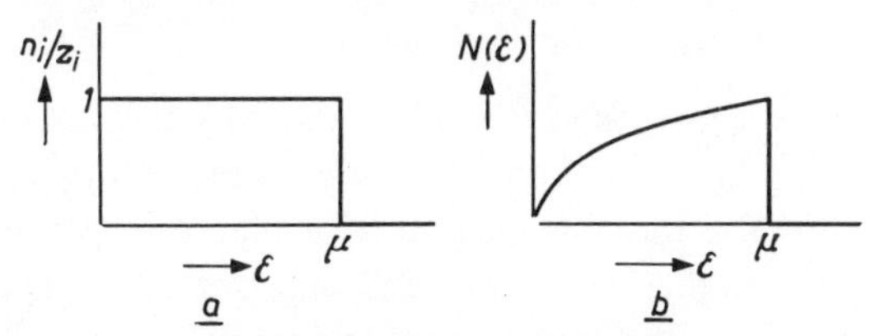

Fig. 61. (a) Occupation of the various energy levels by the "gas" of conduction electrons in a Sommerfeld metal at 0 °K. (b) Energy distribution of these electrons at 0 °K.

The maximum energy $\varepsilon_{max} = \mu$ of the electrons at 0 °K can be calculated without difficulty as follows. At this temperature a sphere in the momentum space with radius p_{max} and volume $4\pi p_{max}^3/3$ is occupied. Since the N electrons present occupy $N/2$ cells of volume h^3/V in the momentum space at 0 °K, we have

$$\frac{4\pi p_{max}^3}{3} = \frac{Nh^3}{2V}$$

whence

$$\varepsilon_{max} = \frac{h^2}{8m}\left(\frac{3N}{\pi V}\right)^{2/3} \tag{4.16.5}$$

where N/V is the number of electrons per unit volume.

The total energy U_0 of the N electrons at 0 °K is given by

$$U_0 = \int_0^{\varepsilon_{max}} \varepsilon N(\varepsilon) \mathrm{d}\varepsilon$$

Substituting (4.16.4) and (4.16.5) one finds

$$U_0 = \tfrac{3}{5} N\varepsilon_{max} = \frac{3h^2}{40m} N \left(\frac{3N}{\pi V}\right)^{2/3} \tag{4.16.6}$$

The average energy of the electrons is thus

$$\varepsilon_{av} = \tfrac{3}{5}\,\varepsilon_{max} \tag{4.16.7}$$

For sodium, Equations (4.16.5) and (4.16.7) supply the values

$$\left.\begin{aligned} \varepsilon_{max} &= 6.3 \times 10^{-12}\ \mathrm{erg} = 3.9\ \mathrm{eV} \\ \varepsilon_{av} &= 3.8 \times 10^{-12}\ \mathrm{erg} = 2.3\ \mathrm{eV} \end{aligned}\right\} \tag{4.16.8}$$

The specific heat per particle of a normal monatomic gas is $3k/2$ ($k = 1.38 \times 10^{-16}$ erg/deg.). In order to impart to the particles in a gas the same average kinetic energy as the free electrons in sodium possess at 0 °K, it would be necessary to heat the gas to more than 18,000 °K.

Thermodynamic properties of the electron gas

On the basis of the above one can understand why the specific heat of the electron gas is very much smaller than that forecast by classical theory. Let us imagine that the electron gas at 0 °K is placed in contact with a normal ideal gas at a higher temperature. Collisions between electrons and atoms will result in equilibrium at a final temperature T. During the adjustment the energy of the electrons as a whole will increase, that of the atoms decrease. Most of the electrons, however, cannot absorb any energy at all during the collisions, since the kinetic energy of the gas atoms is not large enough to raise them to unoccupied levels. Only those electrons which have energies within an interval kT from ε_{max}, approximately, can take up thermal energy. At room temperature kT amounts to only 1% or less of ε_{max}, so that roughly speaking only 1% of the electrons participates in the energy exchange. The contribution of the electrons to the specific heat is therefore very small. Thus the "apparently insurmountable difficulty", mentioned at the beginning of this section, is finally solved. In agreement with this solution of the problem, the curves in Fig. 61 change very little right up to the melting point of the metal (see Fig. 62).

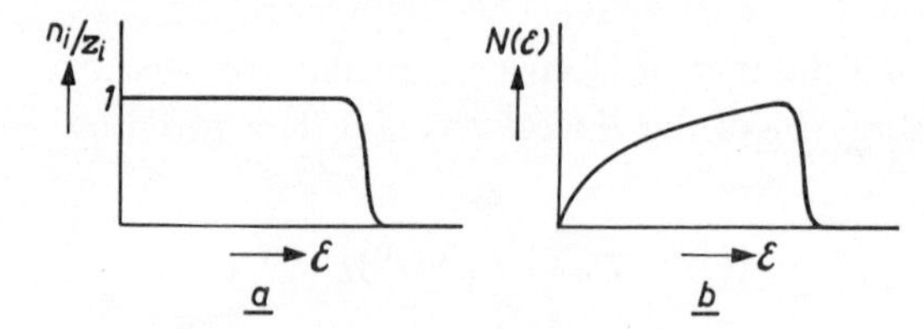

Fig. 62. (a) Occupation of the various levels by the conduction electrons at high temperature. (b) Energy distribution of these electrons at high temperature. Compare these with the corresponding curves for 0 °K in Fig. 61.

The fraction of the electrons which absorbs thermal energy when heated has been seen to be of the order of magnitude of kT/ε_{max}. Since each of the electrons involved absorbs a quantity of energy of the order of kT, the total *thermal* energy of the electrons is proportional to $(kT)^2/\varepsilon_{max}$, while their specific heat is proportional to k^2T/ε_{max}, i.e. to T. At low temperatures the specific heat of the lattice vibrations is proportional to T^3 (see Section 2.14).

The constant of proportionality in the latter case is much larger than that for the electrons. Nevertheless, as the temperature drops, one must finally arrive in a region where the two kinds of specific heat are of the same order of magnitude. Indeed, it has been found possible to measure the electronic specific heat at very low temperatures and to verify the linear dependence on the temperature.

As specified by the third law, the entropy, $S = k \ln m$, of the electron gas is zero at 0 °K. This accords with the fact that the state in which all the levels below ε_{max} are fully occupied and all those above ε_{max} are completely empty can be realized in only one way. As the temperature rises the electron entropy, like the electronic specific heat, increases but little.

The free energy, $F = U - TS$, of the electron gas at 0 °K is given by $F_0 = U_0$, i.e. by (4.16.6).

Since $\mathrm{d}F = -S\mathrm{d}T - p\mathrm{d}V$, the pressure of the gas at 0 °K is given by

$$p_0 = -\frac{\partial F_0}{\partial V} = -\frac{\partial U_0}{\partial V}$$

and thus, employing (4.16.6) and (4.16.5), by

$$p_0 = -\frac{3}{5} N \frac{\partial \varepsilon_{max}}{\partial V} = \frac{2}{5}\frac{N}{V}\varepsilon_{max} = \frac{2}{3}\frac{U_0}{V} \tag{4.16.9}$$

$$p_0 = \frac{h^2}{20m}\left(\frac{3}{\pi}\right)^{2/3}\left(\frac{N}{V}\right)^{5/3} \tag{4.16.9a}$$

Using this formula, we find for sodium a zero point pressure of the electrons of 9.1×10^{10} dyne/cm^2 $= 9 \times 10^4$ atm. Despite this high pressure the electrons do not escape from the metal, due to the strong electrostatic attraction between ions and electrons. The separation of the ions adjusts itself so that the pressure of the electron gas, the repulsion between the ions, the repulsion between the electrons and the mutual attraction of ions and electrons just balance each other.

The free enthalpy of the electron gas is given at 0 °K by

$$G_0 = U_0 + p_0 V$$

and thus, introducing (4.16.9) and (4.16.6), by

$$G_0 = \tfrac{5}{3}\, U_0 = N\varepsilon_{max} \tag{4.16.10}$$

We have already seen that ε_{max} is given by the value which μ from Equation (4.16.1) takes at absolute zero temperature:

$$\varepsilon_{max} = \mu_0. \tag{4.16.11}$$

In agreement with (4.16.10) and (4.16.11), μ in Equation (4.16.1) gives the chemical potential per particle. This quantity corresponds, according to the equation, with the energy at which half the quantum states are occupied. One often speaks of Fermi energy or Fermi level. It may be noted that the Fermi level may well coincide with an energy in the forbidden zone. This is the case, for example, in insulators and intrinsic semi-conductors at not too high temperatures. The upper full band can then only donate a few electrons to the first empty band, so that the upper levels in the full band are more than half occupied and the lowest levels in the empty band less than half occupied.

The great importance of the Fermi level lies in the fact that it is a chemical potential. Consequently, if different substances are placed in contact, the electron equilibrium adjusts itself in such a way that the whole system acquires a common Fermi level.

In so far as the above discussions of the thermodynamic properties of the electron gas in a metal were quantitative, they related to a temperature of 0 °K. What has been said about the properties at higher temperatures was of a qualitative nature. The dependence of the various thermodynamic quantities (including the Fermi level) on the temperature can be accurately calculated. For these calculations and their results, the reader is referred to the literature [1]).

---

[1]) See e.g. J. E. Mayer and M. Goeppert-Mayer, Statistical Mechanics, Chapter **16**, New York, 1940.

CHAPTER 5

THE ENTROPY OF MONATOMIC GASES

5.1. INTRODUCTION

We have seen in the preceding chapter that when calculating the energy distribution in a gas composed of atoms or molecules, it makes no appreciable difference whether one applies the statistics of Fermi-Dirac, of Bose-Einstein or of Maxwell-Boltzmann. It is true that when F.D. statistics are valid, the quantum states may not contain more than one particle, while they may contain any number according to the calculations of B.E. or M.B., but in a normal gas the number of quantum states accessible is so much larger than the number of particles that almost all of them will anyway contain only 0 or 1 particle, whichever statistics we apply.

In contrast to the energy distribution, the number of states in which a gas can exist, and thus also the entropy, is not independent of the statistics employed. Each distribution in which all N identical particles are in different states (occupy different cells in phase space) can, in fact, only be realized in one way by F.D. or B.E. statistics, while $N!$ possibilities exist according to M.B. statistics. Therefore, if we apply the (in principle incorrect) Maxwell-Boltzmann statistics to a gas, the error thus introduced must be corrected by dividing the calculated number of microstates by $N!$.

This is a most attractive consequence of the new statistics since, in classical statistics, the occurrence of the factor $N!$ had led to various difficulties (see Section 5.7). Even before the advent of the new statistics it was therefore felt to be essential to discard this factor, although it was impossible to justify this in a completely satisfactory manner. Before the birth of quantum mechanics, several leading figures in the world of physics, including Planck, Einstein, Ehrenfest and Schrödinger, carried on prolonged discussions about the best way to explain away the division by $N!$.

In the following sections we shall first calculate the entropy (and other thermodynamic properties) of a perfect gas whose particles have no internal degrees of freedom, i.e. of a gas possessing only translational energy. Subsequently we shall take into account the fact that the *internal* energy of a particle also may have various values. In this we restrict ourselves to monatomic gases and refer, for diatomic gases, to Chapter 6. However, in order to give a wider picture, Sections 5.9 to 5.11 will anticipate Chapter 6 to some extent.

5.2. THE ENTROPY OF A PERFECT GAS OF STRUCTURELESS PARTICLES

According to Section 4.15, the distribution function of a gas is given by the formula

$$\frac{n_i}{z_i} = \frac{e^{-\varepsilon_i/kT}}{A} \tag{5.2.1}$$

For each allowed energy ε_i of a gas particle, n_i/z_i is a number which is small with respect to unity. It gives the average occupation of the separate quantum state of energy ε_i and is denoted by the symbol n_i'. The total number of particles N is given by

$$N = \sum \frac{n_i}{z_i} = \Sigma\, n_i' \tag{5.2.2}$$

provided the summation is carried out over all the quantum states separately (and not over the *groups* of states, for then $N = \Sigma n_i$). From (5.2.1) and (5.2.2) follows:

$$A = \frac{\Sigma e^{-\varepsilon_i/kT}}{N} \tag{5.2.3}$$

so that

$$n_i' = \frac{N e^{-\varepsilon_i/kT}}{\Sigma e^{-\varepsilon_i/kT}} \tag{5.2.4}$$

in which the summation must extend over all individual quantum states.

It should be remarked that Equation (5.2.4), valid for this system of non-localized, independent particles, corresponds exactly to (2.12.10) which is valid for a system of localized, independent particles. Now, too, the expression

$$Z = \Sigma e^{-\varepsilon_i/kT} \tag{5.2.5}$$

is called the state sum or partition function of the particle. At 20 °C very many terms make an appreciable contribution to the state sum of a gas particle, while in contrast, only a few terms contribute noticeably to the sum for a particle in a solid. The numerical value of the state sum is therefore very much larger for a gas than for a solid.

After these remarks it is self-evident that Equations (2.12.11) and (2.12.12), valid for the total energy and the average energy per particle of an idealized solid, are also applicable to the perfect gas considered in this section. The entropy of this gas can be found via either B.E. or F.D. statistics. We shall choose B.E. statistics and write (4.12.8) in the form:

$$\ln g = \sum n_i \ln\left(\frac{n_i + z_i}{n_i}\right) + \sum z_i \ln\left(\frac{n_i + z_i}{z_i}\right) \tag{5.2.6}$$

The maximum value of g, according to the distribution function of a B.E. gas (Equation (4.12.6)) is reached when

$$\frac{n_i + z_i}{n_i} = Ae^{\varepsilon_i/kT}, \tag{5.2.7}$$

thus, substituting (5.2.3) and (5.2.5), when

$$\frac{n_i + z_i}{n_i} = \frac{Z}{N} e^{\varepsilon_i/kT} \tag{5.2.8}$$

We thus have:

$$\ln g_{\max} = \sum n_i \left(\ln \frac{Z}{N} + \frac{\varepsilon_i}{kT}\right) + \sum z_i \ln\left(1 + \frac{n_i}{z_i}\right) \tag{5.2.9}$$

In this particular case $n_i/z_i \ll 1$ and thus $\ln(1 + n_i/z_i) \simeq n_i/z_i$. The second sum in (5.2.9) then becomes $\Sigma n_i = N$, so that

$$\ln g_{\max} = N\left(\ln \frac{Z}{N} + 1\right) + \frac{U}{kT} \tag{5.2.10}$$

where $U = \Sigma n_i \varepsilon_i$ is the total internal energy of the gas. The entropy of the gas is given by

$$S = k \ln g_{\max} = kN\left(\ln \frac{Z}{N} + 1\right) + \frac{U}{T}. \tag{5.2.11}$$

If we compare this formula with (2.12.14), representing the entropy of an idealized solid, we see that (5.2.11) is obtained by subtracting $k \ln N!$ from (2.12.14). It is obvious that this is due to the indistinguishability of the particles in a gas and to the fact that they are virtually all in different quantum states (see Section 5.1). Nevertheless, the entropy of a gas has a greater numerical value than that for an equivalent quantity of a solid. This is related to the disorder in a gas discussed in Section 3.2 and is expressed in the much larger value of the state sum as compared to that for a solid.

The reader can deduce for himself without difficulty that the use of F.D. statistics leads to the same Equation (5.2.11) for the entropy of a gas as has been found here with the help of B.E. statistics. On the other hand, in agreement with what was said in Section 5.1, the reader will find that application of M.B. statistics gives a value for S which is too great by an amount $k \ln N!$.

5.3. THE STATE SUM OF A PARTICLE

In order to use Equation (5.2.11) to determine the numerical value of the entropy of a perfect monatomic gas, Z must first be expressed in terms of known quantities.

The summation in (5.2.5) is carried out over all stationary states in which a particle of the gas may be found. Accoiding to Section 4.6 these states are characterized by three quantum numbers n_x, n_y and n_z. For each possible combination of these numbers, the corresponding energy is given, from (4.6.9), by

$$\varepsilon_{n_x,n_y,n_z} = \frac{h^2}{8mV^{2/3}}(n_x{}^2 + n_y{}^2 + n_z{}^2) \tag{5.3.1}$$

The sum over the states is thus:

$$Z_{\mathrm{tr}} = \sum_{n_x,n_y,n_z} e^{-\varepsilon_{n_x,n_y,n_z}/kT} \tag{5.3.2}$$

We call this the translational sum Z_{tr}, because account has been taken only of the translational degrees of freedom of the monatomic gas. It is absolutely impossible to calculate all the terms of the sum separately and then to add them together, since the translational levels, as we saw in Section 4.7, lie very close together and are extremely numerous. In this case, the summation can be replaced by an integration.

Combining (5.3.1) and (5.3.2.) we have

$$Z_{\mathrm{tr}} = \sum_{n_x,n_y,n_z} e^{-\lambda(n_x^2+n_y^2+n_z^2)} \tag{5.3.3}$$

where

$$\lambda = \frac{h^2}{8mV^{2/3}kT} \tag{5.3.4}$$

According to (5.3.3) we can write Z_{tr} as the product of three independent sums:

$$Z_{tr} = \sum_{n_x} e^{-\lambda n_x^2} \sum_{n_y} e^{-\lambda n_y^2} \sum_{n_z} e^{-\lambda n_z^2} = \left\{ \sum_{n=1}^{\infty} e^{-\lambda n^2} \right\}^3 \tag{5.3.5}$$

If we convert the sum into an integral, we obtain an integral the value of which is known:

$$Z_{tr} = \left\{ \int_0^{\infty} e^{-\lambda n^2} \, \mathrm{d}n \right\}^3 = \left\{ \frac{1}{2} \left(\frac{\pi}{\lambda} \right)^{1/2} \right\}^3 \tag{5.3.6}$$

If (5.3.4) is substituted in this equation, we obtain:

$$Z_{tr} = \left(\frac{2\pi mkT}{h^2} \right)^{3/2} V \tag{5.3.7}$$

We repeat that this state sum relates to one separate particle which possesses no internal degrees of freedom, only translational degrees of freedom. Atoms and molecules do not, in general, conform to this model. Not only their translational state, but also their internal state, can be changed by supplying energy. At not extremely high temperatures, however, the atoms of the inert gases, for example, behave as though they possess no degrees of internal freedom: the energy required to excite these gases is so large that even at relatively high temperatures kT is small with respect to this energy. Consequently these excited states play no appreciable role in the state sum.

Many other atoms possess internal degrees of freedom which must be reckoned with even at low or moderately low temperatures. We shall return to this point in the last sections of this chapter.

Molecules consisting of two or more atoms always have internal degrees of freedom which play a part even at low temperatures. For this, the reader is referred to Chapter 6, which is devoted wholly to the calculation of the state sum and the thermodynamic quantities of diatomic gases (see also Sections 5.9 and 5.10).

5.4. THERMODYNAMIC FUNCTIONS OF MONATOMIC GASES AND STATE SUM

It is a simple matter to express all the thermodynamic functions of a perfect monatomic gas in terms of the state sum just discussed.

The entropy of 1 mole of gas is given, according to (5.2.11), by

$$s = R \ln \frac{Z}{N_0} + R + \frac{u}{T} \tag{5.4.1}$$

N_0 being Avogadro's number. The Helmholtz free energy, $F = U - TS$, of 1 mole of this perfect gas is thus given by

$$f = -RT \left(\ln \frac{Z}{N_0} + 1\right) \tag{5.4.2}$$

The free enthalpy (Gibbs' free energy), $G = U - TS + pV$, for 1 mole of a perfect gas is

$$\mu = u - Ts + RT = f + RT$$

or, from (5.4.2):

$$\mu = -RT \ln \frac{Z}{N_0} \tag{5.4.3}$$

Since $\mathrm{d}F = \mathrm{d}U - T\mathrm{d}S - S\mathrm{d}T$ and $\mathrm{d}U = T\mathrm{d}S - p\mathrm{d}V$ (see Chapter 1), we have for the partial derivative of F with respect to the temperature:

$$\left(\frac{\partial F}{\partial T}\right)_V = -S \tag{5.4.4}$$

In combination with (5.4.2), this relationship produces a new expression for the entropy of 1 mole of a perfect gas:

$$s = R \left\{ \ln \frac{Z}{N_0} + T \left(\frac{\partial \ln Z}{\partial T}\right)_v + 1 \right\} \tag{5.4.5}$$

The internal energy, $u = f + Ts$, follows from (5.4.2) and (5.4.5):

$$u = RT^2 \left(\frac{\partial \ln Z}{\partial T}\right)_v \tag{5.4.6}$$

As a last example, the molar heat capacity at constant volume will be expressed in terms of the state sum:

$$c_v = \left(\frac{\partial u}{\partial T}\right)_v = 2RT \left(\frac{\partial \ln Z}{\partial T}\right)_v + RT^2 \frac{\partial}{\partial T}\left(\frac{\partial \ln Z}{\partial T}\right)_v \tag{5.4.7}$$

or

$$c_v = -R \frac{\partial}{\partial T}\left(\frac{\partial \ln Z}{\partial(1/T)}\right)_v \tag{5.4.7a}$$

By inserting the value of Z_{tr}, as given by (5.3.7), in the formulae in this section, the values of the various thermodynamic quantities for our monatomic perfect gas can be found directly.

5.5. ENERGY AND ENTROPY

The molar translational energy of a perfect monatomic gas is given, from (5.3.7) and (5.4.6), by

$$u_{tr} = \tfrac{3}{2}\, RT = \tfrac{3}{2}\, N_0kT \tag{5.5.1}$$

The mean translational energy is thus $3kT/2$ per particle, i.e. $kT/2$ per degree of freedom:

$$\bar{\varepsilon}_x = \bar{\varepsilon}_y = \bar{\varepsilon}_z = \tfrac{1}{2}\, kT \tag{5.5.2}$$

The molar entropy of this gas, from (5.3.7) and (5.4.5) is

$$s_{tr} = R \ln \left\{ \frac{ve^{5/2}}{N_0h^3} (2\pi mkT)^{3/2} \right\} \tag{5.5.3}$$

In Section 1.16, purely thermodynamic reasoning supplied the equation

$$s = R \ln v + c_v \ln T + \text{constant} \tag{5.5.4}$$

The additive constant in this equation is no longer unknown but, as seen from (5.5.3), is given by

$$\text{const.} = R \ln \frac{e^{5/2}(2\pi mk)^{3/2}}{N_0h^3} = R \ln \frac{e^{5/2}(2\pi MR)^{3/2}}{N_0{}^4h^3} \tag{5.5.5}$$

where $M = N_0m$ is the atomic weight and $R = N_0k$ the gas constant.

From the well-known relationship in Chapter 1

$$T\mathrm{d}s = \mathrm{d}u + p\mathrm{d}v \tag{5.5.6}$$

it follows that

$$\left(\frac{\partial s}{\partial v}\right)_u = \frac{p}{T} \tag{5.5.7}$$

From (5.5.3) we derive:

$$\left(\frac{\partial s}{\partial v}\right)_u = \frac{R}{v} \tag{5.5.8}$$

Combining (5.5.7) and (5.5.8) we find the well-known equation of state for a perfect gas:

$$pv = RT \tag{5.5.9}$$

With the help of this equation we can also write (5.5.3) in the form:

$$s_{tr} = R \ln \frac{(2\pi M)^{3/2}(ekT)^{5/2}}{ph^3 N_0{}^{3/2}} \tag{5.5.10}$$

If, in this equation, we fill in the numerical values of the physical constants,

$$\begin{aligned} R &= 1.9872 \text{ cal/deg.} \\ k &= 1.3804 \times 10^{-16} \text{ erg/deg.} \\ h &= 6.6252 \times 10^{-27} \text{ erg. sec.} \\ N_0 &= 6.0232 \times 10^{23} \end{aligned}$$

then we obtain for the entropy of a perfect monatomic gas the expression

$$s_{tr} = \tfrac{3}{2} R \ln M + \tfrac{5}{2} R \ln T - R \ln p + 25.165 \text{ cal/degree.mole} \tag{5.5.11}$$

Here p must be expressed in dyne/cm^2. If p is expressed in atm (1 atm $=$ 1.01325×10^6 dyne/cm^2), then (5.5.11) is transformed to

$$s_{tr} = \tfrac{3}{2} R \ln M + \tfrac{5}{2} R \ln T - R \ln p - 2.316 \text{ cal/degree.mole} \tag{5.5.12}$$

One is often primarily interested in the "standard entropy", which is the entropy s^0 at 25 °C (298.15 °K) and 1 atm. We find:

$$s^0_{tr} = \tfrac{3}{2} R \ln M + 25.990 \text{ cal/degree.mole} \tag{5.5.13}$$

The description "standard state" and the superscript 0 are often used in relation to a gas to indicate, as above, the perfect gas state and a pressure of 1 atm, but *not* to fix a particular temperature (see Section 3.4). If necessary, a suffix is then used to indicate the temperature, e.g. s^0_{1000} is the entropy of 1 mole of a gas regarded as perfect at a pressure of 1 atm and a temperature of 1000 °K.

As an example of the use of (5.5.12), we may calculate the molar entropy of gaseous mercury at 630 °K and 1 atm. Inserting in the equation $M =$ 200.61, $T = 630$, $p = 1$, we find

$$s = 45.51 \text{ cal/degree}$$

Using the experimental values of the specific heats of solid and liquid mercury and the heat of fusion and heat of evaporation of this element, the molar entropy can also be calculated with the help of (2.17.2). One then finds

$$s = 45.4 \text{ cal/degree}$$

Thus we find with the help of (2.17.2) almost exactly the same value for the entropy as from (5.5.12). When it is remembered how completely different are the ways in which these two values of the entropy have been found, it will be appreciated what a fine result this is! One value, in fact, has been calculated by means of specific heats, heats of fusion and evaporation, the other by counting the numbei of ways in which the macroscopic state of the gas can be realized, this latter being done without a thought for heats of fusion or evaporation or for any "previous history" whatsoever at low temperatures.

In other words, one value is obtained in a way which requires no knowledge of the existence of atoms, but only of the results of calorimetric measurements; in contrast, the other is found in a way which regards gaseous mercury as a collection of atoms which exert virtually no forces on each other and which requires no knowledge of the existence of liquid or solid mercury.

Strictly speaking we must correct the value of the calorimetric entropy for the transition from the actual to the perfect gas state. The agreement between the calorimetric and the statistical entropy is then even better. For the method of calculating this correction, we refer to Section 6.13.

5.6. INSENSITIVITY OF THE FORMULA $S = k \ln g$

In the foregoing sections we have calculated the entropy from the number of ways $g_{\max}$, in which the most probable distribution can be achieved. When calculating the entropy of an idealized solid we have already shown in Chapter 2 that the result does not change noticeably if the number of states of the most probable distribution is replaced by the *total* number of states, m, which is possible for the system at a given energy U. We can even go a step further and replace $g_{\max}$ by the number q of *all* the states which have an energy $E \leqslant U$. In this section it will be shown that in this case, too, we find the same value foı the entropy.

In order to determine q it is more advantageous to consider a phase space with $6N$ dimensions than one with 6 dimensions. A point in this $6N$-dimensional phase space gives, by projection on the first 6 axes the positional and momentum co-ordinates x_1, x_2, x_3, p_1, p_2, p_3 of the first atom, by projection on the second 6 axes the co-ordinates of the second atom, and so on. One point in this phase space thus gives the $6N$ co-ordinates of *all* N atoms. According to the quantum conditions (4.6.10) not all of the states are allowed. Each allowed state of the system corresponds to a cell of magnitude h^{3N} in the $6N$-dimensional phase space.

This method of representation, in which each atom has its own position and momentum axes, naturally involves individualization of the atoms. For gases in normal conditions we can, as above, nullify this effect by dividing the number of micro-states by $N!$.

For our monatomic, perfect gas at not too high temperatures the internal energy U is simply the total translational energy of the atoms

$$\frac{p_1^2 + p_2^2 + \dots + (p_{3N})^2}{2m} = U$$

or

$$p_1^2 + p_2^2 + \dots + (p_{3N})^2 = 2\,mU \tag{5.6.1}$$

If we had only two atoms in a one-dimensional space (two atoms moving along the x axis), (5.6.1) would become

$$p_1^2 + p_2^2 = 2\,mU$$

This is the equation of a circle (a two-dimensional sphere) of radius $(2mU)^{1/2}$.

For three atoms in a one-dimensional space or 1 atom in a three-dimensional space we obtain

$$p_1^2 + p_2^2 + p_3^2 = 2\,mU$$

This is the equation of a normal (three-dimensional) sphere of radius $(2mU)^{1/2}$, i.e. the equation of the surface $U =$ const. in the three-dimensional momentum space.

In the same way (5.6.1) is the equation of a $3N$-dimensional "hyper-sphere" in the $3N$-dimensional momentum space of the N atoms. The radius of this sphere is given by $(2mU)^{1/2}$, the volume by $C(2mU)^{3N/2}$. If N is very large C is given approximately by

$$\ln C = \frac{3N}{2}\left(\ln \frac{2\pi}{3N} + 1\right) \tag{5.6.2}$$

In the $3N$-dimensional momentum space the whole volume inside the above-mentioned "hyper-sphere" is open to our system because $E \leqslant U$. In the $3N$-dimensional configuration space a volume V^N is available. The volume in the $6N$-dimensional phase space is thus

$$Q = CV^N\,(2mU)^{3N/2} \tag{5.6.3}$$

The required number of micro-states is thus given by

$$q = \frac{CV^N\,(2\,mU)^{3N/2}}{h^{3N}\,N!} \tag{5.6.4}$$

If we now calculate $k \ln q$ by means of (5.6.4), (5.6.2) and (5.5.1), we obtain for 1 mole of gas a result which is in complete agreement with (5.5.3). Counting too many micro-states (q instead of g_{max}) is just compensated by making use of the approximations (5.6.2) and $\ln N! = N \ln N - N$.

5.7. THE PARADOX OF GIBBS

The division of the number of "classical" micro-states by $N!$, necessitated by the indistinguishability of identical particles, in order to obtain the real number is closely connected with the so-called "Gibbs' paradox" (see Section 1.17).

If we do not divide by $N!$, then according to the preceding sections (see e.g. Equation (5.6.4)) the entropy of N molecules in a volume V at constant temperature is given by

$$S_1 = k \ln V^N + K = kN \ln V + K.$$

that of $2N$ molecules in a volume $2V$ by

$$S_2 = k \ln (2V)^{2N} + 2K = 2\,kN \ln 2V + 2K = 2\,S_1 + 2\,kN \ln 2.$$

The total entropy would thus increase by $2kN \ln 2$ when a partition between two volumes V of the same gas, each containing N molecules, is removed.

If, however, the number of micro-states is divided by $N!$ (Equation (5.6.4)), we find

$$S_1 = k \ln \frac{V^N}{N!} + K$$

$$S_2 = k \ln \frac{(2V)^{2N}}{(2N)!} + 2K$$

and thus, employing formula $\ln N! = N \ln N - N$:

$$S_2 = 2S_1.$$

Thus in reality the total entropy does not change when a partition is removed in the above way from between two volumes of *the same* gas.

If, however, the two gases are different, interchanging two different molecules will lead to a new state, so that in that case an extra factor $(2N)!/N!N!$ occurs in the number of micro-states after mixing. An extra term

$$\Delta S = k \{\ln (2N)! - 2 \ln N!\} = 2kN \ln 2$$

thus appears in the entropy. If we start with a total quantity of 1 mole, i.e. $N_0 = 0.6 \times 10^{24}$ molecules, of which $N_0 x$ are identical but different from the remaining $N_0(1-x)$ mutually identical molecules, then mixing introduces into the number of micro-states an extra factor

$$\frac{N_0!}{(N_0 x)! \{N_0 (1-x)\}!}$$

corresponding to an entropy of mixing

$$\Delta S = k \{-N_0 x \ln x - N_0 (1-x) \ln (1-x)\}$$

or, since $kN_0 = R$ (the gas constant):

$$\Delta S = -R \{x \ln x + (1-x) \ln (1-x)\}$$

Thus we find here once more the entropy of mixing given by Equation (1.17.2b), Chapter 1, but now deduced by a statistical method.

When two quantities of different gases are mixed by removing a partition the entropy thus increases, while this is not the case when the gases mixed are identical. In Gibbs' time this seemed paradoxical as a result of the following consideration. Suppose that the properties of the two sorts of molecules are allowed to approach each other gradually, then it will not be possible to say at which moment they have become so similar that the entropy no longer increases when they are mixed (cf. Section 1.17).

The paradox has been resolved by modern atomic theory since this theory shows that a gradual transition from one molecule to another is inconceivable. The properties of matter change discontinuously. It is always possible to make the distinction: two molecules are identical or they are different. That an entropy of mixing occurs in one case and not in the other, can thus no longer be regarded as paradoxical.

5.8. THE STATE SUM OF A SYSTEM OF PARTICLES

In the early sections of this chapter it was shown how the thermodynamic functions of a perfect gas can be calculated from the state sum of one gas particle. At first sight it may seem strange that the thermodynamic functions may be related to the properties of one particle since they relate essentially to the properties of macroscopic systems. In the following we shall investigate the conditions which must be fulfilled by a system of many particles so that a relationship may be established between the thermodynamic properties and the individual properties of a single particle.

In the state sum defined by (5.2.5), the energies of one particle occur and consequently this sum only exists if one can refer to the energy levels of separate particles. This is the case only if the particles are wholly independent of one another. In the gases, liquids and solids with which one is dealing in reality, a mutual potential energy occurs which cannot be divided between the particles, but which belongs to the system as a whole. It is just as impossible to speak of the energies of the separate particles in an arbitrary macrosystem as it is to speak of the energy levels of the separate atoms in a molecule composed of two or more atoms.

If one wishes to relate the macroscopic properties of, for example, a gas or crystal to its energy levels, then one is obliged to regard the whole gas or crystal as one quantum-mechanical system with energy levels U_0, U_1, U_2, . . . ,U_i, . . . , in which U_0 is the energy of the system at absolute zero temperature. In principle, these levels and their multiplicity could be found by writing the Schrödinger equation for the whole system and then solving it.

Starting with the classical statistical considerations introduced by Gibbs (see Chapter 2), we suppose the existence of a large number of identical copies of the macroscopic system being studied, all of which are in thermal contact with each other and possess a certain quantity of energy in common. This "ensemble of systems" might consist, for example, of a large number of containers with the same quantity of the same gas or a large number of crystals of identical dimensions and composition. The interaction between the systems is supposed to be only just sufficient to ensure thermal contact; it has no effect upon the properties of the separate systems. In other words, this artifice employed by Gibbs introduces a "super-system" (the ensemble of systems) consisting of "super-molecules", each independent of the other. Each "super-molecule" is one of the systems of the ensemble and therefore a macroscopic quantity of material. Each is situated, as it were, in a constant temperature bath, consisting of the $(N - 1)$ other systems and consequently behaves as the one really existing system would do if it were immersed in a bath at constant temperature [1]).

In analogy to the behaviour of a gas consisting of independent molecules, the independent systems will distribute themselves in groups n_0, n_1, n_2, . . . over the allowed energies U_0, U_1, U_2, . . . in such a way that the number of realization possibilities, and thus the entropy, will have the maximum value consistent with the total energy $\Sigma n_i U_i$ of the ensemble. When calculating this number of realization possibilities, it should be remembered that we are now dealing with macroscopic systems which can be labelled. In other words, the systems are distinguishable and the calculations in Section 4.14

[1]) E. Schrödinger, Statistical Thermodynamics, University Press, Cambridge, 1948.

are applicable to the ensemble. The number of realization possibilities for the distribution n_0, n_1, n_2, ... is given by Equation (4.14.1) or (4.14.1a), in which z_i now represents the number of states available to a system of energy U_i.

Reasoning as in Sections 4.14 and 5.2, it is not difficult to find that the most probable distribution of the systems over the energies U_0, U_1, ... is given by an equation analogous to (5.2.4). The quantity

$$\mathcal{Z} = \sum e^{-U_i/kT} \tag{5.8.1}$$

occurring in the denominator of the new expression is called the state sum of the system. The summation must be extended to cover all the states in which the system can exist.

It is also a simple matter to show that Equation (5.2.11), with a few modifications, gives the entropy of the ensemble under consideration. The modifications consist, as will be obvious, of replacing Z by $\mathcal{Z}$ and U by the total energy NU of the ensemble, in which U is the mean energy per system. Since the systems are distinguishable, one must also add a quantity $k \ln N!$ to the entropy. With these modifications, Equation (5.2.11) becomes, for an ensemble of systems:

$$S_{\text{ens}} = k\, N \ln \mathcal{Z} + \frac{NU}{T} \tag{5.8.2}$$

It should be remembered here that for each separate system there is only a very small probability that its energy will differ appreciably from the mean value U. The entropy and the free energy per system are therefore given by

$$S = k \ln \mathcal{Z} + \frac{U}{T} \tag{5.8.3}$$

$$F = -\, kT \ln \mathcal{Z} \tag{5.8.4}$$

The state sum $\mathcal{Z}$ of the whole system is the fundamental quantity which should be used as a starting point when calculating thermodynamic quantities. Unfortunately, it is not possible to do this exactly, since the energies U_i cannot be calculated accurately. Here we come up against one of the central problems of statistical thermodynamics, that of finding satisfactory approximations for the unknown energies U_i. All the methods employed for this purpose, consist essentially of trying to visualize the macroscopic system under discussion as being composed, in one way or another, of mutually independent sub-systems, the interaction between which is limited to what is strictly necessary for the interchange of energy. If this is successful, then each energy U_i can be written as the sum of the energies of the sub-systems:

$$U_i = \sum \varepsilon_j(i) \tag{5.8.5}$$

in which j is the number of the independent sub-systems. In this case Equation (5.8.1) becomes

$$\mathcal{Z} = \sum e^{-\varepsilon_1/kT} \cdot e^{-\varepsilon_2/kT} \ldots e^{-\varepsilon_N/kT} \tag{5.8.6}$$

in which ε_1 is the energy of the first sub-system, ε_2 that of the second, and so on.

Equation (5.8.6) can also be written:

$$\mathcal{Z} = \sum_1 e^{-\varepsilon_1/kT} \cdot \sum_2 e^{-\varepsilon_2/kT} \ldots \sum_N e^{-\varepsilon_N/kT} \tag{5.8.7}$$

in which each sum is the state sum of a separate sub-system.
Thus

$$\mathcal{Z} = \left\{ \sum e^{-\varepsilon/kT} \right\}^N = Z^N \tag{5.8.8}$$

This relationship shows why, in simplified cases, it is possible to express the thermodynamic properties of a system in terms of the state sum of one sub-system. The perfect gas already discussed is the simplest example of a system built up of independent sub-systems (atoms or molecules). It should, however, be remarked that in the case of a gas a correction must be applied to (5.8.8). The derivation of this equation was carried out as if the particles were recognizable individuals, i.e. as if Maxwell-Boltzmann statistics could be applied to a gas. As we already know from the above, this can be corrected by dividing by $N!$. Consequently, we obtain for a perfect gas:

$$\mathcal{Z} = \frac{Z^N}{N!} \tag{5.8.9}$$

As Debye has shown, the sub-division of a system into mutually independent sub-systems is also possible in many cases where the interaction between the constituent atoms of the system is so strong that satisfactory results cannot be obtained by regarding these atoms as the independent sub-systems (see Section 2.14). This is especially true for crystals, where the thermal vibrations consist of strongly correlated movements of the constituent atoms. In Debye's theory the independent sub-systems are obtained by resolving the above movements into mutually independent modes of vibration.

In this case, thus, the sub-systems are not mass-particles which may or may not be localized, but are non-localizable abstract oscillators which can be distinguished from one another by means of their different frequencies. The various independent oscillations are very numerous and the energy U_i of the system can be divided between them in a great number of different ways.

The systems considered in this and the next (last) chapter are all perfect gases. For the sake of simplicity we shall therefore continue to use the term state sum to denote the state sum of one atom or molecule. It should always be kept in mind, however, that for an arbitrary system this does not exist and that, strictly speaking, only $\mathcal{Z}$ fully deserves the designation state sum.

It is perhaps worth while to conclude this section with an example employing small numbers to show that one is not justified in applying Equation (5.8.8) to a gas, since in fact, this amounts to applying M.B. statistics.

We imagine that we have a system consisting of only two identical, independent particles 1 and 2, each of which can only occur in two single states a and b, corresponding to the energies ε_a and ε_b. It is clear that the state sum of one particle is then given by

$$Z = e^{-\varepsilon_a/kT} + e^{-\varepsilon_b/kT} \tag{5.8.10}$$

Application of Equation (5.8.8) would give, for the system of two particles:

$$\mathcal{Z} = Z^2 = e^{-(\varepsilon_a+\varepsilon_a)/kT} + 2e^{-(\varepsilon_a+\varepsilon_b)/kT} + e^{-(\varepsilon_b+\varepsilon_b)/kT} \tag{5.8.11}$$

This does, indeed, correspond to the four states which are possible according to Maxwell-Boltzmann statistics, viz.: (1) both particles have energy ε_a; (2) particle 1 has energy ε_a and particle 2 energy ε_b; (3) particle 2 has energy ε_a and particle 1 energy ε_b; (4) both particles have energy ε_b.

If the system is subject to Bose-Einstein statistics then, in fact, it has only three states, because the second and third states are, in reality, only one state. If it is subject to Fermi-Dirac statistics, then it has only one state because now, furthermore, the first and fourth states are excluded by the Pauli principle (cf. Section 4.9).

From all that has been said we know that the three types of statistics merge together when the number of available states is so large with respect to the number of particles that it only rarely happens that a state contains more than one particle. The only correction which then still has to be made to the M.B. statistics is the division by $N!$ (see Equation (5.8.9)).

5.9. THE STATE SUM OF A PARTICLE WITH INTERNAL DEGREES OF FREEDOM

In general, the molecules of a gas, especially diatomic and polyatomic molecules, not only move in straight lines (the translational motion discussed earlier) but, due to their mutual collisions, they rotate, and at the same time perform oscillations. Finally, there may be energy exchanges between them by electron transitions. The latter also applies to monatomic gas molecules but often occurs only at extremely high temperatures. In the equilibrium state the energy is divided between translational, rotational, vibrational and electron energy in such a way that the number of realization possibilities is a maximum.

The energy levels for the translational motion of a polyatomic molecule are the same as those for a particle without internal degrees of freedom. They are found from (5.3.1) by inserting the mass of the molecule and the volume of the container. For *each* allowed translational movement of the molecule, characterized by the three translational quantum numbers n_x, n_y, n_z, it can exist in various electron states, for each electron state in different vibrational states and for each vibrational state in a large number of rotational states. Each electron-vibrational-rotational state is characterized by a number of internal quantum numbers (see Chapter 6).

The total energy of a gas molecule is always equal to the sum of the translational and the electron-vibrational-rotational energy:

$$\varepsilon = \varepsilon_{\text{tr}} + \varepsilon_{\text{evr}} \tag{5.9.1}$$

As in the preceding chapter, we divide the translational states into groups of z_i states with approximately equal energy ε_i. For each of the z_i translational states corresponding to the translational energy ε_i a molecule can be in any of the g_m internal states corresponding to its internal energy ε_m. The total number of states z_{im} belonging to the energy

$$\varepsilon_{im} = \varepsilon_i + \varepsilon_m \tag{5.9.1a}$$

is thus given by

$$z_{im} = z_i g_m \tag{5.9.2}$$

The suffixes "tr" in ε_i and "evr" in ε_m have been omitted in Equation (5.9.1a), because the suffixes i and m are sufficient indication of the nature of the energy.

The statistical considerations from the preceding chapter can again be applied to a system of molecules with internal degrees of freedom. We turn our attention to the permissible values, ε_{im}, of the *total* energy of a molecule and divide the molecules, in a manner analogous to that used in Sections 4.12, 4.13 and 4.14, into groups n_{im} among these energies. At the same time, of course, the following conditions must be satisfied:

$$\sum_{i,m} n_{im} = N \tag{5.9.3}$$

$$\sum_{i,m} \varepsilon_{im} n_{im} = U \tag{5.9.4}$$

In order to find the most probable distribution we can repeat, almost word for word, the reasoning which led in the sections quoted to the Bose-Einstein, Fermi-Dirac and Maxwell-Boltzmann distributions. Thus we obtain the same formulae in which only the suffix i is replaced by im. Now, too, the Maxwell-Boltzmann distribution rule will suffice and we obtain, from (5.2.4):

$$n'_{im} = \frac{Ne^{-\varepsilon_{im}/kT}}{\sum_{i,m} e^{-\varepsilon_{im}/kT}} \tag{5.9.5}$$

In this equation n'_{im} gives the population of the individual states, each characterized by a particular combination of the translational quantum numbers and a particular combination of the internal quantum numbers. In the state sum

$$Z = \sum_{i,m} e^{-\varepsilon_{im}/kT} \tag{5.9.6}$$

the summation, according to Section 5.2, must be carried out over all these separate states in which a molecule can exist.

The state sum can, if desired, also be written in the form

$$Z = \sum_{i,m} z_{im} e^{-\varepsilon_{im}/kT} \tag{5.9.7}$$

It is obvious that this equation will give the same result as (5.9.6) provided that this time the summation is carried out not over the *separate* quantum states but over the *groups* of states (each characterized by a particular value of i and a particular value of m, and thus by a particular energy).

Making use of (5.9.1a), the state sum in the form (5.9.6) may also be written:

$$Z = \sum_{i,m} e^{-\varepsilon_i/kT} \cdot e^{-\varepsilon_m/kT}$$

or

$$Z = \sum_{i} e^{-\varepsilon_i/kT} \cdot \sum_{m} e^{-\varepsilon_m/kT} \tag{5.9.8}$$

In (5.9.8) the first sum is the translational state sum and the second the electron-vibrational-rotational state sum. Starting with Z in the form (5.9.7) and employing (5.9.1a) and (5.9.2), we obtain

$$Z = \sum_{i,m} z_i e^{-\varepsilon_i/kT} \cdot g_m e^{-\varepsilon_m/kT}$$

or

$$Z = \sum_{i} z_i e^{-\varepsilon_i/kT} \cdot \sum_{m} g_m e^{-\varepsilon_m/kT} \tag{5.9.9}$$

Equations (5.9.8) and (5.9.9) once more have precisely the same significance, since in the first case the summation is performed over the separate states, in the second over the energy levels, which in reality consist of z_i or g_m coincident levels.

According to the above we may write in general

$$Z = Z_{tr} \cdot Z_{evr} \tag{5.9.10}$$

As an approximation, ε_{evr} can be written as the sum of an electronic, a vibrational and a rotational energy:

$$\varepsilon_{evr} = \varepsilon_{el} + \varepsilon_{vi} + \varepsilon_{ro} \tag{5.9.11}$$

Consequently it is possible to calculate an approximate value for Z_{evr} by writing this state sum as a product of an electronic, a vibrational and a rotational state sum:

$$Z_{evr} = Z_{el} Z_{vi} Z_{ro} \tag{5.9.12}$$

Equations (5.9.11) and (5.9.12) are only valid by approximation since, for example, the rotation and the vibration influence one another, so that in fact interaction terms occur in the energy.

5.10. THERMODYNAMIC FUNCTIONS OF POLYATOMIC GASES AND THE STATE SUM

According to the considerations of the preceding section Equation (5.4.1), giving the entropy of one mole of a perfect gas, is valid also for a gas the particles of which possess internal degrees of freedom. The state sum appearing in this formula is now the complete state sum, given by (5.9.10). The same is true for all the equations derived in Section 5.4 from that for the entropy.

They are valid not only for a gas which only possesses translational degrees of freedom, but also for more complicated gases, provided that in the latter case Z is taken to denote the product of Z_{tr} and Z_{evr}.

Thus we find for 1 mole of gas:

$$f = -RT\left(\ln\frac{Z_{tr}}{N_0} + 1\right) - RT\ln Z_{evr} \tag{5.10.1}$$

$$\mu = -RT\ln\frac{Z_{tr}}{N_0} - RT\ln Z_{evr} \tag{5.10.2}$$

$$s = R\left\{\ln\frac{Z_{tr}}{N_0} + T\left(\frac{\partial \ln Z_{tr}}{\partial T}\right)_v + 1\right\} + R\left\{\ln Z_{evr} + T\left(\frac{\partial \ln Z_{evr}}{\partial T}\right)_v\right\} \tag{5.10.3}$$

$$u = RT^2\left(\frac{\partial \ln Z_{tr}}{\partial T}\right)_v + RT^2\left(\frac{\partial \ln Z_{evr}}{\partial T}\right)_v \tag{5.10.4}$$

$$c_v = -R\frac{\partial}{\partial T}\left(\frac{\partial \ln Z_{tr}}{\partial(1/T)}\right)_v - R\frac{\partial}{\partial T}\left(\frac{\partial \ln Z_{evr}}{\partial(1/T)}\right)_v \tag{5.10.5}$$

Comparison of these formulae with those in Section 5.4 shows that we can always split the value of each thermodynamic function into a translational portion and an electron-vibrational-rotational portion.

We shall now rewrite the formulae for the evr-portions separately, at the same time keeping in mind that for our perfect gas Z_{evr}, in contrast to Z_{tr}, is independent of pressure and volume:

$$f_{evr} = \mu_{evr} = -RT\ln Z_{evr} \tag{5.10.6}$$

$$s_{evr} = R\left(\ln Z_{evr} + T\frac{d\ln Z_{evr}}{dT}\right) \tag{5.10.7}$$

$$u_{evr} = RT^2\frac{d\ln Z_{evr}}{dT} \tag{5.10.8}$$

$$c_{v(evr)} = -R\frac{d}{dT}\left(\frac{d\ln Z_{evr}}{d(1/T)}\right) \tag{5.10.9}$$

To avoid misunderstanding, it should be noted that in the American literature the term state sum or partition function or sum over states is frequently used exclusively to denote the internal state sum Z_{evr}.

5.11. APPROXIMATE AND EXACT VALUES OF THE INTERNAL STATE SUM

From the foregoing it appears that the calculation of the translational state sum, and with it the translational portion of the various thermodynamic functions, is carried out in the same way for all gases. Thus, for example, we can always use Equation (5.5.12) to calculate the translational entropy of an arbitrary gas (assumed to be perfect). However, to calculate the total entropy, free energy, etc., we must also know Z_{evr}.

Approximate values of Z_{evr} are calculated by means of Equation (5.9.12). If we substitute this in Equations (5.10.6) to (5.10.9), we see that the evr-part of each thermodynamic function can be written, by approximation, as the sum of an electronic, a vibrational and a rotational part. For the electronic entropy, for example, we obtain by this approximation

$$s_{el} = R\left(\ln Z_{el} + T\frac{\mathrm{d}\ln Z_{el}}{\mathrm{d}T}\right) \tag{5.11.1}$$

It is possible, however, to calculate *exact* values of Z_{evr} (and thus the evr-portion of thermodynamic functions) by using experimentally (spectroscopically) determined values of the energy levels of the molecules. The summation is then carried out term by term. This method has been used principally by Giauque and his collaborators. Both methods of calculation are discussed in Chapter 6 for diatomic molecules. In this chapter we shall restrict ourselves to monatomic molecules. In this case, the internal degrees of freedom only relate to the various allowed electron states, so that the calculation of Z is always "exact" (see the following sections).

When applying the exact method, Equations (5.10.6) to (5.10.9) are preferably written in the form

$$f_{evr} = \mu_{evr} = -RT\ln\sum e^{-\varepsilon/kT} \tag{5.11.2}$$

$$s_{evr} = R\left\{\ln\sum e^{-\varepsilon/kT} + \frac{1}{kT}\frac{\sum\varepsilon e^{-\varepsilon/kT}}{\sum e^{-\varepsilon/kT}}\right\}. \tag{5.11.3}$$

$$u_{evr} = N\frac{\sum\varepsilon e^{-\varepsilon/kT}}{\sum e^{-\varepsilon/kT}} \tag{5.11.4}$$

$$c_{v(evr)} = \frac{R}{(kT)^2}\left\{\frac{\sum\varepsilon^2 e^{-\varepsilon/kT}}{\sum e^{-\varepsilon/kT}} - \left(\frac{\sum\varepsilon e^{-\varepsilon/kT}}{\sum e^{-\varepsilon/kT}}\right)^2\right\} \tag{5.11.5}$$

In all these formulae ε has been used instead of ε_m for the sake of convenience. The various summations, according to the discussions in Section 5.9, extend over all the quantum states in which the atom or molecule may be found. If one prefers to carry out the summation over the allowed values of ε, then a factor g or g_m must be added under the summation symbol to indicate how many states correspond to the energy ε_m.

5.12. THE STATES OF AN H ATOM

Up to now we have treated the monatomic gases as consisting of particles without internal degrees of freedom. For the inert gases, the alkaline-earth metals and Zn, Cd and Hg this is entirely justified, since a great deal of energy is required to bring the atoms of these elements into excited states. So long as the temperature is not extremely high, these monatomic gases thus possess only translational entropy and no electron-entropy. Many other monatomic gases, however, possess a not inconsiderable electron-entropy even at low temperatures. This is calculated via the portion of the state sum which relates to the various electron states.

In discussing these states we shall restrict ourselves at first to the hydrogen atom, although it will appear that, in this case as well, the excited states lie so high that they play no appreciable part in the state sum. It is, however, the simplest case for theoretical treatment because the "electron cloud" around the nucleus here consists of only one electron.

The many different states in which an H atom may exist are based, in the pictorial model of Bohr and Sommerfeld, on the various orbits which the electron can describe. In this model the electron is allowed to move around the nucleus in elliptical orbits according to Kepler's laws. From the infinitely large number of orbits which are possible in classical mechanics, a limited number are selected with the help of three Sommerfeld conditions of the type (4.5.1).

For each condition according to (4.5.1) there is a corresponding quantum number. The first condition divides the permitted orbits into groups which are distinguished by different values of the *principal quantum number n*. In all the elliptical orbits with the same principal quantum number, the major axis has the same length. This length is proportional to n^2. The energy of the H atom is also determined wholly by n and calculation shows that

$$\varepsilon_n = -\frac{2\pi^2 e^4 m}{h^2}\frac{1}{n^2}, \tag{5.12.1}$$

if the ionized state ($n = \infty$) is chosen as the zero level of the energy.

Sommerfeld's second condition introduces the *azimuthal quantum number l*. In the Bohr-Sommerfeld model, l was related directly to the angular momentum of the motion about the nucleus and to the excentricity of the elliptical path.

The angular momentum of the motion of a particle about a fixed centre of attraction is a vector, perpendicular to the plane of the orbit and of constant absolute value given by $mvr \sin \phi$, where r is the length of the line joining the particle to the centre and ϕ is the angle between this line and the direction of the velocity v. For a series of ellipses, all belonging to the same n, the above-mentioned absolute value is given directly, with the exception of a constant factor, by the area of the ellipse. Since for constant n all ellipses have the same length of major axis, the angular momentum becomes smaller as the ellipse becomes more excentric.

Calculations showed that l can only have the values 0, 1, 2, . . ., $(n - 1)$, so that

$$\begin{array}{rcl} l = 0 & \text{for} & n = 1 \\ l = 0 \text{ or } l = 1 & \text{for} & n = 2 \\ & \text{and so on.} & \end{array} \tag{5.12.2}$$

This meant that in the model the electron in the ground state ($n = 1$) could only move in a circular orbit [1]). In the first excited state ($n = 2$) an ellipse was possible besides the circle, for $n = 3$ there were two ellipses possible as well as the circle, and so on (the highest value of l always related to the circular orbit).

The vector, perpendicular to the plane of the orbit, which represents the angular momentum of the orbit, is indicated by the symbol $\mathbf{l}$, and its absolute value by $|\mathbf{l}|$. Sommerfeld's conditions gave for this the discrete values

$$|\mathbf{l}| = \frac{h}{2\pi} l \tag{5.12.3}$$

where l can assume the values stated above.

The third condition of Sommerfeld introduces the *magnetic quantum number m*. In a magnetic field $\mathbf{l}$ can only be orientated in directions such that the

---

[1]) For $l = 0$ the angular momentum is zero and this would correspond to motion along a straight line. The electron would thus collide with the nucleus in the pictorial model. In Sommerfeld's original theory $k = l + 1$ was the azimuthal quantum number and the collision was avoided by excluding the value $k = 0$. In Equation (5.14.12), based on quantum mechanics, $\sqrt{l(l + 1)}$ occurs in place of l (see Section 5.14).

component in the direction of the field, expressed in units of $h/2\pi$, has one of the $(2l + 1)$ values

$$m_l = l;\ (l-1);\ (l-2);\ \ldots;\ -l \tag{5.12.4}$$

It is clear that the two extreme values correspond to **l** being parallel or anti-parallel to the field. The intermediate values differ from each other by one unit.

Roughly speaking, the principal quantum number n gave information about the average distance of the electron from the nucleus, while the azimuthal quantum number l gave information about the shape of the orbit and the magnetic quantum number m about the orientation of the plane of the orbit in space.

For a constant value of n, l can, according to (5.12.2), take n different values, while for each value of l, m can assume $(2l + 1)$ different values according to (5.12.4). The total number of orbits corresponding to a particular principal quantum number n (and thus according to (5.12.1) to a particular energy of the H atom) is therefore

$$\sum_{l=0}^{l=(n-1)} (2l + 1) = n^2 \tag{5.12.5}$$

Each level is thus n^2-fold degenerate (i.e. consists, in fact, of n^2 coincident levels). The degeneracy is partly resolved by taking into account the dependence, in accordance with relativity theory, of the mass of the electron on its velocity. If this is done, it appears that the energy not only depends on n, but also to a very slight extent on l. Each energy level is thus split into a series of n levels, very close together, each with a different value of l.

The remaining degeneracy can be resolved by introducing a disturbance in the form of a magnetic field. At each value (allowed by (5.12.4) and remaining constant) of the projection of **l** on the direction of the field a precession takes place about this direction in the pictorial model.

The study of spectra finally led to the realization that each of the above-mentioned n^2 "states" is double, i.e. consists in fact of two states. In the pictorial model this is based on the rotation of the electrons about their own axes, so that, besides the angular momentum of the orbit they also have their own *angular momentum* or *spin* (cf. Section 4.13).

In analogy to (5.12.3), one can write for this spin:

$$|\mathbf{s}| = \frac{h}{2\pi}\, s. \tag{5.12.6}$$

where s denotes the quantum number for the electron's own rotation and can only assume the value $s = \frac{1}{2}$.

The rotating charge of the electron also causes a magnetic moment. In a magnetic field the spin of an electron can only take a position such that the component in the field direction, expressed in units $h/2\pi$, has a magnitude $+\frac{1}{2}$ or $-\frac{1}{2}$. This is in agreement with (5.12.4), since here the series

$$m_s = s;\ (s-1);\ (s-2);\ \ldots;\ -s \tag{5.12.4a}$$

only permits the values $+\frac{1}{2}$ and $-\frac{1}{2}$ .The spin of an electron is thus either parallel or anti-parallel to the field.

As a result of its movement around the nucleus, the electron moves in a magnetic field which is perpendicular to the plane of its orbit. The spin will orientate itself either parallel or anti-parallel to this field. In the first case the electron's own angular momentum must be added to its orbital angular momentum, in the other, it must be subtracted from it. Therefore, for the quantum number j of the total angular momentum we obtain

$$j = l \pm \tfrac{1}{2} \tag{5.12.7}$$

If there is no magnetic field present, i.e. if $l = 0$, then the two states are physically indistinguishable and we must simply write

$$j = s = \tfrac{1}{2}. \tag{5.12.8}$$

To the two values of j, given by (5.12.7), correspond two energy levels, which are only separated by a very small interval. The magnitude of this splitting is given by the energy which would be required to turn the spin from the parallel to the anti-parallel position with respect to the orbital magnetic field.

5.13. THE INTERNAL STATE SUM OF AN H ATOM

If we neglect the minute separation of the energy levels due to spin and relativity correction, we can say that each level (5.12.1), as explained in the preceding section, corresponds to $2n^2$ states of the H atom. In neglecting these separations we are making an error of, at most, a few thousandths of a percent in the values of the energy.

A more serious error is made by treating the nucleus of the H atom as a fixed centre, while in the model discussed we must assume that the electron and the nucleus rotate about their common centre of gravity. The correction in this case is very simple. It is only necessary to replace m in (5.12.1) by the so-called reduced mass m_r, given by

$$m_r = \frac{m_1}{1 + m_1/m_2} \tag{5.13.1}$$

where m_1 is the mass of the electron and m_2 that of the nucleus (the proton). This problem will be dealt with at more length in the discussion of the rotation of diatomic molecules in Chapter 6.

With Equation (5.12.1), thus modified, we can immediately calculate the distances of the energy levels from the ground level. If these are denoted by $\varepsilon_{12}(= \varepsilon_2 - \varepsilon_1)$, $\varepsilon_{13}(= \varepsilon_3 - \varepsilon_1)$, etc., the state sum of the H atom, due to the $2n^2$-fold degeneracy of each level, is given by

$$Z = 2 + 8e^{-\varepsilon_{12}/kT} + 18e^{-\varepsilon_{13}/kT} + 32e^{-\varepsilon_{14}/kT} + \dots \tag{5.13.2}$$

In the state sum (5.13.2) the zero level of the energy coincides with the ground level of the H atom, while in Section 5.12 the ionized state was chosen as zero level. If the choice of a particular zero level for the energy results in a state sum Z, then the choice of a new zero level lying an amount ε_0 below the original one, gives a state sum $Ze^{-\varepsilon_0/kT}$. Equations (5.10.7) and (5.11.1), however, show that both state sums lead to the same value of the entropy. According to the equations in Section 5.10, the value of the specific heat is also independent of the choice of the zero level of the energy. The same is not true, however, for the values of u, f and μ: they are dependent on the choice of the zero energy level.

It is easily calculated that

$$e^{-\varepsilon_{12}/kT} = e^{-118360/T}$$

Clearly, then, even at a temperature of, say, 10,000 °K, the excited states of the H atom can be completely neglected in the state sum. In all circumstances which are of importance to the chemist, we may thus simply write for the state sum of the H atom:

$$Z = 2$$

This value is based, as has been shown, on the two orientation possibilities for the spin of the one electron.

The electronic entropy of atomic hydrogen is thus $R \ln 2$ per gram-atom according to Equation (5.11.1). In Chapter 6 we shall discover that atomic hydrogen also possesses an equal quantity of nuclear spin entropy.

5.14. EIGENVALUES AND EIGENFUNCTIONS OF AN H ATOM

It has already been pointed out in Chapter 4 that the electron orbits mentioned in the preceding sections cannot be attributed with any physical significance. It is therefore essential to see what remains of the quantitative results of the last two sections if the problem is tackled in the modern manner, starting with Schrödinger's Equation (4.4.1).

We start again by neglecting the motion of the nucleus in the first approximation. If, in this consideration of the H atom, the proton is chosen as the centre of the system of co-ordinates, then we must write for the potential V in (4.4.1):

$$V = -\frac{e^2}{r} \tag{5.14.1}$$

where r is the distance from the nucleus. For the Schrödinger equation we thus obtain:

$$\frac{h^2}{8\pi^2 m}\nabla^2\psi + \left(\varepsilon + \frac{e^2}{r}\right)\psi = 0 \tag{5.14.2}$$

The search for solutions to this equation is a purely mathematical problem, i.e. a problem which involves no difficulties of a non-mathematical nature. It will therefore be sufficient to give merely a rough description of the results.

The eigenvalués ε are found to be exactly the same as those given by Equation (5.12.1). For each eigenvalue ε_n we find one appropriate spherically symmetrical eigenfunction, i.e. a solution ψ to (5.14.2) which is only dependent on r:

$$\psi_n = \mathrm{f}_n(r) \tag{5.14.3}$$

There are, however, also solutions which are not spherically symmetrical, i.e. which depend on the co-ordinates x, y and z separately. For the lowest value of the energy ε_1 this is not the case. It is only associated with a spherically symmetrical function which, with the exception of a constant, is given by

$$\psi_1 = \mathrm{f}_1(r) = e^{-r/a} \tag{5.14.4}$$

The second eigenvalue ε_2, however, is not only associated with a spherically symmetrical function, but also with three non-spherically-symmetrical functions:

$$\psi_x = x \,.\, \mathrm{f}(r) \tag{5.14.5}$$
$$\psi_y = y \,.\, \mathrm{f}(r) \tag{5.14.6}$$
$$\psi_z = z \,.\, \mathrm{f}(r) \tag{5.14.7}$$

where f(r) must naturally satisfy certain conditions. It is not surprising that three non-spherically-symmetrical functions occur together here, because if the first of the three, viz. (5.14.5), is a solution, then (5.14.6) and (5.14.7) must necessarily also be solutions. This is so because x, y and z appear in an equivalent manner in the wave equation, since

$$r = (x^2 + y^2 + z^2)^{1/2}$$

The states described by a spherically symmetrical eigenfunction (5.14.3) are called s-states, while those of the type (5.14.5) are called p-states. There are also p-states belonging to the higher levels ε_3, ε_4, etc.

The third eigenvalue ε_3 possesses, in addition to an s-state and a three-fold p-state, a fivefold d-state. The d-functions are dependent on x, y and z in a more complicated manner than the p-functions (5.14.5), (5.14.6) and (5.14.7).

And so we obtain for the first three energy levels, characterized by the number n, the following scheme:

n	s-functions	p-functions	d-functions	total number of states
1	1			1
2	1	3		4
3	1	3	5	9

Thus we have found again the n^2 states in Equation (5.12.5).

The "azimuthal quantum number" l, introduced in Section 5.12, could only take the value zero when $n = 1$, for $n = 2$ the values 0 and 1, and so on. For each value of l there are $2l + 1$ corresponding states, each of which is separately characterized by a particular value of the "magnetic quantum number" m.

It can be seen that this is in complete agreement with the above scheme, provided that one denotes a spherically symmetrical or s-function by $l = 0$, a p-function by $l = 1$, a d-function by $l = 2$, and so on. In this way, the quantum numbers have here degenerated to mere letters characterizing the various solutions to the Schrödinger equation of the system.

In the rigorous mathematical treatment the azimuthal quantum number l is introduced in a more satisfactory manner than in the rough sketch given above. In the rigorous treatment polar co-ordinates r, θ, ϕ are used. In this case, it is found to be possible to write the wave function ψ as the product of two functions, one of which is dependent only on the distance r and the other only on the angles θ and ϕ:

$$\psi = R(r) \cdot Y(\theta,\phi). \tag{5.14.8}$$

By substituting (5.14.8) in the wave equation one obtains two differential equations which can be solved separately. As a result of this separation of variables the same constant C appears in both equations (but with different sign) and the angle-dependent equation

$$\frac{1}{Y \sin\theta}\frac{\partial}{\partial\theta}\left(\sin\theta\frac{\partial Y}{\partial\theta}\right) + \frac{1}{Y \sin^2\theta}\frac{\partial^2 Y}{\partial\phi^2} = -C \tag{5.14.9}$$

is found to have satisfactory solutions only when

$$C = l(l+1). \tag{5.14.10}$$

where $l = 0,1,2,\ldots$. To each value of l correspond $(2l+1)$ eigenfunctions $Y_{lm}(\theta,\phi)$, each of which is characterized by one of the values $m = l; (l-1); \ldots; -l$ of the "magnetic quantum number". When $l = 0$, the solution is a constant and we thus obtain pure spherically symmetrical functions which were denoted earlier by the symbol s.

The second differential equation obtained by the separation of the variables

$$\frac{r}{R}\frac{d^2(rR)}{dr^2} + \frac{8\pi^2 m}{h^2} r^2 \left(\varepsilon + \frac{e^2}{r}\right) = C \tag{5.14.11}$$

only has satisfactory solutions for the eigenvalues ε_n given by (5.12.1), representing the energy levels of the system. In this way the principal quantum number n is introduced.

Due to the appearance of $C = l(l+1)$ in (5.14.11), the eigenfunctions $R(r)$ not only depend on n, but at the same time on l.

We shall not discuss the mathematical problem further, since we only wished to give the reader a rough impression of the way in which the quantum numbers make their appearance in the quantum-mechanical treatment.

We shall not deal with the significance of the concept of "angular momentum" in quantum mechanics. For our purpose it is sufficient to remark that the azimuthal quantum number l, as in the old theory, is very closely connected with this "angular momentum". Equation (5.12.3) must now be replaced by

$$|\mathbf{l}| = \frac{h}{2\pi}\sqrt{l(l+1)}. \tag{5.14.12}$$

In a similar way, Equation (5.12.6) becomes

$$|\mathbf{s}| = \frac{h}{2\pi} \sqrt{s(s+1)} \qquad (5.14.13)$$

On the other hand, Equation (5.12.4), which relates to the component of the angular momentum in the direction of the magnetic field, remains unchanged.

In the coming sections we shall continue to make use of the pictorial language of the vector model, as derived from the old equations. For example, we shall continue to speak of parallel and anti-parallel spin, but shall accompany them with the correct equations.

For all the solutions ψ to (5.14.2), $|\psi|^2 d\omega$ gives, as always, the probability of finding the electron in the volume element $d\omega$. For each state we can thus calculate a "probability cloud", which gives for each point the probability of finding the electron there. If, for example, we consider the ground state described by (5.14.4) and employ polar co-ordinates, we see that

$$4\pi r^2 \cdot e^{-2r/a} \cdot dr$$

is the probability of encountering the electron between two spherical surfaces with radii r and $r + dr$. The chance of finding the electron at a distance r from the nucleus, is thus determined, with the exception of a constant, by the function $r^2 e^{-2r/a}$. This probability is very small for very small distances, due to the factor r^2, and at the same time very small for large distances, due to the factor $e^{-2r/a}$. The maximum, as can be easily verified, lies at $r = a$.

Calculations show that $a = 0.53 \times 10^{-8}$ cm and this corresponds exactly to the radius of the first circular orbit in Bohr's old theory.

According to the above, a graph of the function $r^2 e^{-2r/a}$ ($= r^2\psi^2$ for $n = 1$; $l = 0$) will show rough agreement with the ψ^2 curve for $n = 1$ in Fig. 54. In a similar manner the $r^2\psi^2$ curve for $n = 2$; $l = 0$ will show rough agreement with the ψ^2 curve for $n = 2$ in Fig. 54. The two maxima, nevertheless, no longer have the same height; the maximum which lies closest to the nucleus is much lower than the other. Analogous considerations apply for $n = 3$ (see Fig. 63).

A correction for the slight motion of the nucleus can be made in this case in a manner exactly analogous to that in the old theory (Section 5.13).

The application of a relativity correction (Section 5.12) becomes superfluous when the relativistic wave mechanics of Dirac is used. Also the spin with all its properties appears then as a logical consequence without being necessary to introduce it as a separate hypothesis. A more detailed treatment of this point is, however, quite beyond the scope of this book.

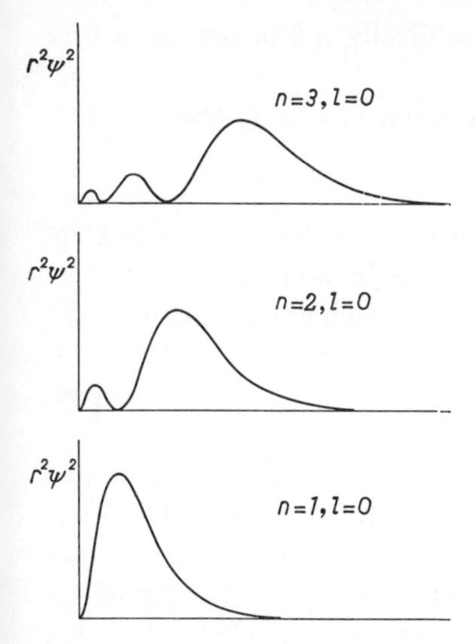

Fig. 63. The function $r^2\psi^2$ for three states of the hydrogen atom. The distance from the nucleus is plotted horizontally on an arbitrary scale.

5.15. ATOMS WITH MORE ELECTRONS

If we take into account the doubling of all states by the spin, it follows from the preceding section that the principal quantum number $n = 1$ corresponds to two s-functions, $n = 2$ to two s-functions and six p-functions and so on. This applies to an H atom, but it is obvious that the same is true for a system consisting of a more highly charged nucleus and one electron, thus for He^{+}, Li^{++}, Be^{+++}, B^{++++}, etc.

At low temperatures the electron will always be found in one of the s-states with $n = 1$, since this makes the energy a minimum. If we add more electrons to the system (e.g. B^{++++}) we must take into account not only the tendency towards a minimum value of the energy, but also the Pauli exclusion principle, according to which it is impossible for two electrons in one atom to be in the same state.

The second electron will occupy the second s-state with $n = 1$. This will in future be referred to in short as a 1s-state. Any further electrons will be forced, however, to occupy the states corresponding to $n = 2$, which are less energetically favourable. From the following will appear that in doing this, it is energetically the least unfavourable to occupy first the two 2s-states, which the third and fourth electrons will thus do. Finally, a fifth electron will be accommodated in one of the 2p-states.

The electronic configuration obtained is denoted by the symbol

$$1s^2\,2s^2\,2p$$

which means two electrons in a 1s-state, two in a 2s-state and one in a 2p-state or, more concisely, two 1s-, two 2s-and one 2p-electron.

The two 1s-electrons have three quantum numbers in common (cf. Sections 5.12 and 5.14). For both of them, $n = 1$, $l = 0$, $m_l = 0$. They differ, however, in spin quantum number ($m_s = +\frac{1}{2}$ and $-\frac{1}{2}$), i.e. they have opposed spins. The same is true for the two 2s-electrons.

If we wish to indicate the quantum numbers of the 2p-electron, we find that the state of the boron atom is not uniquely defined by the symbol $1s^2 2s^2 2p$. For the 2p-electron $l = 1$ and according to Equation (5.12.7) this may correspond to either a parallel or an anti-parallel position of the spin of this same electron. In the next section we shall discuss the symbols which are used to distinguish these two states from one another.

We have already seen that only two s wave functions correspond to each value of n. Consequently, an atom contains a maximum of two 1s-electrons, two 2s-electrons, two 3s-electrons, and so on. Groups of this kind $1s^2$, $2s^2$, $3s^2$, etc., with the maximum number of electrons are called "saturated shells" or "filled shells".

The expression "filled shells" is often used in a slightly different sense, viz. for cases where all the electrons permitted for a given value of n are present, i.e. for $1s^2$; for $2s^2 2p^6$; etc.

When $n = 1$ there are no p wave functions, i.e. there are no 1p-electrons. For all values of n greater than 1 there are six p-functions, so that an atom may contain a maximum of six 2p-electrons, six 3p-electrons, etc. The symbols $2p^6$, $3p^6$, $4p^6$, etc. thus indicate filled shells of p-electrons.

When all shells are filled with electrons we obtain the following scheme for the first three values of n:

n	symbol	total number of electrons
1	$1s^2$	2
2	$2s^2 2p^6$	8
3	$3s^2 3p^6 3d^{10}$	18

Nothing has yet been said about the very important mutual repulsion of electrons in an atom with more than one electron. In the imaginary process described, by which more and more electrons are added to a highly charged nucleus, this repulsion makes it no longer justified to calculate the energy levels, available to the next electron to be added, by means of a simple equation of the type (5.12.1). The problem of calculating these levels becomes more complicated as more electrons are already grouped around the nucleus.

Valuable approximate results can be obtained by regarding the electrons already present around the nucleus as a static, spherically symmetrical "cloud" of negative electricity. This "cloud" corresponds to a central field, i.e. a field in which the force is always directed towards the centre and is dependent only on the distance, not on the angle. This field, together with the Coulomb field of the nucleus, gives a resultant field in which the potential depends on the distance r from the centre in a different manner than in a Coulomb field.

For points outside the "cloud" the field is the same as if all the negative charge were concentrated at the centre. For such points, therefore, it is as though the nuclear charge were reduced by the amount of the negative charge already present. If the "cloud" consists of a electrons, it appears as if the nuclear charge were reduced from Ze (Z being the atomic number) to $(Z - a)e$. The newly added electron is thus attracted outside the "cloud" by a force $(Z - a)e^2/r^2$.

However, if the electron penetrates into the interior of the cloud, the screening action becomes smaller as the electron approaches the nucleus. In fact, it is well known that the field strength in a spherically symmetrical "cloud" at a distance r from the centre is determined entirely by the portion of the "cloud" inside a sphere of radius r. It is as though the negative charge outside this sphere were not present at all. The force exerted on the penetrating electron thus approaches Ze^2/r^2 for very small values of r.

The manner in which the field inside the "cloud" changes as a function of r is naturally dependent on the way in which the electric charge is distributed among the various spherical shells. Each distribution corresponds to a particular potential field $V(r)$. By substituting this function in the Schrödinger equation, the energy levels of the system can be calculated. For this non-Coulomb type of field the energy levels are found to be not only dependent on n (as for a Coulomb field when the small relativity correction is neglected), but also on l. For the same value of n, the energy is greater as l is greater.

The latter can also be seen immediately from the model of Bohr and Sommerfeld. As l increases, the excentricity of the orbit decreases and the electron dives less deeply into the central part of the "cloud", where the attraction of the nucleus is greatest.

The energy thus increases with both increasing values of n (l constant) and increasing values of l (n constant) i.e. in the sequence s, p, d, etc. It is now also clear why in the example of boron just discussed the two 2s-states were occupied first and then one of the 2p-states.

5.16. COUPLING OF THE ANGULAR MOMENTA

We have seen in the preceding section that a symbol of the type $1s^22s^22p$ is in general, not sufficient to describe the state of an atom unambiguously because it gives no information about the coupling between the angular momenta.

In most atoms the interaction of the electrons is such that the orbital angular momenta $\mathbf{l}_1$, $\mathbf{l}_2$, $\mathbf{l}_3$, ... of the separate electrons must be regarded as being strongly coupled to each other as are the spins, $\mathbf{s}_1$, $\mathbf{s}_2$, $\mathbf{s}_3$, ... In this case we speak of Russell-Saunders coupling or LS coupling and obtain a resultant of all the $\mathbf{l}_i$'s, the resultant orbital angular momentum $\mathbf{L}$, and of all the $\mathbf{s}_i$'s, the resultant spin $\mathbf{S}$.

The resultant of, for example, $\mathbf{l}_1$ and $\mathbf{l}_2$ is obtained with the help of the vector parallelogram. This resultant $\mathbf{L}$ is also quantized, i.e. the quantum number L associated with $\mathbf{L}$ according to the relationship

$$|\mathbf{L}| = \frac{h}{2\pi}\sqrt{L(L+1)} \tag{5.16.1}$$

can only take the values

$$L = (l_1 + l_2);\ \ (l_1 + l_2 - 1);\ \ (l_1 + l_2 - 2);\ \ \ldots;\ \ |l_1 - l_2| \tag{5.16.2}$$

Since l_1 and l_2 are whole numbers, L also will always be a whole number. The vectors $\mathbf{l}_1$ and $\mathbf{l}_2$ can thus only make very special angles with one another.

If there are three vectors $\mathbf{l}_1$, $\mathbf{l}_2$, $\mathbf{l}_3$, one first finds the resultants of two of these vectors and then those of every possible resultant and the third vector.

It can be shown that the resultant orbital angular momentum of filled electron shells is equal to zero. For an alkali atom with only one outer electron, L for the atom is thus equal to l for this one electron. For an alkaline-earth atom with two outer electrons with individual quantum numbers l_1 and l_2 the possible values of L for the whole atom are given by Equation (5.16.2).

In a manner analogous to that described above for $\mathbf{L}$, the absolute value of the resultant spin is given by

$$|\mathbf{S}| = \frac{h}{2\pi}\sqrt{S(S+1)} \tag{5.16.3}$$

If there are only two spin vectors $\mathbf{s}_1$ and $\mathbf{s}_2$, the spin quantum number S is subject to (5.16.2), which in this case reads:

$$S = (s_1 + s_2);\ (s_1 + s_2 - 1);\ \ldots;\ |s_1 - s_2| \tag{5.16.2a}$$

Since $s_1 = s_2 = \frac{1}{2}$, S can thus only have the values $S = 1$ and $S = 0$. If $S = 1$, we speak of similarly directed or parallel spins and indicate this by the symbol $\uparrow\uparrow$. If $S = 0$ we speak of opposed spins or anti-parallel spins (symbol $\uparrow\downarrow$).

If there are more than two spin vectors the values of S will be whole numbers for an even number of electrons and whole numbers plus $\frac{1}{2}$ for an odd number. For filled shells S is equal to zero.

The *total angular momentum* $\mathbf{J}$ of the electrons surrounding the nucleus is the resultant of $\mathbf{L}$ and $\mathbf{S}$. The absolute value of $\mathbf{J}$ is given by

$$|\mathbf{J}| = \frac{h}{2\pi} \sqrt{J(J+1)} \tag{5.16.4}$$

with

$$J = (L + S);\ (L + S - 1);\ \ldots;\ |L - S| \tag{5.16.5}$$

In the cases where Russell-Saunders coupling can be assumed, the total angular momentum is thus given by the formulae:

$$\left.\begin{array}{l} \mathbf{l}_1 + \mathbf{l}_2 + \mathbf{l}_3 + \ldots = \mathbf{L} \\ \mathbf{s}_1 + \mathbf{s}_2 + \mathbf{s}_3 + \ldots = \mathbf{S} \end{array}\right\} \quad \mathbf{L} + \mathbf{S} = \mathbf{J}. \tag{5.16.6}$$

in which the addition is not algebraic but vectorial.

With this type of coupling the interaction between the orbital angular momenta on one hand and the spin momenta on the other hand is small by comparison with the strong mutual interaction of the orbital angular momenta and with the similarly strong mutual interaction of the spin momenta of the various electrons. Consequently, states with different values of L or different values of S, for instance, the states L_1S_1, L_2S_1, L_1S_2, L_2S_2, etc., differ widely in energy, while states with different values of J (for constant values of L and S) lie comparatively close together in energy. Thus we have a number of energy levels LS, which lie relatively far apart, but all of which are subdivided into closely spaced components with different values of J. Such a subdivided state is called a *multiplet* and the number of components into which it is divided, the *multiplicity* of the particular state LS.

As can be seen from the foregoing, the multiplicity is determined by (5.16.5). This equation shows that the number of values of J and thus the multiplicity of the particular state of the atom, is given by $2S + 1$, provided that $L \geqslant S$. If, for example, $S = \frac{1}{2}$, then from (5.16.5):

$$J = L + \tfrac{1}{2} \quad \text{or} \quad J = L - \tfrac{1}{2}$$

so that the multiplicity is two. If $S = 1$, then according to (5.16.5):

$$J = L + 1 \quad \text{or} \quad J = L \quad \text{or} \quad J = L - 1$$

corresponding to a multiplicity of three. The components of one multiplet differ in orientation with respect to each other of the relatively weakly coupled vectors **L** and **S**.

When $L = 0, 1, 2, 3, \ldots$ one speaks respectively of S, P, D, F states of the atom. Before this capital letter a superscript of the value $2S + 1$ is written, where S is the quantum number of the spin resultant, while after the letter the suffix J, the quantum number of the total angular momentum, is added.

We have followed the prevailing custom of using the same capital letter for the state $L = 0$ and for the quantum number of the resultant spin, but to avoid confusion we set S for the resultant spin and S for $L = 0$. It should further be remembered that for a single electron one uses small letters, thus s, p, d, f, etc, to indicate the size of the azimuthal quantum number l.

When we say that the chlorine atom has a ^{2}P state as ground state then, from the above, the number 2 means that $2S + 1 = 2$, thus $S = \frac{1}{2}$, while the letter P means that $L = 1$. According to (5.16.5) J can then have two different values, viz. $J = 1 + \frac{1}{2} = \frac{3}{2}$ and $J = 1 - \frac{1}{2} = \frac{1}{2}$. The ground state of the chlorine atom is thus a doublet, i.e. it consists of two states, the states $^2P_{3/2}$ and $^2P_{1/2}$. In the first state **L** and **S** are roughly speaking in the same direction, in the second state in opposite directions. According to the spectrum, the second state lies 1.750×10^{-13} erg higher than the first. If the temperature at which kT is equal to the energy difference between the two states, is called the *characteristic temperature* of electron excitation, then for a gas consisting of chlorine *atoms* this temperature lies at $1.750 \times 10^{-13}/k = 1268$ °K.

We should further like to point out that the "state" $^2P_{3/2}$, because $J = \frac{3}{2}$, really consists of $2J + 1 = 4$ states. In the pictorial model these are the four orientations of **J** which, according to the earlier-discussed direction quantization, are permitted in a magnetic field, the components in the field direction being given in units of $h/2\pi$ by $m_J = \frac{3}{2}, \frac{1}{2}, -\frac{1}{2}$ and $-\frac{3}{2}$ (cf. Equation (5.12.4)). These four states have a slightly different energy in the field. Without a field, one is dealing with four coincident levels (or, if one prefers, one fourfold degenerate level). In the same way, the "state" $^2P_{1/2}$, since $J = \frac{1}{2}$, consists of $2J + 1 = 2$ states ($m_J = \frac{1}{2}$ or $-\frac{1}{2}$).

It is perhaps useful to say a little more here about the word "state" which, as is demonstrated by the foregoing, is used with several different meanings. Strictly speaking, we may only speak of a particular state insofar as this is described by one particular eigenfunction (see the preceding chapter).

However, it is often found convenient to group together all the eigenfunctions belonging to the same eigenvalue and to call them one state with a statistical weight, which is given by the number of eigenfunctions. Thus if it is said that a "state" has a statistical weight g, this means that this "state" is in reality a combination of g states. The use (or abuse) of the word state is carried even further when (as above) one speaks, for example, of the ^2P state of a chlorine atom, for this consists of two, energetically widely separated "states" $^2P_{3/2}$ and $^2P_{1/2}$, each of which consists of four or two states respectively.

5.17. THE STATE SUM OF ANY ATOM

We can calculate the state sum of any atom immediately, if the energy levels are known and if it is known how many states (eigenfunctions) correspond to each level. For these data, the reader is referred to a well-known book of reference [1]).

In this field, the unit of energy chiefly employed is the cm^{-1}, corresponding to $hc = 1.9862 \times 10^{-16}$ erg/atom (h being Planck's constant and c the velocity of light). Each exponent ε/kT in the state sum changes through the use of this unit into $\varepsilon hc/kT$, if k is expressed in erg/degree. We can also write for this $1.4388\ \varepsilon/T$, in which $1.4388\ \varepsilon$ is called the characteristic temperature of excitation of the level in question (see preceding section).

We shall demonstrate the calculation of the state sum of an atom with a couple of examples and begin with the chlorine atom just discussed.

As we saw in the preceding section, the ground state of the chlorine atom is a ^2P state, consisting of the components $^2P_{3/2}$ and $^2P_{1/2}$. According to Landolt-Börnstein, the energy difference between these two states is 881 cm^{-1} corresponding to a characteristic temperature of $1.4388 \times 881 = 1268$ °K. Since the $^2P_{3/2}$ state is the lowest state here, the electronic state sum is given by

$$Z_{el} = 4 + 2e^{-1268/T} \tag{5.17.1}$$

due to the quadruple weight of the $^2P_{3/2}$ state and the double weight of the $^2P_{1/2}$ state.

According to Equation (5.11.1) the electronic entropy of 1 mole (gram-

[1]) Landolt-Börnstein, Zahlenwerte und Funktionen aus Physik, Chemie, Astronomie, Geophysik und Technik, 1. Band 1. Teil: Atome und Ionen, Springer-Verlag, 6. Auflage, Berlin 1950.

atom) of atomic chlorine at very low temperatures is thus $R \ln 4$, while at high temperatures it approaches $R \ln 6$. For intermediate temperatures, Equation (5.11.1) gives the correct value.

At extremely high temperatures, however, the state sum can become considerably larger than 6. Equation (5.17.1), in fact, only gives a small fraction of the state sum.

The electron configuration of the chlorine atom is

$$1s^2\ 2s^2\ 2p^6\ 3s^2\ 3p^5$$

and in this configuration the chlorine atom can indeed only occur in the ^2P states discussed. There are, however, excited states in which the electron configuration is changed. The least energy is required for the transfer of one of the 3p-electrons into a 4s-state, producing the configuration

$$1s^2\ 2s^2\ 2p^6\ 3s^2\ 3p^4\ 4s.$$

This configuration occurs in a ^4P state with the components $^4P_{5/2}$, $^4P_{3/2}$ and $^4P_{1/2}$ and in a ^2P state with the components $^2P_{3/2}$ and $^2P_{1/2}$. The lowest of these states, viz. $^4P_{5/2}$, has an energy of as much as 71954 cm^{-1} above the ground state. The term in the state sum which corresponds to it thus has a value of

$$6e^{-103530/T}$$

and even at a temperature of, say, 10 000 °K it will play no appreciable part.

As a second example, we shall consider the states of the iron atom. The ground state of this atom is a ^5D state, which means, as we saw in the preceding section, that the quantum number of the electron spin $S = 2$ and the quantum number of the orbital angular momentum of the electrons $L = 2$ also. The quantum number J of the total angular momentum of the electrons can thus assume the values $J = 4, 3, 2, 1$ or 0. The statistical weights of these states are given by $2J + 1$ and are thus respectively $g = 9, 7, 5, 3$ and 1. If, as usual, we choose the lowest state as the zero level of the energy, then the energies of these five states are 0, 415.9, 704.0, 888.1, and 978.1 cm^{-1} and the characteristic temperatures 0, 598, 1013, 1278, and 1407 °K.
The electronic state sum of the iron atom is thus:

$$Z_{el}^{(1)} = 9 + 7e^{-598/T} + 5e^{-1013/T} + 3e^{-1278/T} + e^{-1407/T}. \qquad (5.17.2)$$

Just as for chlorine, this is naturally only a small part of the total electronic state sum. We have considered only five levels, but Landolt-Börnstein mentions several hundreds of levels for the unionized iron atom alone.

The state which follows directly upon the ^5D state mentioned is a ^5F state, also with five components. Because $S = 2$ and $L = 3$, J can assume the values 5, 4, 3, 2 and 1, corresponding to statistical weights 11, 9, 7, 5 and 3. The energies are 6928.3, 7376.8, 7728.7, 7985.8, 8154.7 cm^{-1}, corresponding to the characteristic temperatures 9969, 10614, 11120, 11490, and 11733 °K. This therefore makes a second contribution to the state sum:

$$Z_{\text{el}}^{(2)} = 11e^{-9969/T} + 9e^{-10614/T} + 7e^{-11120/T} + \\ + 5e^{-11490/T} + 3e^{-11733/T} \tag{5.17.3}$$

It can be seen that these terms begin to play a noticeable part in the state sum, even at fairly low temperatures (above about 1000 °K). If it is required to calculate the state sum of the Fe atom for a temperature of, say, 5000 °K, then very many more terms must be included than are given by (5.17.2) and (5.17.3).

Table 9 gives the statistical weights and the characteristic temperatures of the lowest electron levels of various atoms. In compiling the table, all energy

TABLE 9

STATISTICAL WEIGHTS AND CHARACTERISTIC TEMPERATURES OF THE LOWEST ELECTRON LEVELS OF VARIOUS ATOMS

Element	Statistical weights					Characteristic temperatures				
He, Ne, Ar, Kr, X, Rn	1					0				
Be, Mg, Ca, Sr, Ba	1					0				
Zn, Cd, Hg	1					0				
Pb	1					0				
H, Li, Na, K, Rb, Cs	2					0				
Cu, Ag, Au	2					0				
N, P, As, Sb, Bi	4					0				
Mn	6					0				
Cr, Mo	7					0				
Al	2	4				0	160.5			
F	4	2				0	581.3			
Cl	4	2				0	1268			
Br	4	2				0	5302			
C	1	3	5			0	21.3	60.9		
Si	1	3	5			0	111.0	321.3		
Sn	1	3	5			0	2434	4932		
O	5	3	1			0	227.5	326.3		
S	5	3	1			0	573	824		
Fe	9	7	5	3	1	0	598.5	1013	1278	1407

TABLE 10

VALUES OF SOME IMPORTANT PHYSICAL CONSTANTS

Gas constant	R	$= 1.9872$ cal/degree
Boltzmann's constant	k	$= 1.3804 \times 10^{-16}$ erg/degree
Planck's constant	h	$= 6.6252 \times 10^{-27}$ erg.sec
Velocity of light	c	$= 2.9979 \times 10^{10}$ cm/sec
Avogadro's number (chem.)	N_0	$= 6.0232 \times 10^{23}$
Avogadro's number (phys.)	N_0	$= 6.0247 \times 10^{23}$
Charge on electron	e	$= 4.8029 \times 10^{-10}$ e.s.u.
Mass of electron	m_e	$= 9.1085 \times 10^{-28}$ g
Mass of proton	m_p	$= 1.6724 \times 10^{-24}$ g
Mass of neutron	m_n	$= 1.6747 \times 10^{-24}$ g

levels corresponding to characteristic temperatures greater than 9000 °K have been omitted.

Finally, Table 11 gives the exact values, calculated from Table 9 and Equations (5.5.13) and (5.11.1) of the statistical entropies of monatomic gases in the perfect state at 298.15 °K and 1 atm. The values given do not include the nuclear spin entropy and the isotopic mixing entropy, which will be discussed in Chapter 6.

Naturally, most of the elements appearing in Table 11 are not stable in the monatomic gaseous form at 298.15 °K and 1 atm; they will spontaneously condense or combine to diatomic molecules. At high temperatures, however, one may possibly have to deal with chemical reactions, in which these elements occur in an appreciable concentration in the atomic form and for thermodynamic calculations it is then often convenient to be able to use the entropies given in Table 11 for the "standard state" (25 °C and 1 atm; cf. Section 5.5).

TABLE 11

EXACT VALUES OF THE STATISTICAL ENTROPIES IN CAL/DEG.MOLE OF SOME ELEMENTS IN THE MONATOMIC PERFECT GASEOUS STATE AT 298.15 °K AND 1 ATM, EXCLUDING THE NUCLEAR SPIN ENTROPY AND THE ISOTOPIC MIXING ENTROPY

Gas	Translational entropy	Electronic entropy	Total entropy
H	26.01	1.38	27.39
Li	31.77	1.38	33.15
Na	35.34	1.38	36.72
K	36.92	1.38	38.30
Rb	39.25	1.38	40.63
Cs	40.57	1.38	41.95
Cu	38.37	1.38	39.75
Ag	39.94	1.38	41.32
Au	41.74	1.38	43.12
Be	32.54	0.00	32.54
Mg	35.50	0.00	35.50
Ca	36.99	0.00	36.99
Sr	39.32	0.00	39.32
Ba	40.66	0.00	40.66
Zn	38.45	0.00	38.45
Cd	40.07	0.00	40.07
Hg	41.79	0.00	41.79
Al	35.81	3.49	39.30
C	33.40	4.36	37.76
Si	35.93	4.19	40.12
Sn	40.23	0.02	40.25
Pb	41.89	0.00	41.89
N	33.86	2.75	36.61
P	36.22	2.75	38.97
As	38.86	2.75	41.61
Sb	40.30	2.75	43.05
Bi	41.92	2.75	44.67
Cr	37.76	3.87	41.63
Mo	39.59	3.87	43.46
O	34.26	4.22	38.48
S	36.33	3.76	40.09
Mn	37.93	3.56	41.49
F	34.77	3.15	37.92
Cl	36.63	2.83	39.46
Br	39.05	2.75	41.80
I	40.43	2.75	43.18
Fe	37.98	5.13	43.11
He	30.12	0.00	30.12
Ne	34.95	0.00	34.95
Ar	36.98	0.00	36.98
Kr	39.19	0.00	39.19
X	40.53	0.00	40.53

CHAPTER 6

THE ENTROPY OF DIATOMIC GASES

6.1. INTRODUCTION

The calculation by statistical means of the entropy of diatomic gases, as in the case of monatomic gases, consists chiefly of the calculation of the state sum. Since all the other thermodynamic functions can also be expressed in terms of the state sum (see Chapter 5), the calculation of the entropy from this sum must be regarded primarily as an example of the calculation of thermodynamic functions by statistical means. We might also have chosen as our example the free energy, the calculation of which, according to the equations in Sections 5.10 and 5.11, is somewhat easier, or the specific heat, for which the calculation is rather more complicated than that for the entropy.

In this chapter we shall show how approximate (but nevertheless fairly accurate) values of the state sum of diatomic gases can be calculated with the help of idealized models of molecules. The main points of the idealization are, that in treating the rotation of the molecules they are regarded as being rigid, while in treating the vibration, they are regarded as harmonic oscillators. We shall also show how *exact* values of the state sum can be calculated on the basis of the energy levels of the molecules derived from spectra. Finally, we shall discuss the calculation of chemical equilibria between gases, from the statistically determined thermodynamic quantities.

6.2. THE ROTATION OF RIGID MOLECULES

It has already been pointed out in Chapters 2 and 5 that a polyatomic gas in general possesses not only translational, but also rotational and vibrational energy.

The translational energy of a molecule is given by

$$\varepsilon_{tr} = \tfrac{1}{2}mv^2 = \frac{p^2}{2m}\,. \tag{6.2.1}$$

From this we can easily find the rotational energy of a rigid molecule rotating about a particular axis. The mass of the atoms, of which the molecule is composed, is concentrated almost wholly in the nuclei, which are so small that they may be regarded as mass-points. The masses of these points will be called m_1, m_2, m_3, ... and their distances from the axis r_1, r_2, r_3, ... At an angular velocity ω (expressed in radians) the mass-points have velocities $r_1\omega$, $r_2\omega$, $r_3\omega$, ... The kinetic energy of the rotation is thus given, according to (6.2.1), by

$$\varepsilon_{ro} = \tfrac{1}{2}\,(m_1 r_1^2 + m_2 r_2^2 + \ldots)\omega^2$$

or

$$\varepsilon_{ro} = \tfrac{1}{2}\, I\omega^2 \tag{6.2.2}$$

where

$$I = \sum m_i r_i^2. \tag{6.2.3}$$

We shall restrict ourselves to diatomic molecules, so that this becomes simply:

$$I = m_1 r_1^2 + m_2 r_2^2 \tag{6.2.3a}$$

I is called the *moment of inertia* with respect to the chosen axis. Comparison of Equations (6.2.2) and (6.2.1) shows that, in the expression for the rotational energy, the moment of inertia I has taken the place of the mass m and the angular velocity ω that of the linear velocity v. The momentum mv is replaced by the *angular momentum* $mvr = I\omega$, which will also be denoted by the symbol p. In complete analogy with (6.2.1), (6.2.2) may now be written:

$$\varepsilon_{ro} = \tfrac{1}{2}\, I\omega^2 = \frac{p^2}{2I}. \tag{6.2.2a}$$

According to the quantum theory, the energy can only assume discrete values. With the aid of the Schrödinger equation one finds that these are given for a diatomic molecule by

$$\varepsilon_{ro} = \frac{h^2}{8\pi^2 I} J(J+1) \tag{6.2.4}$$

where J is a positive integer (including zero), the rotational quantum number:

$$J = 0,\ 1,\ 2,\ 3,\ \ldots$$

One finds that $2J + 1$ eigenfunctions correspond to each of the eigenvalues of the Schrödinger equation as given by (6.2.4). At each value of the energy permitted by quantum mechanics the molecule can thus be found in $2J + 1$ different states. This is often indicated by saying that the energy levels in this particular case are degenerate to the $(2J + 1)$th degree or that the statistical weight of the levels is $(2J + 1)$. If we wish, we may also regard the energy levels as being always single and say that at every value of J there is a group of $2J + 1$ coincident levels (cf. the two preceding chapters).

In the pictorial model each value of the energy corresponds to a particular angular velocity of the molecule and each of the $2J + 1$ states, belonging to a discrete value of the energy, to a particular orientation of the axis of rotation (and thus of the vector representing the angular momentum, which coincides with this axis) in space.

For the permissible values of the angular momentum we have

$$p = \frac{h}{2\pi} \sqrt{J(J + 1)} \tag{6.2.5}$$

in accordance with (6.2.2a) and (6.2.4).

Equation (6.2.4) makes it obvious why no account had to be taken of rotation in monatomic gases (Chapter 5). In fact, in monatomic molecules virtually the whole mass is concentrated in one atomic nucleus, so that the moment of inertia is exceedingly small. Consequently, the rotational energy quanta, according to (6.2.4), are very large as compared with those of diatomic or polyatomic molecules and also very large with respect to the average translational energy of the molecules which was seen in the previous chapter to be of the order of magnitude of kT. As a result, the rotation plays no part in the mutual exchange of energy in monatomic gases.

For diatomic molecules, the moment of inertia of the rotation about the line joining the nuclei is of the same order of magnitude as that of the single atoms. Thus this rotation, too, will not be excited at normally attainable temperatures. It is quite a different matter with the rotation of these molecules about an axis perpendicular to the line joining the nuclei.

The moment of inertia of this rotation is found as follows. From mechanics it is known that the rotational axis perpendicular to the line joining the nuclei must pass through the centre of gravity of the molecule. If the distances of the two atoms from the centre of gravity are r_1 and r_2 and their masses m_1 and m_2, then

$$r_1 + r_2 = r_0 \tag{6.2.6}$$

$$m_1 r_1 = m_2 r_2 \tag{6.2.7}$$

where r_0 is the distance between the two nuclei. From (6.2.6) and (6.2.7) we obtain the two equations:

$$r_1 = \frac{m_2}{m_1 + m_2} r_0. \tag{6.2.8}$$

$$r_2 = \frac{m_1}{m_1 + m_2} r_0 \tag{6.2.9}$$

The moment of inertia is given, according to (6.2.3a), by

$$I = m_1 r_1^2 + m_2 r_2^2$$

or, substituting (6.2.8) and (6.2.9):

$$I = \frac{m_1 m_2^2 + m_2 m_1^2}{(m_1 + m_2)^2} r_0^2 = \frac{m_1 m_2}{m_1 + m_2} r_0^2 = m_r r_0^2 \tag{6.2.10}$$

where

$$m_r = \frac{m_1 m_2}{m_1 + m_2} \quad \text{or} \quad \frac{1}{m_r} = \frac{1}{m_1} + \frac{1}{m_2} \tag{6.2.11}$$

The moment of inertia of a diatomic molecule with respect to an axis perpendicular to the line joining the nuclei is seen from Equation (6.2.10) to be the same as that of one single mass m_r at a distance r_0 from this axis. The mass m_r is referred to as the *reduced mass*.

Henceforth, where the moment of inertia of a diatomic molecule is mentioned, this will always be taken to refer to the moment of inertia with respect to an axis perpendicular to the line joining the nuclei, as given by Equation (6.2.10). We shall calculate this moment of inertia for an HCl molecule, where the distance between hydrogen nucleus and chlorine nucleus $r_0 = 1.284 \times 10^{-8}$ cm. According to (6.2.11), the reciprocal of the reduced mass is given by

$$\frac{1}{m_r} = \frac{6.0232 \times 10^{23}}{1.0080} + \frac{6.0232 \times 10^{23}}{35.457} = 6.145 \times 10^{23} \text{ gram}^{-1}$$

Thus the moment of inertia, from (6.2.10), is

$$I_0 = m_r r_0^2 = \frac{(1.284 \times 10^{-8})^2}{6.145 \times 10^{23}} = 2.682 \times 10^{-40} \text{ g.cm}^2$$

The amount of energy required to transfer the HCl molecule from the rotation state $J = 0$ to the state $J = 1$, is given from (6.2.4) by

$$\varepsilon_1 - \varepsilon_0 = \frac{h^2}{4\pi^2 I} = 4.145 \times 10^{-15} \text{ erg}$$

Even at 300 °K the value of kT is already about 10 times as large. The temperature at which kT is equal to $h^2/4\pi^2 I$ is called by definition the *characteristic temperature of the rotation:*

$$\theta_{ro} = \frac{h^2}{4\pi^2 k I_0}. \tag{6.2.12}$$

For HCl, this temperature is very low:

$$\theta_{ro}^{HCl} = 30\ °K.$$

Thus at room temperature HCl molecules not only perform their translational motion, but at the same time they rotate with a variable number of quanta about an axis perpendicular to the line joining the nuclei. If the number of *degrees of freedom* is defined as the number of co-ordinates which is necessary and sufficient to describe the motion, then a HCl molecule, besides possessing the three degrees of freedom of the translational motion (three co-ordinates x, y, z are required to describe this motion), also has two rotational degrees of freedom. To describe the rotation of any arbitrary (non-linear) molecule, one requires to know three quantities, e.g. data about the rotation around three mutually perpendicular axes. In diatomic molecules one of these degrees of freedom can be excluded, as was shown above, viz. that for the rotation about the line joining the nuclei.

In the previous chapter we saw that for the translational motion, the mean energy per molecule was $3kT/2$, thus $kT/2$ per degree of freedom. According to the *law of equipartition of energy*, the same is true for the rotational degrees of freedom, provided that the temperature lies far above the characteristic temperature and provided we do not relinquish the model of the rigid rotator. At medium temperatures, i.e. temperatures so low that vibration still takes virtually no part (see following section), but high enough for the equipartition value of the rotational energy to have been reached, the energy per mole for diatomic gases is thus approximately $5RT/2$, the molar heat capacity at constant volume $5R/2$ and the molar heat capacity at constant pressure $7R/2$ (see Equation (1.9.2)).

To conclude this section we may make a few remarks on the calculation of the moment of inertia and the characteristic temperature of HCl.

(1) In reality the moment of inertia I is not calculated, as was done in this section, from the nuclear separation r. The reverse is the case: I is deduced from the spectrum and r is then calculated from I.

(2) The value r_0 used in the calculations for the nuclear separation in the molecule (assumed to be rigid) is *not* the equilibrium separation in the pictorial model (i.e. not the separation of stationary nuclei), but a mean

separation for the lowest vibration state. In fact, we shall see in the following section, that the state in which the nuclei are at rest with respect to each other, does not exist. Even at the lowest temperatures, long before vibration states are excited by mutual collisions, the nuclei perform their zero-point vibration and the molecules possess a certain minimum vibrational energy. This zero-point vibration is associated with an effective moment of inertia I_0, which is slightly larger than the moment of inertia I_e in the fictitious equilibrium state. I_0 corresponds to the distance used r_0, which is slightly larger than the "equilibrium separation" r_e. Both I_e and r_e only have significance in the pictorial model, not in reality. Both quantities refer to the non-existent state without zero-point vibration.

(3) The calculation of the moment of inertia and the characteristic temperature for HCl was carried out as though gaseous HCl only contained one sort of molecule. In reality, HCl is a mixture of 75% HCl^{35} and 25% HCl^{37}. (The percentage of deuterium in hydrogen is so small that the presence of DCl^{35} and DCl^{37} can be neglected). Therefore for a more accurate calculation, the reduced masses and the moments of inertia of HCl^{35} and HCl^{37} must be found separately. With $r_0 = 1.284 \times 10^{-8}$ cm for both kinds of molecules (the separations are equal because the mass has practically no effect on the forces operating), we find the following, only very slightly different values for the moments of inertia:

$$I_0(HCl^{35}) = 2.681 \times 10^{-40} \text{ g.cm}^2$$
$$I_0(HCl^{37}) = 2.685 \times 10^{-40} \text{ g.cm}^2.$$

In calculating these moments of inertia we made use of the atomic weights $Cl^{35} = 34.9788$ and $Cl^{37} = 36.9777$ [1]). As is usual for isotopes, the weights are on the physical scale, based on an atomic weight of 16.00000 for the most frequently occurring oxygen isotope O^{16}. On the other hand, the atomic weight 35.457 used in this section for natural chlorine was expressed, as is customary, on the chemical scale, based on an atomic weight of 16.00000 for natural oxygen (mixture of O^{16} with small quantities of the rare isotopes O^{17} and O^{18}). To calculate the reciprocal mass of Cl^{35} and Cl^{37} (required for the calculation of the moments of inertia) it was therefore necessary to make use of an "Avogadro's number" which is greater than that used earlier in this section. In Table 10 (page 248) was stated:

N_0 (physical) = number of atoms in 16.00000 g of $O^{16} = 6.0247 \times 10^{23}$
N_0 (chemical) = number of atoms in 16.00000 g of natural oxygen $= 6.0232 \times 10^{23}$

[1]) W. Riezler, Einführung in die Kernphysik, Vierte Auflage, Berlin, 1950.

6.3. HARMONICALLY VIBRATING MOLECULES

In the previous section we treated a diatomic molecule as a rigid body, in which the atomic nuclei are separated by a fixed distance r_0. As we know, this picture is not true to fact. By supplying sufficient energy, it is even possible to split the molecule into its constituent atoms. Smaller quantities of energy can be absorbed by the molecule as vibrational energy. The vibration takes place along the line joining the two nuclei.

As they move along this connecting line, the two atoms are acted upon by attractive and repulsive forces. At one particular distance, the equilibrium distance r_e, these forces just balance each other. At distances greater than r_e the attractive forces are predominant, at distances smaller than r_e, the repulsive forces. In the pictorial model the atoms would be able to remain at rest with respect to each other for an unlimited period of time at the distance r_e. As we shall see later in this section, and as was stated at the end of the previous section, the pictorial model fails here.

If we start from the equilibrium distance r_e, work must be performed to enlarge or diminish this distance. The potential energy P is a function of r. It is a minimum when $r = r_e$, where the resultant force $K = - \mathrm{d}P/\mathrm{d}r = 0$. Experiments have shown that it increases very rapidly as r becomes smaller than r_e and that it approaches asymptotically to a certain, finite value as r becomes very large, i.e. as dissociation occurs.

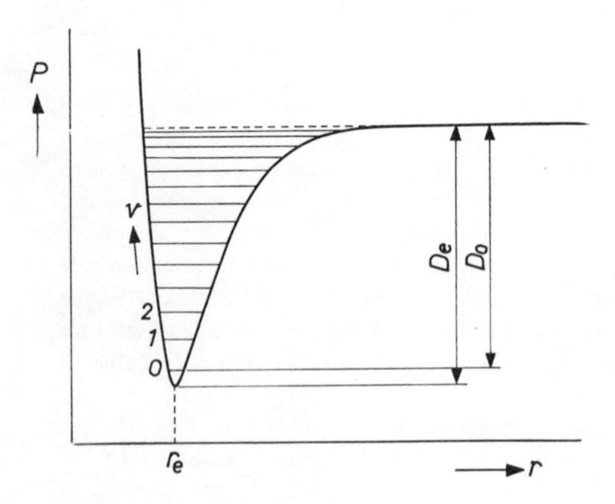

Fig. 64. Schematic representation of the potential curve and the energy levels of a vibrating diatomic molecule.

Fig. 64 gives a schematic representation of the general shape of the potential curve. The energy needed to dissociate the molecule (starting from the inachievable state in which the atoms are at rest at a distance of r_e from each other) is denoted by the symbol D_e. A fairly good empirical approximation to the curve is given by Morse's formula

$$P = C + D_e \left\{ e^{-2a(r-r_e)} - 2e^{-a(r-r_e)} \right\} \tag{6.3.1}$$

where C is an arbitrary constant because there is always an arbitrary additive constant in the energy.

From (6.3.1) we can find directly the expression for the force acting between the two atoms:

$$K = -\frac{\mathrm{d}P}{\mathrm{d}r} = 2aD_e \left\{ e^{-2a(r-r_e)} - e^{-a(r-r_e)} \right\}. \tag{6.3.2}$$

This formula can be considerably simplified if we only concern ourselves with oscillations of small amplitude. To do this, we expand the force in a MacLaurin series of the well known form:

$$\mathrm{F}(x) = \mathrm{F}(0) + x\mathrm{F}'(0) + \frac{x^2}{2!}\mathrm{F}''(0) + \frac{x^3}{3!}\mathrm{F}'''(0) + \ldots \tag{6.3.3}$$

With $x = r - r_e$, we thus obtain from (6.3.2):

$$\mathrm{F}(x) = K = -2a^2D_e(r - r_e) + 3a^3D_e(r - r_e)^2 - \ldots \tag{6.3.4}$$

For very small amplitudes, i.e. small values of $(r - r_e)$, we can neglect the quadratic and higher terms and find:

$$K = -2a^2D_e(r - r_e) \tag{6.3.5}$$

For very small values of the amplitude, the force is thus proportional to the deviation from the equilibrium position. This is the well-known case of a *harmonic oscillator.*

If the masses of the two atoms are m_1 and m_2 and if we call their (variable) distances from the centre of gravity r_1 and r_2 (so that always $r_1 + r_2 = r$), then the equations of motion for the two harmonically oscillating atoms read:

$$K = m_1 \frac{\mathrm{d}^2 r_1}{\mathrm{d}t^2} = -2a^2D_e(r - r_e) \tag{6.3.6}$$

$$K = m_2 \frac{\mathrm{d}^2 r_2}{\mathrm{d}t^2} = -2a^2D_e(r - r_e) \tag{6.3.7}$$

Dividing (6.3.6) by m_1 and (6.3.7) by m_2 and adding:

$$\frac{\mathrm{d}^2 (r_1 + r_2)}{\mathrm{d}t^2} = -2a^2D_e(r - r_e)\left(\frac{1}{m_1} + \frac{1}{m_2}\right)$$

or, substituting (6.2.11):

$$\frac{d^2r}{dt^2} = -\frac{2a^2D_e}{m_r}(r - r_e) \tag{6.3.8}$$

For d^2r/dt^2 we may also write $d^2(r - r_e)/dt^2$. If we again denote $r - r_e$ by the symbol x, we must then look for the solution of the differential equation

$$\frac{d^2x}{dt^2} = -\frac{2a^2D_e}{m_r}x \tag{6.3.9}$$

It can be easily verified that this solution is given by

$$x = x_0 \sin 2\pi\nu(t - t_0), \tag{6.3.10}$$

where

$$\nu = \frac{1}{2\pi}\left(\frac{2a^2D_e}{m_r}\right)^{\frac{1}{2}} \tag{6.3.11}$$

The vibration of the two atoms with respect to one another is seen from (6.3.9) to be the same as that of one particle of mass m_r under the action of a restoring force $2a^2D_ex$. The frequency of the vibration is given by (6.3.11).

The potential energy of the harmonic oscillator is given by

$$P = -\int K dx = 2a^2D_e\int x dx, \tag{6.3.12}$$

and thus, neglecting an additive constant, by

$$P = a^2D_ex^2 \tag{6.3.13}$$

This formula could have been found directly by expanding P (Equation (6.3.1)) in a MacLaurin series and neglecting the higher terms.

By combining (6.3.11) and (6.3.13) we obtain:

$$P = \tfrac{1}{2}(2\pi\nu)^2m_rx^2 \tag{6.3.14}$$

The total energy of the harmonic oscillator is equal to the sum of the kinetic and the potential energy:

$$\varepsilon_v = \frac{p_x{}^2}{2m_r} + \tfrac{1}{2}(2\pi\nu)^2m_rx^2 \tag{6.3.15}$$

The quantum theory prescribes that this energy also can only assume discreet values. With the aid of the Schrödinger equation, one finds that these are given by

$$\varepsilon_v = (v + \tfrac{1}{2})\,h\nu \tag{6.3.16}$$

where v is the vibrational quantum number which can only have the values 0, 1, 2, 3, . . . We find here that each value of v is associated with only one eigenfunction (cf. Chapter 2).

It is clear from (6.3.16) that the vibrationless state is not allowed by quantum mechanics. Even in the lowest state ($v = 0$) the molecule possesses the vibrational energy $\varepsilon_0 = \frac{1}{2}h\nu$, the zero-point energy.

According to (6.3.13) the potential curve of the harmonic oscillator is a parabola. A potential curve of this kind is illustrated in Fig. 65. The figure also shows several energy levels which, from (6.3.16), are equally spaced. To avoid confusion, we should point out that the horizontal lines indicate the total energy, while the parabola only gives the potential energy.

The parabolic potential curve permits no dissociation. We already know, from the foregoing, that the potential curve in reality has the shape sketched in Fig. 64. The fact is that the parabola is valid only for very small vibrational

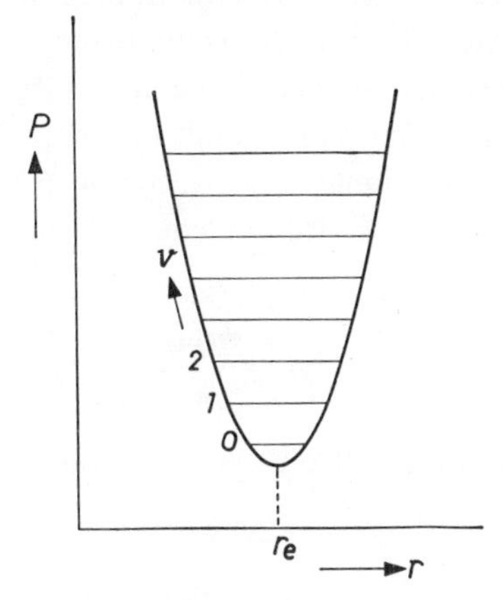

Fig. 65. Potential curve and energy levels of a harmonic oscillator.

amplitudes, and indeed small amplitudes were explicitly assumed when Equation (6.3.4) was replaced by Equation (6.3.5). In reality the energy levels are not separated by equal distances; with increasing quantum number, as shown in Fig. 64, they lie closer to one another.

The frequency of vibration of the harmonic oscillator is the same in each energy state. For the anharmonic oscillator, which will be discussed further in Section 6.11, this is not the case. One must then distinguish between the frequency of the oscillation with infinitely small amplitude which could occur in the classic model and which, in agreement with the foregoing, will be indicated by the suffix e, and the frequency of the zero-point oscillation which will be indicated by the suffix 0. These differ, not only from each other, but also from the continually diminishing frequencies of the higher vibrational states. Regarding the molecule as a harmonic oscillator (which, strictly speaking, is only justified for infinitely small amplitudes) thus amounts to putting all the frequencies equal to ν_e.

Most tables do not give the true frequency ν_e in sec^{-1}, but the "wave number" ω_e, expressed in cm^{-1}. These are connected by the relationship

$$\nu_e = c\omega_e \tag{6.3.17}$$

where c is the velocity of light. Inserting (6.3.17) in (6.3.16) we may write:

$$\varepsilon_v = hc\,(v + \tfrac{1}{2})\,\omega_e \tag{6.3.18}$$

As long as kT is small as compared with $h\nu_e = hc\omega_e$, vibrational quanta will hardly ever be excited by collisions. All the molecules will then perform only their zero-point oscillation, so that the vibrational energy is independent of the temperature, i.e. makes no contribution to the specific heat. However, if kT is large compared with $hc\omega_e$, the molecules will be able to absorb one or more vibrational quanta as a result of their collisions.

Analogous to the rotational case (Section 6.2), we can define a *characteristic temperature of the vibration* as that temperature for which kT is equal to $hc\omega_e$:

$$\theta_{\text{vi}} = \frac{hc\omega_e}{k}. \tag{6.3.19}$$

For example, for HCl we find from the spectrum:

$$\omega_e^{\text{HCl}} = 2989 \text{ cm}^{-1}$$

so that from (6.3.19):

$$\theta_{\text{vi}}^{\text{HCl}} = 4300\ ^\circ\text{K}$$

Under normal circumstances we may thus regard HCl as a gas, the molecules of which cannot absorb or give up vibrational energy.

For temperatures which are much higher than the characteristic temperature, the average vibrational energy per molecule (above the zero-point energy) is kT, of which $kT/2$ can be attributed to kinetic energy and $kT/2$ to potential energy. In the sense discussed in the previous section, the oscillation along a straight line is motion with only one degree of freedom. Now it is often required to include the vibrational motion when applying the *law of equipartition of energy*, according to which the energy, at sufficiently high temperatures, will be $kT/2$ per degree of freedom. One must then take a slightly different view of the concept of degree of freedom and define the number of degrees of freedom as the number of terms occurring in the expression for the energy that are quadratic in the positional or momentum co-ordinates. According to Equation (6.3.15), a diatomic molecule then has *two* vibrational degrees of freedom. We shall omit the proof of the equipartition law because it is of little importance within the scope of this book. Furthermore, even at high temperatures it does not apply rigidly, because this would only be the case if the moment of inertia were invariable, if the oscillations were purely harmonic and if the rotation and vibration did not interact.

Table 12 gives the characteristic temperatures of the rotation for some important diatomic molecules and also the values already given in Table 5 (Section 2.15) for the characteristic temperatures of the vibration. It can be seen from the table that the characteristic temperatures for the vibration lie far above those for the rotation. All the characteristic temperatures of rotation which are included lie below room temperature, all those for vibration above it.

Strictly speaking, when stating the oscillation frequency and characteristic temperatures of HCl, we should have made a distinction between HCl^{35} and HCl^{37}. However, we have already seen in the previous section that the moments of inertia (and thus also the characteristic temperatures of rotation) of HCl^{35} and HCl^{37} can be considered equal to a first approximation. The same applies to the vibration frequencies and thus to the characteristic temperatures of vibration. To illustrate this, we shall use Equation (6.3.11) to calculate the vibration frequency of HCl^{37} from that of HCl^{35}, which is 2989.74 cm^{-1}. According to the equation quoted, the frequency for isotopic molecules is inversely proportional to $m_r^{1/2}$. Thus:

$$\omega_e(HCl^{37}) = 2989.74\left(\frac{m_r(HCl^{35})}{m_r(HCl^{37})}\right)^{\frac{1}{2}} = 2987.47 \text{ cm}^{-1}$$

The only isotopic molecules for which the moments of inertia, vibration frequencies and characteristic temperatures differ widely, and which are thus mentioned separately in the table, are H^1H^1, H^2H^2 and H^1H^2. The reason for the great differences is naturally to be found in the fact that here the heavy isotope is no less than twice as heavy as the light one because the nucleus of the light isotope consists of a single proton, while that of the heavy isotope consists of a proton plus a neutron. The well-known use of a separate name and symbol for the heavy isotope (deuterium, D) serves to stress this great difference in properties.

TABLE 12

CHARACTERISTIC TEMPERATURES OF ROTATION AND VIBRATION FOR SOME DIATOMIC MOLECULES

Molecule	θ_{ro} (°K)	θ_{v1} (°K)
I_2	0.11	310
Br_2	0.23	465
Cl_2	0.69	810
O_2	4.2	2270
N_2	5.8	3400
D_2	87.5	4490
H_2	175	6340
NO	4.9	2740
CO	5.6	3120
HI	19	3320
HBr	24	3810
HCl	30	4300
HF	40	5960
HD	131	5500

6.4. ELECTRONIC STATES OF MOLECULES

Just like atoms, the diatomic and polyatomic molecules can attain higher electronic states as a result of collisions. In most cases, however, the energy quanta required to achieve the transition from the lowest to the second lowest state are very much greater than the vibration quanta, so that below 2000 °K, or even much higher temperatures, it is not necessary to take into consideration the occurrence of molecules in higher electronic states. Nevertheless, there are some cases in which the characteristic temperature for electronic excitation lies much lower, so that one or more higher states play a part in the state sum. As in atoms, this will most often occur in those cases where these higher states form a *multiplet* together with the lowest state.

In fact, just as one may speak of the multiplicity of the electronic state of an *atom*, thereby indicating the number of states which are only distinguishable from one another by the different relative orientations of the spin resultant (see Chapter 5), so one may also speak of the multiplicity of the electronic state of a *molecule*, which is similarly determined by the resultant electron spin.

Just as in the case of an atom with *LS* coupling, where the important quantities were the resultant **L** of the orbital angular momenta and the resultant **S** of the spins (the number of different ways in which they could be combined to form the resultant **J** gave the required multiplicity), so in a diatomic molecule, in the coupling of the most immediate interest, the important factors are the component **Λ** of the resultant orbital angular momentum along the line joining the nuclei and the spin resultant **S**.

When Λ, the quantum number corresponding to the vector **Λ**, is equal to 0, 1, 2, . . . , the electronic states of the molecule are called Σ, Π, Δ, . . . states (by analogy with the notation S, P, D, . . . for atoms when $L = 0, 1, 2, \ldots$).

The orbital motion of the electrons generates, in general, a magnetic field in the direction of the line joining the nuclei. In this field, the spin resultant can only orientate itself in such a way that the component **Σ** in the direction of the field, expressed in units of $h/2\pi$, according to (5.12.4) has one of the values

$$\Sigma = S, S-1, \ldots, -S \tag{6.4.1}$$

The spin resultant thus has $2S + 1$ different orientation possibilities.

The quantum number Σ of the component of the spin resultant along the line joining the nuclei, must not be confused with the symbol Σ for the state where $\Lambda = 0$, just as confusion must be avoided between the quantum number S of the spin resultant in an atom and the symbol S for the state where $L = 0$.

While for the *atoms* the total angular momentum of the electrons was obtained by vectorial addition (according to the rules of the quantum theory) of **L** and **S**, in the case of *molecules* the total angular momentum of the electrons about the line joining the nuclei is obtained by adding **Λ** and **Σ**. In this case, however, the addition is purely algebraic because **Λ** and **Σ** both lie in the line joining the nuclei. The number of permitted values of the sum is determined by the number of permitted values of **Σ**, thus by the value of **S** or the corresponding quantum number S.

We have to deal almost exclusively with cases where $S = 0, \frac{1}{2}$ or 1. In the case $S = 0$, i.e. in the case where there is no resultant spin vector, the quantum number Ω of the resultant angular momentum of the electrons about the line joining the nuclei, is naturally given exclusively by the value of Λ. In the case $S = \frac{1}{2}$, according to (6.4.1), Σ can only have the values $+\frac{1}{2}$ and $-\frac{1}{2}$, so that Ω can have the values $\Lambda + \frac{1}{2}$ and $\Lambda - \frac{1}{2}$. These different values are associated with somewhat different energies. Thus the spin splits what would otherwise be an energetically single state into a doublet. This multiplicity (given by $2S + 1$) is, again, denoted by a superscript written before the state symbol. For example, if $\Lambda = 1$, then for $S = \frac{1}{2}$ we are dealing with a $^2\Pi$ state. The value of $\Omega = \Lambda + \Sigma$ is added as a suffix after the state symbol. The above doublet Π state thus consists of the components $^2\Pi_{1/2}$ and $^2\Pi_{3/2}$. Fig. 66 gives the vector diagrams for this case.

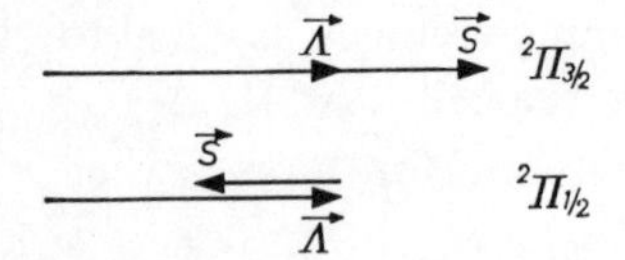

Fig. 66. Vector diagrams of a $^2\Pi$ state. In this and the following figure, a vector is indicated by a symbol surmounted by an arrow in contrast to the symbol in bold type without an arrow which is used in the text.

When $S = 1$, according to (6.4.1), Σ can only have the values 1, 0 and -1. The quantum number Ω can thus only have the values $\Lambda + 1$ or Λ or $\Lambda - 1$. Again, for these $2S + 1 = 3$ values there are slightly different values of the energy. For example, when $\Lambda = 2$, and $S = 1$ we are dealing with a triplet Δ state, consisting of the components $^3\Delta_1$, $^3\Delta_2$ and $^3\Delta_3$ (see Fig. 67).

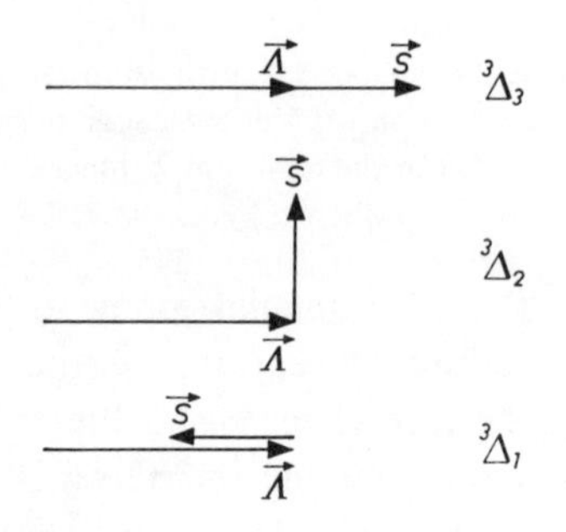

Fig. 67. Vector diagrams of a $^3\Delta$ state.

It is clear from the examples discussed that, in general, the following is valid: if the electrons of the diatomic molecule in question have a resultant spin, multiplicity of the energy levels will occur, because the spin vector can be oriented in $2S + 1$ different ways with respect to the line joining the nuclei, which corresponds to a splitting of each energy level into $2S + 1$ different levels.

The levels can only be split, however, if the component along the line joining the nuclei, of the resultant *orbital* angular momentum of the electrons is *not* equal to zero. For if $\Lambda = 0$, i.e. if we are dealing with a Σ state, there is no magnetic field in the direction of the line joining the nuclei, since such a field is only generated by the orbital motion of the electrons provided that $\Lambda \neq 0$. It is then no longer possible for the spin vector to orient itself with respect to the axis, since experience has shown that the spin, in contrast to the orbital angular momentum, is not affected by an electrical field. The strong electrical field between the nuclei is thus unable to cancel the $(2S+1)$-fold degeneracy of the Σ state.

However, if the molecule rotates, the axis of rotation (perpendicular to the line joining the nuclei) gives the spin vector the possibility of orientation, by which the $(2S+1)$-fold degeneracy is cancelled (see following section).

Finally, we may point out that all states for which $\Lambda \neq 0$ are twofold degenerate. In the pictorial model this corresponds to the fact that the vector $\mathbf{\Lambda}$ can point in either direction. In these two states, the direction of rotation of the electrons about the line joining the nuclei is different. This degeneracy, according to equation $S = k \ln g$, also leads to a larger value of the entropy.

6.5. COUPLING BETWEEN THE MOLECULAR ROTATION AND THE ELECTRONIC MOTION

The ground state of most diatomic molecules is a $^1\Sigma$ state, i.e. a state in which both the resultant electron spin and the resultant orbital angular momentum are equal to zero. The total angular momentum of the molecule, which we shall consistently denote by $\mathbf{J}$, is in this case identical with that of the rotation of the two atomic nuclei about their common centre of gravity (see Section 6.2).

In other states than $^1\Sigma$ there are various possibilities of coupling between the electron spin, the orbital angular momentum of the electrons and the angular momentum of rotation of the nuclei. One can distinguish five different possibilities, of which only two are of importance to us.

The first of these two cases will be involved in the calculation of the entropy of NO. In the ground state of this molecule, $\Lambda = 1$ and $S = \frac{1}{2}$ (see Section 6.10). In the sense of the preceding section the electronic motion is very closely coupled to the line joining the nuclei, while the coupling between the angular momentum of rotation of the nuclei and the angular momentum of the electronic motion is weak. The two, mutually perpendicular, angular momenta yield the total angular momentum $\mathbf{J}$. The corresponding quantum number J can never be smaller than the quantum number of the angular momentum of the electronic motion $\Omega = \Lambda + \Sigma$.

According to quantum mechanics, the absolute value $p = |\mathbf{J}|$ of the total angular momentum can again only assume the values given by Equation (6.2.5), in which J, for the various rotation states of the molecule, has the values $J = \Omega, \Omega + 1, \Omega + 2, \ldots$.

According to the preceding section, and since $\Lambda = 1$ and $S = \frac{1}{2}$, the NO molecule may be found in two different electronic states, characterized by $\Omega = \frac{1}{2}$ and $\Omega = \frac{3}{2}$. Thus in this case there will be two series of rotational levels, those belonging to the state $\Omega = \frac{1}{2}$ having the quantum numbers $J = \frac{1}{2}, \frac{3}{2}, \frac{5}{2}, \ldots$ and to the state $\Omega = \frac{3}{2}$, having the quantum numbers $J = \frac{3}{2}, \frac{5}{2}, \frac{7}{2}, \ldots$

Another type of coupling is evident in the oxygen molecule. In this molecule, $\Lambda = 0$ and $S = 1$. Since $\Lambda = 0$, the resultant spin is not coupled to the line joining the nuclei. The angular momentum of the rotation of the molecule as a whole, together with the spin produces the resultant **J**. The corresponding quantum number J can, from Equation (5.16.2), assume the values

$$J = (K + S), (K + S - 1), \ldots, |K - S|, \tag{6.5.1}$$

where K is the quantum number of the nuclear rotation and S that of the spin. For O_2 thus, since $S = 1$, J has the values

$$J = K + 1, \quad K \quad \text{and} \quad K - 1.$$

Each rotational level K is hereby split into three levels with slightly different energies.

6.6. THE ROTATIONAL ENTROPY OF HETERONUCLEAR MOLECULES

According to Section 5.11, the rotational entropy (insofar as it can be separately considered) is given by

$$s_{ro} = R\left(\ln Z_{ro} + T\frac{d \ln Z_{ro}}{dT}\right). \tag{6.6.1}$$

The state sum for the rotation Z_{ro}, which occurs here, can be written, from Section 5.9, in the form

$$Z_{ro} = \sum g_{ro} e^{-\varepsilon_{ro}/kT}. \tag{6.6.2}$$

If the diatomic molecule is regarded, to a first approximation, as being rigid then, from Section 6.2:

$$\varepsilon_{ro} = \frac{h^2}{8\pi^2 I} J(J+1) \tag{6.6.3}$$

and

$$g_{ro} = 2J + 1\,. \tag{6.6.4}$$

The state sum thus becomes:

$$Z_{ro} = \sum_{J=0}^{\infty} (2J+1)\, e^{-J(J+1)h^2/8\pi^2 IkT}\,. \tag{6.6.5}$$

For temperatures at which kT is very large compared to the rotation quanta or, in other words, at which $T \gg \theta_{ro}$, the terms of the sum differ so little from each other that the summation can be replaced by an integration:

$$Z_{ro} = \int_0^{\infty} (2J+1)\, e^{-J(J+1)h^2/8\pi^2 IkT}\, \mathrm{d}J \text{ (for } T \gg \theta_{ro}). \tag{6.6.6}$$

Now

$$(2J+1) = \frac{\mathrm{d}J(J+1)}{\mathrm{d}J},$$

and

$$\int^{\infty} e^{-ax}\, \mathrm{d}x = \frac{1}{a}, \tag{6.6.7}$$

so that (6.6.6) becomes:

$$Z_{ro} = \frac{8\pi^2 IkT}{h^2} \quad \text{(for } T \gg \theta_{ro}). \tag{6.6.8}$$

Substituting (6.6.8) in (6.6.1) yields for the entropy:

$$s_{ro} = R \ln I + R \ln T + R \ln \frac{8\pi^2 k}{h^2} + R \quad \text{(for } T \gg \theta_{ro}). \tag{6.6.9}$$

Inserting the numerical values of the physical constants (see Table 10, page 248) we obtain:

$$s_{ro} = R \ln I + R \ln T + 177.67 \text{ cal/degree.mole (for } T \gg \theta_{ro}) \tag{6.6.10}$$

The "standard value" of the rotational entropy, i.e. the value s^0_{ro} of s_{ro} at 25 °C (298.15 °K) is thus given for $298 \gg \theta_{ro}$ by

$$s^0_{ro} = R \ln I + 188.99 \text{ cal/degree.mole} \tag{6.6.11}$$

For most diatomic molecules, Table 12 shows that the rotational characteristic temperatures lie so low with respect to 298 °K that the standard rotational entropy can be calculated by means of the simple Equation (6.6.11). On the other hand, it can be seen from the same table that the vibrational characteristic temperatures are so high for most diatomic molecules that the vibration need not be taken into consideration at all at 298 °K.

Of these diatomic molecules with relatively low characteristic temperature for rotation and relatively high characteristic temperature for the vibration, the majority have a $^1\Sigma$ state as electronic ground state, which means, according to Section 6.4, that both the resultant orbital angular momentum of the electrons about the line joining the nuclei and the resultant electron spin are equal to zero. This singlet Σ state occurs, for example, in HF, HCl, HBr, HI and CO. Higher electronic states are here unattainable at normal temperatures, so that all the molecules are to be found at the lowest electronic energy level, which is not degenerate. If we choose this state as the energetic zero-level, the electronic state sum is simply given by

$$Z_{el} = 1.e^{-0/kT} = 1$$

(In the same way, leaving the vibration out of consideration means that the energy is reckoned from the state in which the molecules perform their zero-point oscillation, so that $Z_{vi} = 1$).

The standard entropy of the gases in question (HF, HCl, HBr, HI and CO) can be calculated, as shown above, by simply adding together the translational entropy, given by Equation (5.5.13), and the rotational entropy, given by Equation (6.6.11):

$$s^0 = s^0_{tr} + s^0_{ro} = \tfrac{3}{2} R \ln M + R \ln I + 214.98 \tag{6.6.12}$$

In this equation, I must be given the value of I_0, i.e. the value of the effective moment of inertia of the molecule when performing its zero-point oscillation. In this way, one finds the values given in Table 13 for the standard entropies of the gases under discussion.

These entropy values, calculated in such a simple manner, are already accurate to within a few units in the last decimal place (cf. Table 26).

It must be most emphatically stated that the equations derived in this section can *not* be used to calculate the entropies of molecules with two similar nuclei (homonuclear molecules), since for these we must first introduce a symmetry factor based on the indistinguishability of identical particles. Before we can do this, we must first discuss the nuclear spin entropy, which has been left out of consideration up to now (see the following section).

TABLE 13

MOLECULAR WEIGHT M, MOMENT OF INERTIA I_0 AND APPROXIMATE VALUE OF THE STANDARD ENTROPY s^0 OF SOME HETERONUCLEAR MOLECULES

Molecule	M	$I_0 \times 10^{40}$ g.cm^2	s^0 cal/degree.mole
HF	20.01	1.362	41.49
HCl	36.47	2.681	44.63
HBr	80.92	3.348	47.44
HI	127.93	4.333	49.33
CO	28.01	14.56	47.21

For gases such as I_2, Br_2 and Cl_2 whose vibrational characteristic temperatures do not lie far above room temperature, one must also take into account the occurrence of higher vibrational states at room temperature, thus of a certain amount of vibrational entropy. This will be calculated in Section 6.9.

Furthermore, there are also homonuclear and heteronuclear molecules with a resultant electron spin, as a result of which, at not too low temperatures, the molecules are divided between various electronic states and therefore also possess a certain amount of electronic entropy. For the treatment of this subject, the reader is referred to Section 6.10.

6.7. NUCLEAR SPIN ENTROPY

In general, atomic nuclei also possess their own angular momentum or spin **i**. As a result of this, as described in Chapter 5, directional quantization will occur in a magnetic field. For each of the $\rho = 2i + 1$ possible orientations (where i is the quantum number of the nuclear spin), there is one corresponding nuclear spin eigenfunction. In the absence of an external field, all these ρ eigenfunctions correspond to the same value of the energy.

If we are dealing with a heteronuclear diatomic molecule, composed of a nucleus 1 with ρ_1 eigenfunctions and a nucleus 2 with ρ_2 eigenfunctions, then each eigenfunction of 1 can be combined with each eigenfunction of 2, so that the system of two nuclei can be found in $\rho_1\rho_2$ different nuclear spin states, each of which is described by a product function of the form $\psi_j(1)\psi'_k(2)$. In this product $\psi_j(1)$ is one of the ρ_1 eigenfunctions which describe the states of the first nucleus and $\psi'_k(2)$ one of the ρ_2 which describe the states of the second nucleus.

These $\rho_1\rho_2$ different nuclear spin states cause each rotational-vibrational-electronic state to be $\rho_1\rho_2$-fold. The energy differences between the $\rho_1\rho_2$ different nuclear spin states are so minute (see also the following section) that this can be accounted for in the state sum by a constant factor $Z_{sp} = \rho_1\rho_2$. The nuclear spin entropy can thus be expressed by an equation of the type (5.11.1) which, as a result of the above, takes the following simple form:

$$s_{sp} = R \ln \rho_1\rho_2,$$

where R is the gas constant.

No symmetry requirements are here involved, for the two different nuclei 1 and 2 are always distinguishable.

If, however, the molecules are homonuclear, i.e. if the two nuclei 1 and 2 are identical, then according to Chapter 4, a symmetry condition must be imposed on the wave functions. This symmetry condition is to the effect that for homonuclear, diatomic molecules, the actually occurring stationary nuclear states of the molecule in general are *not* described by product functions of the type $\psi_j(1)\psi_k(2)$, but by the linear combinations

$$\psi_j(1)\psi_k(2) + \psi_j(2)\psi_k(1) \tag{6.7.1}$$

or

$$\psi_j(1)\psi_k(2) - \psi_j(2)\psi_k(1)\,. \tag{6.7.2}$$

Here, $\psi_j(i)$ and $\psi_k(i)$ are each one of the eigenfunctions of each of the two similar nuclei. The first combination is symmetrical, the second anti-symmetrical with respect to the nuclei.

The number of anti-symmetrical nuclear spin eigenfunctions, i.e. the number of ways in which two indistinguishable particles can be divided between ρ states, with the restriction that they may never be in the same state, is given from Section 2.5 by

$$\frac{\rho!}{2!\,(\rho-2)!} = \tfrac{1}{2}\,\rho\,(\rho-1). \tag{6.7.3}$$

The number of symmetrical nuclear spin eigenfunctions, i.e. the number of ways in which we can divide two indistinguishable particles between ρ

states such that they may also be found in the same state, is given from Section 2.13 by

$$\frac{(2+\rho-1)!}{2!\,(\rho-1)!} = \tfrac{1}{2}\,\rho\,(\rho+1). \qquad (6.7.4)$$

We saw in Chapter 4 that for the elementary material particles (protons, neutrons, electrons), the wave function must always be anti-symmetrical with respect to the particles, i.e. it must change sign when two particles are interchanged. Applied to the simplest diatomic molecule, viz. H_2, this means that the wave function must change sign when the two nuclei are interchanged. This interchange not only involves interchanging the spin co-ordinates, but also the position co-ordinates of the protons. The condition that the wave function must be anti-symmetrical with respect to the nuclei, cannot then apply to the nuclear spin function, but to the *total* wave function. The nuclear spin function may be either symmetrical or anti-symmetrical in the nuclei, provided the total wave function is anti-symmetrical. This total wave function can be written, to a first approximation, as the product of the eigenfunctions which correspond to the separate degrees of freedom. Besides the nuclear spin eigenfunction, this product also contains the eigenfunctions of the translation, the rotation and the vibration and, furthermore, the electronic eigenfunction.

The translational motion is that of the centre of gravity and the eigenfunction which corresponds to it does not contain the co-ordinates of the nuclei, in other words, it is symmetrical with respect to the nuclei. The rotational eigenfunctions for which $J = 0, 2, 4, \ldots$ are symmetrical with respect to the nuclei, those for which $J = 1, 3, 5, \ldots$ are anti-symmetrical with respect to the nuclei. For a rotator of which the axis of rotation has a fixed direction, this is easily seen. For in that case the rotational eigenfunction ψ^{ro} has the form

$$\psi^{ro} = \sin J\phi\,,$$

where ϕ is the angle of rotation about the axis. Interchanging the two nuclei means that ϕ becomes $\phi + \pi$. This leaves the sign unchanged when J is even, while it changes sign when J is odd. The mathematical treatment of the rotator with a free axis is not so simple. The symmetry properties of the ψ^{ro} function, however, remain the same. Consequently, the rotational eigenfunction for even values of J is symmetrical and for odd values of J is anti-symmetrical with respect to the nuclei.

In the same way, for the vibrational eigenfunctions, those with an even v are symmetrical while those with odd v are anti-symmetrical with respect to the nuclei. The eigenfunction of the electrons, finally, can be either symmetrical or anti-symmetrical with respect to the nuclear co-ordinates; for the ground state it is symmetrical for nearly all diatomic molecules.

We shall restrict ourselves here to a consideration of hydrogen at not too high temperatures, i.e. at temperatures at which virtually all the molecules are in the lowest vibrational and the lowest electronic state, both of which are symmetrical with respect to the nuclei. In the total wave function, written as the product of the separate eigenfunctions, we do not need to take into account the eigenfunctions of translation, vibration and the electrons, since they can not change the symmetry character of the product.

The symmetry character of the total eigenfunction of H_2, according to the foregoing, is determined by the product of the rotational eigenfunction ψ^{ro} and the nuclear spin eigenfunction ψ^{sp}. The anti-symmetry condition which must be imposed upon the total wave function, is satisfied by both the combination

$$\psi^{ro}_{anti}\ \psi^{sp}_{symm} \tag{6.7.5}$$

and the combination

$$\psi^{ro}_{symm}\ \psi^{sp}_{anti} \tag{6.7.6}$$

The quantum number of the spin of a proton has the value $i = \frac{1}{2}$. The number of possible orientations in a magnetic field (the number of nuclear spin wave functions) is thus $\rho = 2i + 1 = 2$. For the combination of two protons we have, from (6.7.3), $\frac{1}{2}\rho(\rho - 1) = 1$ anti-symmetrical nuclear spin eigenfunction and, from (6.7.4), $\frac{1}{2}\rho(\rho + 1) = 3$ symmetrical nuclear spin eigenfunctions. In the pictorial vector model, the anti-symmetrical function is described as the state in which the nuclear spins are anti-parallel oriented ($\uparrow\downarrow$) and the three symmetrical functions as the three states which are possible with a parallel orientation ($\uparrow\uparrow$) of the nuclear spins. (The quantum number of the resultant nuclear spin in the latter case is $I = 1$, which corresponds to $2I + 1 = 3$ possible orientations in an external field). As always, the pictorial representation must not be taken too literally here.

According to (6.7.5), the rotational eigenfunctions with $J = 1, 3, 5, \ldots$, which are anti-symmetrical with respect to the nuclei, can only occur with one of the $\frac{1}{2}\rho(\rho + 1) = 3$ symmetrical nuclear spin eigenfunctions, while according to (6.7.6), the rotational eigenfunctions with $J = 0, 2, 4, \ldots$, which are symmetrical with respect to the nuclei, can only occur with the $\frac{1}{2}\rho(\rho - 1) = 1$ anti-symmetrical nuclear spin eigenfunction. In other words: the H_2 molecules with a parallel orientation of the nuclear spins can only exist in rotational states which are characterized by odd quantum numbers, while the H_2 molecules with an anti-parallel orientation of the nuclear spins can only rotate with even quantum numbers.

Although at first sight it might be expected that a continual transition from one nuclear orientation to the other takes place within the H_2 molecule, this is *not* the case. From theoretical considerations also, it follows that the probability of such transitions is very small. Thus in practice this means that normal hydrogen consists of a mixture of two different "modifications". One speaks of *parahydrogen* with anti-parallel orientation of the spins and even rotation quantum numbers and of *orthohydrogen* with parallel orientation of the spins and odd rotation quantum numbers:

$$H_2 \begin{cases} \text{para} \;\; \uparrow\downarrow & J = 0,\ 2,\ 4,\ \ldots \\ \text{ortho} \;\; \uparrow\uparrow & J = 1,\ 3,\ 5,\ \ldots \end{cases}$$

The triple weight of the ortho-state (three symmetrical nuclear spin eigenfunctions) results in an equilibrium state in which there are three times as many ortho molecules as para molecules present in normal hydrogen, provided that the temperature is so high that kT is large compared with the rotation quanta, i.e. at room temperature and higher temperatures. For in this case the preference for the lowest rotational state, which is a para-state, no longer plays a part: the molecules are distributed among a fairly large number of rotational states, of which those with odd J (the ortho states) have been invested by the nuclear spin with a weight 3, while the para states have remained single.

In other words, half of the levels have become triple, due to the nuclear spin, while the other half has remained single. On an average, the number of states must therefore be multiplied, as a result of the nuclear spin, by a factor $(3 + 1)/2 = 2$ at high temperatures. This corresponds to an entropy term $R \ln 2$.

For practical reasons (see also the following section), this term $R \ln 2$ is not denoted by the name nuclear spin entropy. The nuclear spin entropy of homonuclear molecules at high temperatures is defined in the same way as in the case of heteronuclear molecules, viz. as

$$s_{\mathrm{sp}} = R \ln \rho_1 \rho_2$$

where $\rho_1 = \rho_2 = 2$ in the case of H_2. According to this definition, the nuclear spin entropy of H_2 is $R \ln 4$. In reality, due to the nuclear spin, only an entropy of

$$R \ln 2 = R \ln \rho_1 \rho_2 - R \ln 2$$

occurs. In other words, the chosen definition necessitates the introduction of an extra entropy term $-R \ln 2$, which is based on the symmetry of the molecule and does not occur in the case of heteronuclear molecules. Indeed, if no symmetry condition existed all the levels in the case under consideration ($\rho_1 = \rho_2 = 2$) would have been quadrupled by the nuclear spin, while now they are alternately single and triple. In the next section we shall deal with this question in a more precise manner.

If we cool normal hydrogen to very low temperatures then, as a result of the above-mentioned stability of the two forms, the proportion ortho: para = 3 : 1 will continue to hold, but all the para molecules will fall back to their lowest rotation level, viz. the level $J = 0$ and all the ortho molecules to the lowest ortho level, the level $J = 1$. At very low temperatures, however, this can never be the equilibrium state, for at 0 °K all the molecules must be in the state with the lowest energy, i.e. in the state for which $J = 0$. Finally, therefore, at decreasing temperature, all hydrogen *in the equilibrium state* must be transformed into para hydrogen, so that the nuclear spin entropy falls to zero.

This zero value of the nuclear spin entropy, in the equilibrium state, is virtually reached at about 20 °K. If we consider the reaction $H_2 \rightleftarrows 2\,H$ at this temperature, it is clear that it is accompanied by an important change in the nuclear spin entropy. For while H_2 at 20 °K in the equilibrium state is virtually wholly in the para state, monatomic hydrogen is still equally divided between the $2i + 1 = 2$ nuclear spin states ($\uparrow$ and $\downarrow$), i.e. it has a statistical weight 2, which corresponds to a nuclear spin entropy of $2R \ln 2$ (*two* gram-atoms). The occurrence of the reaction thus changes the nuclear spin entropy by an amount $2R \ln 2 = 2.75$ entropy units.

But we must immediately add, that equilibrium hydrogen is the only known example of a material whose nuclear spin entropy decreases with falling temperature within the range which is easily accessible to experiment. In other materials, the nuclear spin entropy only begins to decrease at very much lower temperatures. Consequently, with the exception of the example just discussed, the nuclear spin entropy does not change by chemical reactions. Equilibrium conditions can be calculated without taking the nuclear spin entropy into account. Entropy tables therefore commonly omit to include the nuclear spin contribution. This clearly demonstrates that the designation "absolute entropies", so often used for these tabulated values, must not be taken too absolutely.

It will be clear from the foregoing, that the cancellation of the nuclear spin entropy on each side of chemical reaction equations is based on the facts that (a) both the number and the nature of the nuclei remains unchanged in these reactions, and (b) at not too low temperatures the nuclear spin entropy of homonuclear molecules can be defined in the same way as that for heteronuclear molecules. Due to these two factors, each type of nucleus i retains its own nuclear spin entropy $R \ln \rho_i$ per gram-atom, independent of the manner of chemical combination.

Because the nuclei of homonuclear molecules were attributed by definition with a nuclear spin entropy $R \ln \rho_i$ per gram-atom, despite the fact that half the rotational levels are inaccessible to each separate molecule (ortho or para), we were forced to introduce a compensating entropy term, which was seen to have the value $- R \ln 2$ for diatomic homonuclear molecules. In other words, only by introducing this extra entropy term $- R \ln 2$, based on the symmetry of homonuclear molecules, are we able to ignore the nuclear spin entropy in chemical reactions at not too low temperatures.

This can be expressed less fully and therefore less accurately by saying that the nuclear spin entropy can be left out of consideration in chemical reactions because the number and the nature of the nuclei do not change in these reactions, while in contrast, the symmetry term $- R \ln 2$ (for diatomic homonuclear molecules) may not be neglected because the symmetry is not invariable in chemical reactions.

Special attention should be drawn to the fact that the dependence on the temperature of the nuclear spin entropy of equilibrium hydrogen is still noticeable at relatively high temperatures, because the rotation quanta are large, due to the small moment of inertia, and because the nuclear spin states are coupled to the rotational states by the symmetry conditions. To transfer a H_2 molecule at a low temperature from the para state to one of the three ortho states, it must be moved from the lowest to the lowest-but-one rotational level. In other homonuclear molecules, the distance between these two levels is much smaller. In heteronuclear molecules no coupling exists between the nuclear spin states and the rotational states. The energy differences between the $\rho_1\rho_2$ different nuclear spin states are therefore extremely small and the maximum nuclear spin entropy $s_{sp} = R \ln \rho_1\rho_2$ is already reached at extremely low temperatures (far below 1 °K).

6.8. NUCLEAR SPIN ENTROPY AND ROTATIONAL ENTROPY OF HOMONUCLEAR MOLECULES

If, instead of ordinary hydrogen H_2, we consider heavy hydrogen D_2, we are dealing with a homonuclear diatomic molecule with each nucleus (deuteron) consisting of one proton and one neutron. The anti-symmetry condition applies to both protons and neutrons. If the nuclei are interchanged, the total wave function thus changes sign twice, i.e. it is symmetrical with respect to the nuclei.

The electronic eigenfunction of H_2 was symmetrical with respect to the nuclei for the ground state and the same applies to D_2. If we once more restrict ourselves to the ground state of the vibration, thus to not too high temperatures, then the symmetry condition will be satisfied by the combinations

$$\psi^{\mathrm{ro}}_{\mathrm{anti}}\ \psi^{\mathrm{sp}}_{\mathrm{anti}} \tag{6.8.1}$$

and

$$\psi^{\mathrm{ro}}_{\mathrm{symm}}\ \psi^{\mathrm{sp}}_{\mathrm{symm}} \tag{6.8.2}$$

Both the neutron and the proton have a nuclear spin, of which the quantum number is $\frac{1}{2}$. In the deuteron the two spins are "parallel", so that the quantum number of the spin of the deuteron is 1. The number of possible orientations in a magnetic field (the number of nuclear spin wave functions) is thus $\rho = 2i + 1 = 3$. For the combination of two deuterons, i.e. for the D_2 molecule, we have from (6.7.3), $\frac{1}{2}\rho(\rho - 1) = 3$ anti-symmetrical nuclear spin eigenfunctions and from (6.7.4), $\frac{1}{2}\rho(\rho + 1) = 6$ symmetrical nuclear spin eigenfunctions.

The rotational eigenfunctions with odd J, according to (6.8.1), can here only occur together with one of the three anti-symmetrical nuclear spin eigenfunctions, the rotational eigenfunctions with even J, according to (6.8.2), only with one of the six symmetrical nuclear spin eigenfunctions.

Once again, transitions between the symmetrical and the anti-symmetrical states are forbidden (though, as for H_2, not very strictly), so that in practice we are dealing with two modifications. The modification with the greatest statistical weight, i.e. the modification with the symmetrical nuclear spin eigenfunctions, is always called the ortho modification, that with the smallest weight, i.e. that with the anti-symmetrical nuclear spin eigenfunctions, the para modification. The statistical weight of ortho D_2 is thus twice as large as that of para D_2. In contrast to H_2, ortho D_2 rotates with even quantum numbers, para D_2 with odd. In D_2, it is thus the ortho-modification which is the stable form at low temperatures (for the rotational state with $J = 0$ is an ortho state). Due to the statistical weights in the proportion 2 : 1, normal deuterium is a mixture of $\frac{2}{3}$ ortho deuterium and $\frac{1}{3}$ para deuterium.

If we proceed from the discussion of H_2 and D_2 to the general case, then it is clear that the total wave function is symmetrical with respect to the nuclei for homonuclear molecules whose nuclei have an even mass number (consist of an even number of protons plus neutrons), but anti-symmetrical for an odd mass number.

If the mass number is *odd*, then the rotational eigenfunctions with even J (symmetrical in the nuclei) must be combined with one of the $\frac{1}{2}\rho(\rho - 1)$ anti-symmetrical nuclear functions, on the other hand, the rotational eigenfunctions with odd J must be combined with one of the $\frac{1}{2}\rho(\rho + 1)$ symmetrical nuclear functions. The rotational states with even J thus have, for an odd mass number of the nuclei, an extra weight of $\frac{1}{2}\rho(\rho - 1)$ and the rotational states with odd J an extra weight of $\frac{1}{2}\rho(\rho + 1)$.

Conversely, if the mass number is *even*, the rotational states with even J must be combined with one of the $\frac{1}{2}\rho(\rho + 1)$ symmetrical nuclear functions and the rotational states with odd J with one of the $\frac{1}{2}\rho(\rho - 1)$ anti-symmetrical nuclear functions. Here, therefore, the even rotational states get an extra weight of $\frac{1}{2}\rho(\rho + 1)$ and the odd rotational states an extra weight of $\frac{1}{2}\rho(\rho - 1)$.

For the combination of the rotational and the nuclear spin states we therefore obtain, according to Equation (6.6.5), the following state sums:

Heteronuclear molecules:

$$Z_{rs} = \rho_1\rho_2 \sum_{J=0}^{\infty} (2J + 1)e^{-J(J+1)h^2/8\pi^2 IkT} \qquad (6.8.3)$$

Homonuclear molecules with odd mass number:

$$Z_{rs} = \tfrac{1}{2}\,\rho\,(\rho - 1) \sum_{J=0,2,\ldots}^{\infty} \ldots + \tfrac{1}{2}\,\rho\,(\rho + 1) \sum_{J=1,3,\ldots}^{\infty} \ldots \qquad (6.8.4)$$

Homonuclear molecules with even mass number:

$$Z_{rs} = \tfrac{1}{2}\,\rho\,(\rho + 1) \sum_{J=0,2,\ldots}^{\infty} \ldots + \tfrac{1}{2}\,\rho\,(\rho - 1) \sum_{J=1,3\ldots}^{\infty} \ldots \qquad (6.8.5)$$

For the sake of completeness, we have also included here the corresponding state sum of heteronuclear molecules. Naturally, the state sums may only be written in this form so long as the extremely small energy differences between the $\rho_1\rho_2$ nuclear spin states of the heteronuclear molecules may be neglected. The same applies to the energy differences between the $\frac{1}{2}\rho(\rho - 1)$ anti-symmetrical states of the homonuclear molecules and the energy differences between the $\frac{1}{2}\rho(\rho + 1)$ symmetrical states of these molecules. If one wished to calculate the state sum for extremely low temperatures (say 0.001 °K), the Equations (6.8.3), (6.8.4) and (6.8.5) could not be used. For example, in Equation (6.8.3), the nuclear spin states could not then be accounted for by a simple factor $\rho_1\rho_2$, i.e. it would no longer be justified to collect the nuclear-spin-rotational-states together in groups of $\rho_1\rho_2$ states with *nearly* the same energy. On the contrary, just because each group of states in reality does contain very small energy differences, which are then not small compared to kT, one should include a separate term in the state sum for each separate state.

Equations (6.8.4) and (6.8.5) have been derived on the supposition that the electronic ground state of the molecules is symmetrical with respect to the nuclei. In the exceptional cases in which this ground state is anti-symmetrical with respect to the nuclei, Equation (6.8.5) is then valid for the odd mass numbers and Equation (6.8.4) for even mass numbers.

If the nuclei have no spin, then $\rho = 2i + 1 = 1$, so that there are $\frac{1}{2}\rho(\rho - 1) = 0$ anti-symmetrical nuclear functions and $\frac{1}{2}\rho(\rho + 1) = 1$ symmetrical function for the diatomic molecule.

According to Equations (6.8.4) and (6.8.5), half the rotational states then drop out, i.e. the molecule can only rotate with even or only with odd quantum numbers. This becomes evident in the spectrum by the absence of half the lines. This case is found to occur for O_2^{16}. Furthermore, this type of molecule is one of the exceptional cases, in which the electronic function is anti-symmetrical with respect to the nuclei. Despite the even mass number, thus, Equation (6.8.4) must be applied, from which it appears that the oxygen molecules can only rotate with odd quanta.

If the temperature is much higher than the rotational characteristic temperature, i.e. at least twice as large (which is shown by Table 12 to be the case at room temperature for all diatomic molecules, except H_2), then by approximation:

$$\sum_{J=0,2,\ldots}^{\infty} \ldots = \sum_{J=1,3,\ldots}^{\infty} \ldots = \tfrac{1}{2} \sum_{J=0,1,2,\ldots}^{\infty} \ldots \tag{6.8.6}$$

With the aid of this approximation and Equation (6.6.5), both the Equations (6.8.4) and (6.8.5) for homonuclear molecules are converted into

$$Z_{rs} = \frac{\rho^2}{2} Z_{ro} \quad (\text{for } T \gg \theta_{ro}) \tag{6.8.7}$$

while, according to (6.6.5) and (6.8.3), we have for heteronuclear molecules:

$$Z_{rs} = \rho_1\rho_2 Z_{ro}. \tag{6.8.8}$$

In the latter case, Z_{rs} is thus the product of Z_{ro} and the total number of nuclear spin eigenfunctions $\rho_1\rho_2$, while in the former case, Z_{rs} is *half* the product of Z_{ro} and the total number of nuclear spin eigenfunctions $\rho^2 = \frac{1}{2}\rho(\rho - 1) + \frac{1}{2}\rho(\rho + 1)$. The factor 2, by which ρ^2 in (6.8.7) must be divided, originates (as appears from the foregoing) from taking into account the symmetry conditions. In classical statistics, this factor 2 was introduced as the "*symmetry number*" σ. This symmetry number is equal to 1 for a heteronuclear diatomic molecule and equal to 2 for a homonuclear diatomic molecule.

The introduction of the symmetry number makes it possible to combine the last two equations into one which, substituting Equation (6.6.8), becomes

$$Z_{rs} = \frac{8\pi^2 IkT}{h^2}\frac{\rho_1\rho_2}{\sigma} \quad (\text{for } T \gg \theta_{ro}) \tag{6.8.9}$$

As we have already mentioned in the previous section, the nuclear spin entropy is not usually taken into account. This amounts to omitting the factor $\rho_1\rho_2$ in (6.8.9). The symmetry number, however, remains, so that the entropy of a homonuclear gas is smaller than that of a heteronuclear gas by an amount $R \ln 2 = 1.38$ entropy units per mole. This can also be shown in a much simpler way (see Section 2 of the Appendix).

For H_2, D_2 and N_2, the electronic ground state is a $^1\Sigma$ state. Furthermore, the vibrational characteristic temperatures of these gases are very high (see Table 12). At room temperature, thus, both the electronic and the vibrational entropy can be neglected. The standard entropy of these gases consists entirely of translational and rotational entropy, so that the standard entropy can be calculated from the simple Equation (6.6.12), provided that the term 214.98 is replaced by $214.98 - 1.38 = 213.60$. In this way we can calculate the values given in Table 14 for the standard entropy s^0.

TABLE 14

MOLECULAR WEIGHT M, MOMENT OF INERTIA I_0, AND APPROXIMATE VALUE OF THE STANDARD ENTROPY s^0 OF SOME HOMONUCLEAR MOLECULES

Molecule	M	$I_0 \times 10^{40}$ g.cm^2	s^0 cal/degree.mole
H_2	2.016	0.4719	31.17
D_2	4.030	0.9360	34.59
N_2	28.02	13.99	45.75

Even the standard entropy of H_2, calculated in this way, is only 0.04 entropy units too small, despite the fact that for this gas, the characteristic temperature of the rotation is so high that it is really not justifiable to use the approximate Equation (6.8.9) here. At still lower temperatures, Equation (6.8.9) becomes quite useless for calculating the entropy of H_2. One is obliged to return to the original Equation (6.8.4). As we have already seen, Equations (6.8.4) and (6.8.9) supply the entropy including the nuclear spin entropy, while in most tables (as in Table 14) the nuclear spin entropy is excluded.

6.9. THE VIBRATIONAL ENTROPY

If we assume that, to a first approximation, the rotational and vibrational motions are independent of each other, we can write, according to Section 5.11:

$$s_{\text{vi}} = R \ln Z_{\text{vi}} + RT \frac{\mathrm{d} \ln Z_{\text{vi}}}{\mathrm{d}T} \tag{6.9.1}$$

If we assume further that the vibrations are those of a simple harmonic oscillator of wave-number ω_e cm^{-1}, then the energy levels are given, according to Section 6.3, by

$$\varepsilon_{\text{vi}} = hc\,(v + \tfrac{1}{2})\,\omega_e \text{ erg} \tag{6.9.2}$$

If we place the zero-level of the energy at the lowest vibrational state, i.e. at the state with the zero-point vibration, then (6.9.2) becomes

$$\varepsilon_{\text{vi}} = vhc\omega_e \text{ erg} \tag{6.9.3}$$

The state sum for the vibration is then given by

$$Z_{\text{vi}} = e^{-0/kT} + e^{-hc\omega_e/kT} + e^{-2\,hc\omega_e/kT} + \dots \tag{6.9.4}$$

If we introduce into this equation the vibrational characteristic temperature as defined by (6.3.19), we obtain:

$$Z_{\text{vi}} = 1 + e^{-\theta_{\text{vi}}/T} + e^{-2\theta_{\text{vi}}/T} + \dots \tag{6.9.4a}$$

For $x < 1$ we have

$$1 + x + x^2 + \dots = \frac{1}{1-x} \tag{6.9.5}$$

so that

$$Z_{\text{vi}} = \frac{1}{1 - e^{-\theta_{\text{vi}}/T}} \tag{6.9.4b}$$

The reader will recall that this expression for the state sum of a harmonic oscillator has already been derived in Section 2.11, during the discussion on a solid of the Einstein model.

Substitution of (6.9.4b) in (6.9.1) gives:

$$s_{vi} = -R \ln (1 - e^{-\theta_{vi}/T}) + \frac{R\theta_{vi}/T}{e^{\theta_{vi}/T} - 1}. \tag{6.9.6}$$

It is immediately clear from this equation that the vibrational entropy s_{vi} is smaller as θ_{vi}, the vibrational characteristic temperature, is higher. When $\theta_{vi} = T$, the formula supplies for s_{vi} the value $s_{vi} = 2.07$ entropy units, while for $\theta_{vi} = 7T$, $s_{vi} = 0.015$ units and for $\theta_{vi} = 8T$, $s_{vi} = 0.006$ units.

Thus, if the vibrational characteristic temperature is more than 7 or 8 times as large as the temperature for which one wishes to calculate the entropy, the vibrational entropy can be neglected. In the standard entropy, i.e. the entropy at 298.15 °K and 1 atm, the vibrational entropy can therefore be neglected if the vibrational characteristic temperature lies above 2200 °K. This is the case for all the diatomic gases mentioned in Table 12, with the exception of Cl_2, Br_2 and I_2. To calculate the standard entropy of these three homonuclear gases, we must proceed as in the preceding section, after which the value found must be increased by the vibrational entropy, given by Equation (6.9.6). The amount of electronic entropy is negligible because for these gases also, the electronic ground state is a $^1\Sigma$ state.

Table 15 below, contains the values, thus calculated, of the standard entropies of the three gases mentioned. To make comparison possible, the table also includes the values of the translational, the rotational and the vibrational entropies separately.

TABLE 15

APPROXIMATE VALUES OF THE STANDARD ENTROPY IN CAL/DEG.MOLE OF SOME HOMONUCLEAR GASES WITH LOW VIBRATIONAL CHARACTERISTIC TEMPERATURES

Molecule	M	$I_0 \times 10^{40}$	s^0_{tr}	s^0_{ro}	s^0_{vi}	s^0
Cl_2	70.91	115.2	38.69	14.01	0.52	53.22
Br_2	159.83	346.5	41.12	16.20	1.29	58.61
I_2	253.82	750.6	42.49	17.74	2.00	62.23

These values of the standard entropy also are accurate to within only a few units in the last decimal place (cf. Table 26).

If the state sum is not calculated by means of Equation (6.9.4b), but by calculating the terms in (6.9.4a) separately and adding them together, we can then see from the magnitude of the terms, which oscillations make an appreciable contribution to the entropy. It appears that in gaseous chlorine at room temperature, oscillations with 1 and 2 quanta already play a noticeable part, in bromine oscillations with 1, 2, 3 and 4 quanta and in iodine oscillations with 1 to 7 quanta. This can also be deduced directly from the vibrational characteristic temperatures, viz. by finding which multiples of the characteristic temperature supply an amount smaller than about 2200 °K.

6.10. THE ELECTRONIC ENTROPY

In the examples of entropy calculation treated above, we considered exclusively diatomic molecules of which the electronic ground state is a $^1\Sigma$ state. As a result, we did not have to take into account the electronic entropies. As examples of cases where this does have to be done, we shall consider in this section the homonuclear molecule O_2, whose ground state is a $^3\Sigma$ state and the heteronuclear molecule NO, whose ground state is a $^2\Pi$ state.

Oxygen

The oxygen molecule in its ground state has two unpaired electrons, i.e. two electrons with parallel spins ($\uparrow\uparrow$), while the other electrons are paired, i.e. compensate each other two by two. It thus has a resultant electron spin of which the quantum number $S = 1$. On the other hand, there is no resultant orbital angular momentum about the line joining the nuclei, so that the ground state of O_2 is a $^3\Sigma$ state (cf. Sections 6.4 and 6.5). For non-rotating oxygen molecules this would mean that they were in a degenerate state of weight 3. Non-rotating molecules, however, cannot occur in gaseous O_2 (at least, not in $O^{16}O^{16}$) even at the lowest temperatures, since according to Section 6.8, they can only rotate with odd quanta. There is thus always at least one rotation quantum excited.

Due to the rotation, the spin vector is able to orient itself with respect to the axis of rotation, whereby the degeneracy is cancelled. Each rotational energy level is, in other words, split into three components which have been denoted by Mulliken [1]) by the symbols F_1, F_2 and F_3. At F_1 the resultant

---

[1]) R. S. Mulliken, Phys. Rev. **32**, 880 (1928).

spin and the angular momentum of the rotation of the molecule are parallel, at F_2 they are perpendicular to one another and at F_3 anti-parallel.

These states, however, lie energetically very close together. The F_2 level lies about 4.10^{-16} erg higher than the F_1 and F_3 levels, which lie even closer together. The characteristic temperature for the transition to the F_2 level is $4 \cdot 10^{-16}/k = 3$ °K. At temperatures above about 6 °K we can neglect the energy differences between the three components and regard the O_2 molecules as molecules which are in a degenerate state of weight 3. The state sum is then simply three times as large as it would have been without resultant electron spin and for the rotational plus electronic entropy we can write:

$$s_{re} = R \ln 3Z_{ro} + RT \frac{d \ln 3Z_{ro}}{dT} = R \ln Z_{ro} + RT \frac{d \ln Z_{ro}}{dT} + R \ln 3$$

The electronic entropy is thus $R \ln 3$.

Table 16 gives for O_2 the molecular weight M, the moment of inertia I_0 corresponding to the zero-point vibration, the standard values of the translational and rotational entropy, which have been calculated in the normal manner, and finally, the electronic entropy and the standard value of the "total" entropy.

TABLE 16

MOLECULAR WEIGHT M, MOMENT OF INERTIA I_0 IN G.CM^2 AND STANDARD ENTROPY VALUES IN CAL/DEG. MOLE FOR O_2

M	32.00
$I_0 \times 10^{40}$	19.47
s^0_{tr}	36.32
s^0_{ro}	10.48
s^0_{el}	2.18
s^0	48.98

Nitric oxide

In contrast with O_2, NO does have a resultant orbital angular momentum of the electrons about the line joining the nuclei. The quantum number Λ of this vector is equal to 1, so that the ground state of NO is a Π state (see Section 6.4). Further, NO has one unpaired electron. The quantum number of the resultant electron spin is thus $\frac{1}{2}$ and the ground state of NO a doublet state ($^2\Pi$). The two components of this doublet have different directions of the spin vector with respect to $\mathbf{\Lambda}$. The spin vector can align itself either parallel or anti-parallel to $\mathbf{\Lambda}$ ($\Lambda + \Sigma = \frac{3}{2}$ or $\Lambda + \Sigma = \frac{1}{2}$). Between the two components $^2\Pi_{3/2}$ and $^2\Pi_{1/2}$ there is an energy difference of

240×10^{-16} erg. The characteristic temperature for the transition to the higher state is

$$240 \times 10^{-16}/k = 174\ °\text{K}$$

To obtain a rough approximation to the value of the electronic entropy at 298 °K, we shall start by neglecting the energy difference between the two components. Each rotational energy level then represents two states, each of which is also double, because the total angular momentum of the electrons can be in the direction N → O or N ← O. Thus as a rough approximation we are dealing with a fourfold degeneracy and thus with an electronic entropy $R \ln 4 = 2.75$ units. A better approximation is obtained by taking into account the energy difference between $^2\Pi_{1/2}$ and $^2\Pi_{3/2}$. If we place the zero-level of the energy at the lowest state, then the electronic state sum Z_{el} is given by

$$Z_{el} = g_0\, e^{-0/kT} + g_1\, e^{-\varepsilon_1/kT} = 2 + 2\, e^{-174/T} \tag{6.10.1}$$

According to (5.11.1), the electronic entropy is given by

$$s_{el} = R\left(\ln Z_{el} + T\frac{\mathrm{d}\ln Z_{el}}{\mathrm{d}T}\right) \tag{6.10.2}$$

If we substitute (6.10.1) in (6.10.2), we find the following value for the electronic entropy at 298.15 °K:

$$s_{el} = 2.67 \text{ cal/degree.mole}$$

The translational and rotational entropies are calculated in the usual manner. Thus we obtain for NO the values given in Table 17 for the standard entropies.

TABLE 17

MOLECULAR WEIGHT M, MOMENT OF INERTIA I_0 IN G.CM^2 AND STANDARD ENTROPIES IN CAL/DEG.MOLE FOR NO

M	30.01
$I_0 \times 10^{40}$	16.51
s^0_{tr}	36.13
s^0_{ro}	11.53
s^0_{el}	2.67
s^0	50.33

Actually, the electronic motion and the rotation of the molecule may not be considered separately (see Section 6.5). The angular momentum of the electrons is vectorially coupled to the angular momentum of the rotating molecule. The vectorial summation of the electron spin, the orbital angular momentum of the electrons about the line joining the nuclei and the angular momentum of the rotation of the molecule, leads to one single vector **J**, which represents the total angular momentum of the molecule.

Since the electronic angular momentum (from orbit and spin) and the angular momentum of the rotating molecule are perpendicular to one another, the total angular momentum can never be smaller than that of the electrons. The minimum value of the quantum number J of the total angular momentum is thus $\frac{1}{2}$ for the $^2\Pi_{1/2}$ state and $\frac{3}{2}$ for the $^2\Pi_{3/2}$ state. In this way we obtain the energy levels, shown in Fig. 68, for the rotation of NO. The state sum for the combination of electron motion and rotation reads

$$Z_{\text{er}} = 2 \sum_{J=\frac{1}{2},\frac{3}{2},\ldots}^{\infty} (2J+1)\, e^{-J(J+1)h^2/8\pi^2 IkT} + \\ + 2e^{-\varepsilon_1/kT} \sum_{J=\frac{3}{2},\frac{5}{2},\ldots}^{\infty} (2J+1)\, e^{-J(J+1)h^2/8\pi^2 IkT} \tag{6.10.3}$$

where ε_1 is again the energy difference between the two states $^2\Pi_{3/2}$ and $^2\Pi_{1/2}$ (cf. Fig. 68).

When $T \gg \theta_{\text{ro}}$ $(= h^2/4\pi^2 Ik)$, the sums in (6.10.3) may be replaced by integrals. Each integral, according to Section 6.6, has approximately the value

$$\frac{8\pi^2 IkT}{h^2} = \frac{2T}{\theta_{\text{ro}}} \tag{6.10.4}$$

Thus when $T \gg \theta_{\text{ro}}$, Equation (6.10.3) becomes:

$$Z_{\text{er}} = \frac{8\pi^2 IkT}{h^2}(2 + 2e^{-\varepsilon_1/kT}) = \frac{2T}{\theta_{\text{ro}}}(2 + 2e^{-174/T}) \tag{6.10.5}$$

For high temperatures, Z_{er} is nothing else but the product of the quantities which we have defined earlier (see Equations (6.6.8) and (6.10.1)) as the rotational and the electronic state sum:

$$Z_{\text{er}} = Z_{\text{el}} Z_{\text{ro}} \quad (\text{for } T \gg \theta_{\text{ro}}) \tag{6.10.6}$$

This now justifies (in connection with the value of θ_{ro} in Table 12) the earlier calculation of the standard entropy of NO (Table 17). The value **50.33** for the standard entropy in fact deviates less than 0.01 entropy units from the exact value. To calculate the entropy at much lower temperatures, however, Equation (6.10.3) will have to be employed.

6.11. CALCULATION OF THE EXACT VALUES OF THE ENTROPY

In the preceding sections, it was found possible to calculate approximate values for the entropy of diatomic gases by considering the rotational, vibrational and electronic states as being independent of each other and by idealizing the rotation and the vibration.

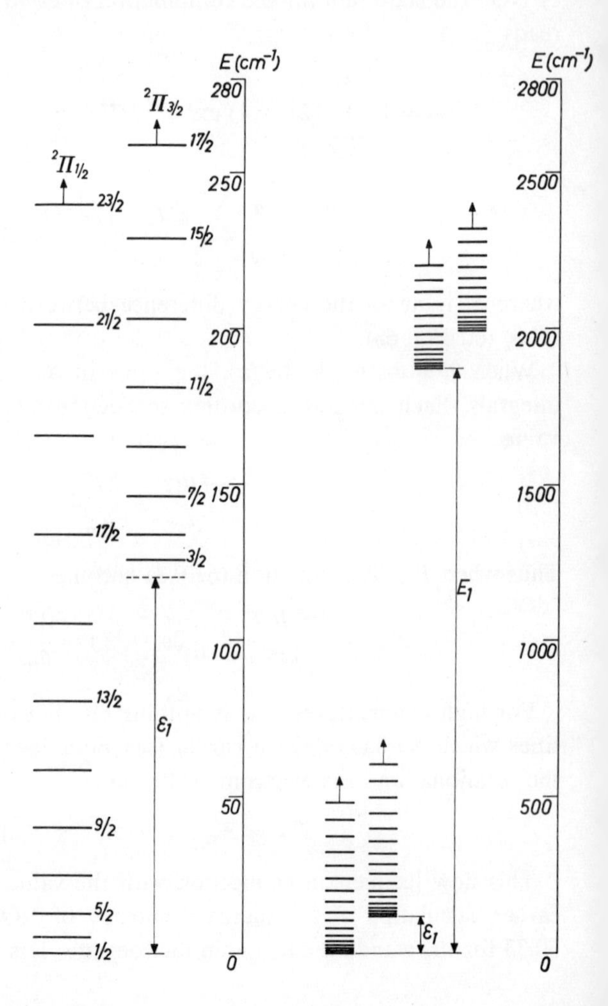

Fig. 68. Energy levels of an NO molecule. The left-hand side of the figure shows the rotational levels for the ground state of the vibration ($v = 0$). The right-hand side shows the same levels on a ten times smaller scale and, furthermore, the vibrational-rotational levels for $v = 1$.

In reality, however, a diatomic molecule is not rigid, so that when it rotates, an expansion will occur under the influence of the centrifugal force. The moment of inertia therefore increases when the molecule rotates with more quanta. Furthermore, as a result of the asymmetrical shape of the potential curve (Fig. 64), the effective moment of inertia is dependent on the vibrational state. It increases when the molecule vibrates with more quanta. The rotation and vibration are therefore dependent on one another; when exact calculations are being made, the rotational-vibrational energy may not be written as the sum of a rotational energy and a vibrational energy which are independent of one another. It is also immediately obvious from the non-parabolic shape of the potential curve that the oscillations are not harmonic. Finally, the rotational-vibrational states are, furthermore, not independent of the electronic states.

Exact values of the entropy can be calculated by means of Equation (5.11.3) which (summing over the energy levels instead of over the separate states) can also be written in the form:

$$s_{\text{evr}} = R \left\{ \ln \sum g e^{-\varepsilon/kT} + \frac{1}{kT} \frac{\sum \varepsilon g e^{-\varepsilon/kT}}{\sum g e^{-\varepsilon/kT}} \right\}. \qquad (6.11.1)$$

The sums occurring in this equation are determined, in the exact calculations, by substituting in the terms of each sum the *experimentally* determined values (deduced from the spectrum) of the energy levels $\varepsilon_1, \varepsilon_2, \ldots$ and by including in the summation all those terms which make an appreciable contribution to the sum. In this method of calculation, no mechanical model is employed, neither that of the rigid rotator, nor that of the harmonic oscillator.

The first attempt to calculate the entropy of a diatomic gas, viz. HCl, in this way, was made in 1926 by Hicks and Mitchell [1]). Their attempt failed because at that time there was still great uncertainty about the statistical weights g. Finally, in 1928, Giauque and Wiebe [2]), making use of the correct statistical weights, succeeded in calculating the exact value of the entropy of HCl. This was the first *exact* calculation of the entropy of a non-monatomic gas.

Hicks and Mitchell made their calculations in such a way that each energy level was deduced separately from the spectrum. They employed no empirical or theoretical formulae to calculate these levels. Their method of calculating the entropy is, in principle, the most accurate. For irregularities (or disturbances) sometimes occur in the otherwise regular arrange-

[1]) H. C. Hicks and A. C. G. Mitchell, J. Amer. Chem. Soc. **48**, 1520 (1926).

[2]) W. F. Giauque and R. Wiebe, J. Amer. Chem. Soc. **50**, 101 (1928).

ment of the rotation lines, a fact which is obscured by the use of a general formula for the energy levels. Experiments have shown, however, that in all the cases which are of interest to us, the irregularities are so small that empirical formulae can be used without loss of accuracy. Giauque and his collaborators, who have been responsible for nearly all the exact calculations of the entropies of diatomic gases, have therefore always made use of these formulae. They will be discussed below.

The energy levels of the rigid rotator are given by Equation (6.2.4). This becomes:

$$\varepsilon_{\mathrm{ro}} = \frac{h}{8\pi^2 cI} J(J+1) \tag{6.11.2}$$

if, as in Section 5.17, we employ the cm^{-1} as energy unit. This unit corresponds to the energy of a radiation quantum with wave number 1 cm^{-1}, so that

$$1\ \mathrm{cm}^{-1} = hc = 1.9862 \times 10^{-16}\ \mathrm{erg/molecule.}$$

When the molecule rotates with more quanta, I becomes larger and therefore

$$B = \frac{h}{8\pi^2 cI} = \frac{27.989 \times 10^{-40}}{I}\ \mathrm{cm}^{-1} \tag{6.11.3}$$

becomes smaller. A very good approximation is obtained by replacing B by

$$B\{1 - uJ(J+1)\} \tag{6.11.4}$$

where u is very small with respect to 1 and in which B relates to the rotationless state. Combining the three last equations, we obtain:

$$\varepsilon_{\mathrm{ro}} = BJ(J+1) - DJ^2(J+1)^2 \tag{6.11.5}$$

Here D is a positive constant which is much smaller than B (and not to be confused with the dissociation energy). It is a measure of the effect of the centrifugal force.

In very exceptional cases, experiments make a further extension of (6.11.5) necessary:

$$\varepsilon_{\mathrm{ro}} = BJ(J+1) - DJ^2(J+1)^2 + HJ^3(J+1)^3 + \ldots \tag{6.11.6}$$

Equations (6.11.5) and (6.11.6) take account of the fact that a diatomic molecule is not a *rigid* rotator. In an analogous manner, account can be taken of the fact that it is not a pure *harmonic* oscillator, viz. by adding to the equation

$$\varepsilon_{\mathrm{vi}} = \omega_e(v + \tfrac{1}{2}) \tag{6.11.7}$$

terms in higher powers of $(v + \frac{1}{2})$:

$$\varepsilon_{\text{vi}} = \omega_e (v + \tfrac{1}{2}) - x_e\omega_e (v + \tfrac{1}{2})^2 + y_e\omega_e (v + \tfrac{1}{2})^3 + \dots \quad (6.11.8)$$

Here, ω_e is the wave-number (in cm^{-1}) which the anharmonic oscillator in the classical model would have for infinitesimal amplitudes. The constant $x_e\omega_e$ is much smaller than ω_e and positive. It is called the "anharmonicity constant" and is only written in the form $x_e\omega_e$ for historical reasons. The constant $y_e\omega_e$ is much smaller even than $x_e\omega_e$ and in most cases it can be neglected.

We saw in Section 6.3 that Equation (6.3.1) by Morse gives a fairly satisfactory approximation to the potential curve of a real molecule. It was expressly put in this form by Morse because, when substituted in the Schrödinger equation, it gives the eigenvalues in the form

$$\varepsilon_{\text{vi}} = \omega_e (v + \tfrac{1}{2}) - x_e\omega_e (v + \tfrac{1}{2})^2 \quad (6.11.9)$$

with two constants. If Morse's equation is used, it is thus pointless to use a better approximation for the vibrational levels than (6.11.9). It can be seen from this equation that, for increasing values of v, the energy levels of the anharmonic oscillator lie closer and closer together. The dissociation limit is reached when the interval between successive levels becomes zero.

No account has been taken above of the fact that the rotation and the vibration are not independent of each other. The effective moment of inertia is dependent on the vibrational state of the molecule. The possibility of introducing an effective moment of inertia for each vibrational state is based on the fact that the vibrational frequency is much greater than that of the rotation. Due to the anharmonicity of the vibrations, the mean distance between the nuclei, and with it the moment of inertia, increases as the molecule vibrates with a higher quantum number. In Equations (6.11.3) to (6.11.6), B therefore decreases with increasing value of the vibrational quantum number v. In order to express this influence of the vibration on the rotational levels, (6.11.6) must be written:

$$\varepsilon_{\text{ro}} = B_{\text{vi}}J(J + 1) - D_{\text{vi}}J^2(J + 1)^2 + H_{\text{vi}}J^3(J + 1)^3 + \dots \quad (6.11.10)$$

where B_{vi} and D_{vi} can be represented approximately by formulae of the form

$$B_{\text{vi}} = B_e - \alpha_e (v + \tfrac{1}{2}) + \gamma_e (v + \tfrac{1}{2})^2 + \dots \quad (6.11.11)$$

$$D_{\text{vi}} = D_e + \beta_e (v + \tfrac{1}{2}) + \dots \quad (6.11.12)$$

The energy levels of the rotating and vibrating molecule, according to (6.11.8) and (6.11.10), are given by

$$\varepsilon_{\text{rv}} = \omega_e(v+\tfrac{1}{2}) - x_e\omega_e(v+\tfrac{1}{2})^2 + \ldots + B_{\text{vi}}J(J+1) - D_{\text{vi}}J^2(J+1)^2 + \ldots \quad (6.11.13)$$

If the molecule is in a higher electronic state, the equation will also include a term which gives the electronic energy with respect to the lowest electronic state. The constants ω_e, $x_e\omega_e$, B_{vi} and D_{vi}, which occur in (6.11.13), have different values for different electronic states.

Since the molecules still perform their zero-point oscillation even at absolute zero temperature, the zero-level for the energy is usually placed at the rotationless zero-point oscillation. This means that when the energy levels are calculated by means of (6.11.13), $(v + \frac{1}{2})$ must be replaced by v and $(v + \frac{1}{2})^2$ by $(v^2 + v)$:

$$\varepsilon'_{\text{rv}} = \omega_e v - x_e\omega_e(v^2 + v) + \ldots + B_{\text{vi}}J(J+1) - D_{\text{vi}}J^2(J+1)^2 + \ldots \quad (6.11.14)$$

In (6.11.14) and the associated Equations (6.11.11) and (6.11.12), the higher terms indicated by dots can nearly always be neglected. Very often, D_{vi}

TABLE 18

VIBRATION AND ROTATION CONSTANTS, TOGETHER WITH DISSOCIATION ENERGY AND NUCLEAR SEPARATION OF SOME DIATOMIC MOLECULES

Molecule	Fundamental state	Vibration		Rotation		D_0 (eV)	r_e (10^{-8} cm)
		ω_e (cm^{-1})	$x_e\omega_e$ (cm^{-1})	B_e (cm^{-1})	α_e (cm^{-1})		
$I^{127}I^{127}$	$^1\Sigma$	214.57	0.6127	0.03736	0.000117	1.5417	2.666_6
$Br^{79}Br^{81}$	$^1\Sigma$	323.2	1.07	0.08091	0.000275	1.971	2.283_6
$Cl^{35}Cl^{35}$	$^1\Sigma$	564.9	4.0	0.2438	0.0017	2.475	1.988
$O^{16}O^{16}$	$^3\Sigma$	1580.36	12.073	1.44567	0.01579	5.080	1.20740
$N^{14}N^{14}$	$^1\Sigma$	2359.61	14.456	2.010	0.0187	7.373	1.094
$H^2H^2(D_2)$	$^1\Sigma$	3118.4_6	64.09_7	30.429	1.0492	4.553_6	0.7416_4
H^1H^1	$^1\Sigma$	4395.2_4	117.99_5	60.809	2.993	4.476_3	0.7416_6
H^1H^2(HD)	$^1\Sigma$	3817.09	94.958	45.655	1.9928	4.511_2	0.7413_6
$N^{14}O^{16}$	$^2\Pi$	1904.03	13.97	1.7046	0.0178	5.29_6	1.1508
$C^{12}O^{16}$	$^1\Sigma$	2170.21	13.461	1.9313_9	0.01748_5	—	1.1281_9
H^1I^{127}	$^1\Sigma$	2309.5_3	39.73	6.551	0.183	3.056_4	1.604_1
H^1Br	$^1\Sigma$	2649.67	45.21	8.473	0.226	3.75_4	1.413_8
H^1Cl^{35}	$^1\Sigma$	2989.74	52.05	10.5909	0.3019	4.430	1.27460
H^1F^{19}	$^1\Sigma$	4138.52	90.069	20.939	0.770_5	6.40	0.9171

and γ_e also can be neglected. In that case, the rotational-vibrational levels for the lowest electronic state of the most important diatomic molecules can be calculated with the aid of the constants ω_e, $x_e\omega_e$, B_e and α_e given in Table 18. They are taken from a book by Herzberg [1]), in which the reader can also find all particulars on the experimental determination and the theoretical background of these quantities.

In this table, as we have seen, ω_e gives the wave-number of the (non-existent) oscillation with infinitely small amplitude, while $x_e\omega_e$ is a measure of the anharmonicity of the vibration (the asymmetry of the potential curve). B_e is a measure of the reciprocal value of the moment of inertia in the (non-existent) vibration-free and rotationless state; α_e is a measure of the effect of the vibration on the moment of inertia. It will be seen from the first column that these constants relate to the most common isotopic molecule. Only in the case of hydrogen are the constants also given for the more rare isotopes D_2 and HD, while the constants for H^1Br do not relate to the compound of hydrogen with a particular isotope of bromine, but with the normal mixture of isotopes. The last two columns list the dissociation energy D_0 in electron volts and the nuclear separation r_e for the unattainable vibration-free state in Ångström units (10^{-8} cm). The values of r_e and I_e can be calculated directly from B_e via (6.11.3):

$$I_e = m_r r_e^2 = \frac{27.989 \times 10^{-40}}{B_e} \tag{6.11.15}$$

In a similar manner, employing (6.11.11), it is possible to calculate the moment of inertia I_0 and the (effective) nuclear separation r_0 for the state in which the molecule executes its zero-point vibration (cf. Section 6.2).

The exact calculation of the entropy of a gas, according to Equation (6.11.1), chiefly amounts to carrying out two summations. The first sum contains the terms $ge^{-\varepsilon/kT}$, the second the terms $ge^{-\varepsilon/kT} \times \varepsilon/kT$ (for the sake of convenience the factor $1/kT$ is included in the summation). In the first exact calculations of the entropy of diatomic gases, these terms were calculated one at a time, making use of an equation of the form (6.11.14) to calculate the energies ε. These exact calculations will be demonstrated in the following sections, taking HCl as an example of a heteronuclear gas and H_2 of a homonuclear gas.

The calculation and summation of the separate terms at high temperatures is very time-consuming and therefore use was soon being made of mathematical methods in order to achieve the desired result quicker and without

[1]) G. Herzberg, Molecular Spectra and Molecular Structure, I. Spectra of Diatomic Molecules, Second Edition, Van Nostrand Co., New York, 1950.

sacrificing accuracy. For these methods the reader is referred to the literature [1]).

From the following it will appear that the exact values of the standard entropy (the entropy at 298.15 °K and 1 atm) only differ very little from the approximate values already calculated. This is mainly due to the fact that at the above-mentioned temperature, the vibrational entropy plays virtually no part. At higher temperatures, the differences between exact and approximate values continue to increase.

The total "statistical entropy" is found by adding together the "spectroscopic entropy" calculated from (6.11.1) and the translational entropy calculated from (5.5.13). It should be noted that this sum is often indicated by the name "spectroscopic entropy". In the following sections it will be compared with the "calorimetric entropy", which is calculated with the help of the third law of thermodynamics (cf. Section 2.17).

6.12. THE SPECTROSCOPIC ENTROPY OF HCl^{35}

We shall demonstrate the exact calculation of entropy with the example of HCl^{35}. Table 19 gives the energy levels for the rotation of the HCl^{35} molecule for the vibrational states $v = 0$ and $v = 1$ of the electronic ground state. They have been calculated from Equations (6.11.14), (6.11.11) and (6.11.12). The constants ω_e, $x_e\omega_e$, B_e and α_e which occur in them are to be found in Table 18. The value of D_e is 0.00053 cm^{-1}. The constants γ_e and β_e can be neglected.

According to the preceding section, the rotational-vibrational entropy at any particular temperature is found by calculating the terms $ge^{-\varepsilon/kT}$ and $ge^{-\varepsilon/kT} \times \varepsilon/kT$ for each energy level. Here

$$g = g_{\mathrm{ro}} \cdot g_{\mathrm{vi}} = 2J + 1\,, \tag{6.12.1}$$

since the weight g_{vi} of the vibrational states is 1 and the weight g_{ro} of the rotational states is $2J + 1$. Consequently, if k is expressed in erg/deg., one obtains:

$$\sum ge^{-\varepsilon/kT} = 1 + 3e^{-20.8777\,hc/kT} + 5e^{-62.6203\,hc/kT} + \ldots + \\ + e^{-2885.64\,hc/kT} + 3e^{-2905.92\,hc/kT} + \ldots \tag{6.12.2}$$

[1]) W. F. Giauque and R. Overstreet, J. Amer. Chem. Soc. **54**, 1731 (1932); H. L. Johnston and C. O. Davis, J. Amer. Chem. Soc. **56**, 271 (1934); L. S. Kassel, J. Chem. Phys. **1**, 576 (1933); Chem. Reviews **18**, 277 (1936); A. R. Gordon and C. Barnes, J. Chem. Phys. **1**, 297 (1933); J. E. Mayer and M. Goeppert-Mayer, Statistical Mechanics, New York, 1940.

In a similar way one can calculate $\Sigma g e^{-\varepsilon/kT} \times \varepsilon/kT$ for each desired temperature. It is obvious that in doing this, the number of terms employed is such that when still more terms are added the sum remains virtually unchanged. For a temperature of 298.15 °K, we have calculated the required number of terms (Table 19). If the sums found in Table 19 are inserted in Equation (6.11.1), we then find for the standard value of the rotational-vibrational entropy of HCl^{35} the value

$$s^0_{\mathrm{rv}} = R\left(\ln 20.226 + \frac{19.926}{20.226}\right) = 7.933 \text{ cal/degree.mole} \qquad (6.12.3)$$

TABLE 19

ENERGY LEVELS ε OF HCl^{35} AND THE CORRESPONDING VALUES OF $g e^{-\varepsilon hc/kT}$ AND $g e^{-\varepsilon hc/kT} \times \varepsilon hc/kT$ FOR $T = 298.15$ °K

vJ	ε (cm^{-1})	$g e^{-\varepsilon hc/kT}$ ($T = 298.15$ °K)	$g e^{-\varepsilon hc/kT} \cdot \varepsilon hc/kT$ ($T = 298.15$ °K)
$v = 0$			
$J = 0$	0.0000	1.0000	0.0000
1	20.8777	2.7125	0.2733
2	62.6203	3.6960	1.1169
3	125.2025	3.8257	2.3114
4	208.5860	3.2893	3.3108
5	312.7200	2.4322	3.6705
6	437.5409	1.5739	3.3231
7	582.9723	0.9001	2.5322
8	748.9253	0.4580	1.6552
9	935.298	0.2083	0.9401
10	1141.976	0.0849	0.4679
11	1368.832	0.0311	0.2056
12	1615.726	0.0103	0.0801
13	1882.506	0.0031	0.0278
14	2169.006	0.0008	0.0086
15	2475.048	0.0002	0.0024
16	2800.441	0.0000	0.0006
...	....	....	....
...	....	....	....
$v = 1$			
$J = 0$	2885.64	0.0000	0.0000
1	2905.92	0.0000	0.0000
...	....	....	....
		20.226	19.926

6.13. THE STANDARD ENTROPY OF NORMAL HCl

Besides the entropy of HCl^{35}, that of HCl^{37} must also be known in order to be able to calculate the entropy of normal HCl. In other words, the calculation of the entropy of HCl requires a knowledge not only of the molecular constants of HCl^{35}, but also of those of HCl^{37}. They can be calculated from one another because the binding forces, and therefore the nuclear separation and the electron configuration, are the same in isotopic molecules.

We have already seen in Section 6.3, that the vibrational frequency of harmonically vibrating isotopic molecules is inversely proportional to the root of the reduced mass:

$$\left.\begin{aligned} \frac{\omega^i}{\omega} &= \left(\frac{m_r}{m_r^i}\right)^{\frac{1}{2}} = \rho \\ \text{or} \qquad \omega^i &= \rho\omega \,. \end{aligned}\right\} \tag{6.13.1}$$

By convention the superscript i relates to the heavier isotope, so that ρ is smaller than 1. In the case of the anharmonic oscillator, ω_e must be replaced by $\rho\omega_e$ and $x_e\omega_e$ by $\rho^2 x_e\omega_e$.

According to (6.11.15), B_e is inversely proportional to m_r, so that the value of B_e for the lighter molecule must be replaced by $\rho^2 B_e$ on transition to the heavier molecule. Calculation also shows that α_e must be replaced by $\rho^3\alpha_e$ and D by $\rho^4 D$.

Table 20 gives the molecular constants of HCl^{37} which have been calculated in the manner indicated, using the atomic weights H = 1.0080, Cl^{35} = 34.9788 and Cl^{37} = 36.9777.

TABLE 20

CALCULATED MOLECULAR CONSTANTS OF HCl^{37} IN CM^{-1}

$\omega_e(HCl^{37})$	=	$\rho\omega_e(HCl^{35})$	=	2987.47
$x_e\omega_e(HCl^{37})$	=	$\rho^2 x_e\omega_e(HCl^{35})$	=	51.97
$B_e(HCl^{37})$	=	$\rho^2 B_e(HCl^{35})$	=	10.5749
$\alpha_e(HCl^{37})$	=	$\rho^3\alpha_e(HCl^{35})$	=	0.3012
$D(HCl^{37})$	=	$\rho^4 D(HCl^{35})$	=	0.00053

With the aid of these constants, the values of $Z = \Sigma g e^{-\varepsilon/kT}$ and of $T\mathrm{d}Z/\mathrm{d}T = \Sigma g e^{-\varepsilon/kT} \times \varepsilon/kT$ (for $T = 298.15$ °K) can be calculated in the same way for HCl^{37} as was done in Table 19 for HCl^{35}. One finds $Z = 20.256$

and $TdZ/dT = 19.956$. The standard value, s^0_{rv} (HCl^{37}), of the rotational-vibrational entropy of HCl^{37} is then found by means of (6.11.1). The result will be found in Table 21, together with the rotational-vibrational entropy of HCl^{35}, found in the preceding section, and the standard values, s^0_{tr}, of the translational entropy, calculated via (5.5.13) for the two gases. Addition gives the value for each gas of the total standard entropy:

$$s^0 = s^0_{rv} + s^0_{tr}$$

The calculated figures, in Table 21 do not include the nuclear spin entropy. As we have already noted in Section 6.7, this is left out of consideration because it does not change in chemical reactions. For the same reason, the entropy of isotopic mixing is not included in the tabulated values of the entropy (the proportion of the isotopes does not change in chemical reactions). Normal HCl consists of 75.4% of HCl^{35} and 24.6% of HCl^{37}, so that the standard entropy of normal HCl can simply be written as the sum:

$$(0.754 \times 44.604) + (0.246 \times 44.768) = 44.644 \text{ cal/degree.mole}$$

TABLE 21

EXACT VALUE OF THE STANDARD ENTROPY OF HCl IN CAL/DEG.MOLE

s^0_{rv} (HCl^{35}) = 7.933	s^0_{rv} (HCl^{37}) = 7.936
s^0_{tr} (HCl^{35}) = 36.671	s^0_{tr} (HCl^{37}) = 36.832
s^0 (HCl^{35}) = 44.604	s^0 (HCl^{37}) = 44.768
s^0(HCl) = 44.644	

We shall now compare this statistical entropy with the calorimetric entropy, which is derived in the usual manner from measurements of specific heats and heats of transition. In the case of HCl we are not only dealing with the transitions from solid to liquid and from liquid to gas, but also with a transition in the solid state [1]). The entropies of transition are obtained by dividing each separate heat of transition by the corresponding absolute temperature. The required data for this are to be found in Table 22.

The increase in the entropy with rising temperature is obtained, in accordance with Section 2.17, by graphical integration of the equation

$$s = \int c_p \, d \ln T. \tag{6.13.2}$$

[1]) N. L. Alpert, Phys. Rev. **75**, 398 (1949), has studied the atomic mechanism of this transition by means of the modern method of nuclear magnetic resonance.

TABLE 22

CHANGES OF STATE OF HCl ACCORDING TO GIAUQUE AND WIEBE (LOC. CIT.)

Change of state	Temperature °K	Heat of transition
Transition in solid HCl	98.36	284.3 cal/mole
Transition solid-liquid	158.91	467.0 ,,
Transition liquid-gas (1 atm)	188.07	3860 ,,

Measurements of specific heat are only available down to a temperature of 16 °K. To find the entropy corresponding to the range 0-16 °K, extrapolation must be employed. This is done with the aid of Debye's theory (Section 2.14), according to which the specific heat at constant volume for a solid at very low temperatures is proportional to T^3. No appreciable error is made if we assume that this is also true for c_p, since $(c_p - c_v)$ is negligibly small for solids at low temperatures. Thus we have

$$c_p = a\,T^3 \tag{6.13.3}$$

If the lowest temperature T_e at which specific heat measurements have been made lies in the range for which this equation is valid (which is the case for HCl), then the missing quantity of entropy is given by

$$s(T_e) - s(0) = \int_0^{T_e} \frac{c_p}{T}\,\mathrm{d}T = \int_0^{T_e} a\,T^2\,\mathrm{d}T = \tfrac{1}{3}\,aT_e^3 = \tfrac{1}{3}\,c_p(T_e) \tag{6.13.4}$$

At 16 °K, c_p (HCl) has a value of a mere 0.90 cal/deg.mole. The missing amount of entropy is therefore 0.30 cal/deg.mole.

Table 23 gives a summary of the calculation of the calorimetric entropy of gaseous HCl at 1 atm at the boiling point.

TABLE 23

THE CALORIMETRIC ENTROPY OF GASEOUS HCl AT 188.1 °K

0-16 °K, extrapolated	0.30 cal/degree.mole
16-98.36 °K, graphical	7.06
transition entropy 284.3/98.36	2.89
98.36-158.91 °K, graphical	5.05
entropy of melting 476.0/158.91	3.00
158.91-188.07 °K, graphical	2.36
entropy of evaporation 3860/188.07	20.52
entropy of gaseous HCl at the boiling point	41.2 ± 0.1

The entropy calculated in this way relates to the real gas, while the statistical entropy relates to the perfect gas state. In order to compare the two entropy values, obtained by such different routes, with each other, we must apply a correction to the calorimetric entropy for the transition from the real to the perfect gas state. To do this, we first imagine that the gas pressure is reduced from 1 to 0 atm, at which pressure perfect and real gases are identical. This is accompanied by an entropy change

$$\Delta s(1) = \int_1^0 \left(\frac{\partial s}{\partial p}\right)_T^{\mathrm{re}} \mathrm{d}p$$

in which the superscript re indicates the real gas. After that, we compress the gas in the perfect state from a pressure 0 to a pressure 1, for which the entropy change amounts to

$$\Delta s(2) = \int_0^1 \left(\frac{\partial s}{\partial p}\right)_T^{\mathrm{per}} \mathrm{d}p$$

The total entropy change is thus given by

$$\Delta s = \int_0^1 \left\{ \left(\frac{\partial s}{\partial p}\right)_T^{\mathrm{per}} - \left(\frac{\partial s}{\partial p}\right)_T^{\mathrm{re}} \right\} \mathrm{d}p \tag{6.13.5}$$

Since

$$\left(\frac{\partial s}{\partial p}\right)_T = -\left(\frac{\partial v}{\partial T}\right)_p$$

(6.13.5) can also be written:

$$\Delta s = \int_0^1 \left\{ \left(\frac{\partial v}{\partial T}\right)_p^{\mathrm{re}} - \left(\frac{\partial v}{\partial T}\right)_p^{\mathrm{per}} \right\} \mathrm{d}p \tag{6.13.6}$$

Here

$$\left(\frac{\partial v}{\partial T}\right)_p^{\mathrm{per}} = \frac{R}{p} \tag{6.13.7}$$

while $(\partial v/\partial T)_p^{\mathrm{re}}$ must be found experimentally. In many cases the experimental results can be satisfactorily represented by the Van der Waals' equation of state or by that of Berthelot. Berthelot's equation, for moderate pressures, can be written in the form

$$pv = RT \left\{ 1 + \frac{9}{128} \frac{p T_k}{T p_k} \left(1 - \frac{6 T_k^2}{T^2}\right) \right\} \tag{6.13.8}$$

where p_k and T_k are the critical pressure and critical temperature. From (6.13.8) it follows that

$$\left(\frac{\partial v}{\partial T}\right)_p = \frac{R}{p} + \frac{27}{32} \frac{R T_k^3}{p_k T^3} \tag{6.13.9}$$

If we substitute (6.13.7) and (6.13.9) in (6.13.6), we obtain:

$$\Delta s = \frac{27}{32} \frac{R T_k^3}{p_k T^3} \int_0^1 \mathrm{d}p = \frac{27}{32} \frac{R T_k^3}{p_k T^3} \tag{6.13.10}$$

For HCl, which obeys Berthelot's equation of state better than that of Van der Waals, $T_k = 349.5$ °K and $p_k = 81.6$ atm. If we substitute these values in (6.13.10), we find with $R = 1.987$ and $T = 188.1$, a value for Δs of 0.13 cal/deg.mole. The entropy of perfect HCl gas at the boiling point is thus $41.2 + 0.13 = 41.3$ cal/deg.mole.

The standard value of the calorimetric entropy of HCl, i.e. the value of perfect gaseous HCl at 298.15 °K and 1 atm is calculated as follows. According to Table 12, the rotational characteristic temperature of HCl lies so low and the vibrational characteristic temperature so high, that the equipartition value of the rotational specific heat is already reached at 188.1 °K, while over the whole range between 188.1 and 298.1 °K, the vibration still plays no perceptible part. For the molar c_p of perfect HCl gas, we can thus reckon, in this temperature range, on a constant value of $7R/2$ (cf. Section 6.2), so that the entropy increase from 188.1 °K to 298.1 °K is given by

$$s_{298.1} - s_{188.1} = c_p \int_{188.1}^{298.1} \frac{\mathrm{d}T}{T} = 6.95 \ln \frac{298.1}{188.1} = 3.2.$$

Thus the calorimetric way of calculating the standard entropy of HCl gives us the value $41.3 + 3.2 = 44.5$ cal/deg.mole, in fine agreement with the value of 44.64 calculated statistically.

In the same way as here described for HCl, the standard entropies of the related gases HBr [1]) and HI [2]) were determined.
For these gases too almost exact agreement was found between the value of the statistical entropy and that of the calorimetric entropy.

---

[1]) W. F. Giauque and R. Wiebe, J. Amer. Chem. Soc. **50**, 2193 (1928).
[2]) W. F. Giauque and R. Wiebe, J. Amer. Chem. Soc. **51**, 1441 (1929).

6.14. COMPARISON OF THE STATISTICAL AND THE CALORIMETRIC ENTROPY

One might perhaps feel surprised that such good agreement was found in the preceding section between the statistical and the calorimetric entropy of HCl, considering that the zero-level of the statistical entropy was chosen in such a comparatively arbitrary manner. We started, in fact, from a zero-level at which the translational-rotational-vibrational-electron weight was equal to unity, but at which the nuclear spin weight was equal to the product $\rho_1\rho_2$ of the spin weights of the separate atoms. The nuclear spin entropy was thus excluded from the statistical entropy. Also the isotopic mixing entropy was excluded, i.e. at the chosen zero-level of the entropy, the isotopic molecules are already homogeneously mixed.

The agreement found between the statistical and the calorimetric entropy thus means that the Debye extrapolation of the specific heat implies an HCl crystal at 0 °K, in which the translational-rotational-vibrational-electronic entropy is equal to zero, but in which the nuclear spin entropy and the isotopic mixing entropy are still present in the crystal. This zero-point entropy is not real, but is due, as was discussed for the more general case in Section 2.17, to the relatively high value of the lowest measuring temperature.

In all heteronuclear molecules, the separations $\Delta\varepsilon$ of the $\rho_1\rho_2$ nuclear spin energy levels are so small that even at the lowest measuring temperature, kT is still large compared with $\Delta\varepsilon$, with the result that the molecules are equally distributed over the various levels even at that lowest temperature. These states will thus not make a contribution to the specific heat. Only at temperatures at which kT is much smaller than $\Delta\varepsilon$, will all the molecules be found in the lowest nuclear spin state, which corresponds to zero nuclear spin entropy.

If the measurements extended to within the immediate neighbourhood of the absolute zero-point, then the emptying of the higher nuclear spin levels would be bound to manifest itself as a maximum or as various maxima in the curve of the specific heat. In that case, $\int c_p \mathrm{d} \ln T$ would also include the nuclear spin entropy. Analogous reasoning holds in relation to the separation of the isotopic molecules and the accompanying entropy of mixing. Extrapolation according to the T^3 relationship from, say, 15 °K, thus leads to neglecting the nuclear spin entropy and the isotopic mixing entropy in the calorimetric entropy, i.e. to a hypothetical state at 0 °K in which these quantities have not been given up but are still present in the crystal as zero-point entropy.

Extrapolation from a temperature between 10 and 20 °K is an advantage, up to a point, in the case of HCl because in this way the exclusion of the nuclear spin entropy and the isotopic mixing entropy is automatically achieved. In what follows, however, we shall meet with a case (normal hydrogen) in which extrapolation from the above-mentioned temperatures obscures an important contribution to the entropy which must not be omitted from the calorimetric entropy. In other words, unless the measurements are extended to much lower temperatures than those mentioned, the value found for the calorimetric entropy of hydrogen will be considerably smaller than that for the spectroscopic entropy (see following section).

Extrapolation from too high a temperature is not the only factor which may lead to finding too low a value for the calorimetric entropy. The same error may be introduced when equilibrium fails to be established at low temperatures. Experiments point to the occurrence of this kind of "frozen" equilibrium in the case of CO, for example [1]). For the discussion of this point, the reader is referred to Section 2.17.

6.15. THE EXACT ENTROPY OF H_2

To demonstrate the calculation of the exact value of the entropy of a homonuclear gas, we choose hydrogen as our example. Since hydrogen nuclei possess their own angular momentum (spin), the molecules can occur, according to Section 6.7, in two different nuclear spin states, the ortho and the para state. We may consider the properties of four different kinds of hydrogen: pure para-hydrogen, pure ortho-hydrogen, the equilibrium mixture and normal hydrogen with the proportion ortho: para = 3 : 1.

In the equilibrium mixture, the even and odd rotation levels are accessible to all H_2 molecules. Under these conditions, the number of molecules present at each rotational level is given, according to (5.2.4), by

$$n = \frac{gNe^{-\varepsilon_J/kT}}{Z} \tag{6.15.1}$$

In this equation g for the even levels, i.e. the para-states, is equal to $2J + 1$, while for the odd levels, i.e. the ortho-states, it is equal to $3(2J + 1)$

[1]) J. O. Clayton and W. F. Giauque, J. Amer. Chem. Soc. **54**, 2610 (1932) and **55**, 5071 (1933).

because of the three possible orientations of the resultant nuclear spin (↑↑). The ratio of the number of para to the number of ortho-molecules in equilibrium hydrogen is thus given by

$$\beta = \frac{e^{-\varepsilon_0/kT} + 5e^{-\varepsilon_2/kT} + 9e^{-\varepsilon_4/kT} + \ldots}{3(3e^{-\varepsilon_1/kT} + 7e^{-\varepsilon_3/kT} + 11e^{-\varepsilon_5/kT} + \ldots)} =$$

$$= \frac{\sum_{J=0,2,\ldots}^{\infty} (2J+1)\, e^{-\varepsilon_J/kT}}{3 \sum_{J=1,3,\ldots}^{\infty} (2J+1)\, e^{-\varepsilon_J/kT}}, \qquad (6.15.2)$$

where both numerator and denominator converge so rapidly that in calculating β at room temperature it is sufficient to take four terms in the numerator and three in the denominator (see Table 25).

From Equations (6.8.6) and (6.15.2) it follows that β reaches a constant value of 1/3 at temperatures which are far above the rotational characteristic temperature as, in fact, we had already seen in Section 6.7. At the standard temperature 298.15 °K, this ratio 1 : 3 has already virtually been reached. (This may be checked from Table 25). At this temperature and above, there is therefore practically no difference between equilibrium hydrogen and normal hydrogen. At very low temperatures, according to (6.15.2), β approaches infinity. At these temperatures, thus, equilibrium hydrogen becomes identical with para-hydrogen, which also had already been noted in Section 6.7.

The rotational-nuclear-spin state sum for equilibrium hydrogen at temperatures where only the lowest vibrational state ($v = 0$) plays a role, is given by

$$Z_{rs}(0) = \sum_{J=0,2,\ldots}^{\infty} (2J+1)\, e^{-\varepsilon_J/kT} + 3 \sum_{J=1,3,\ldots}^{\infty} (2J+1)\, e^{-\varepsilon_J/kT} \qquad (6.15.3)$$

in which it is now possible to use a considerably better approximation equation for ε_J than in Section 6.8. This is of especial importance because the expansion effect (the influence of the centrifugal force) is abnormally large in hydrogen. For very accurate calculations, Equation (6.11.10) is here even used with three constants to calculate the energy levels for the rotation:

$$\varepsilon_J = B_{vi} J(J+1) - D_{vi} J^2 (J+1)^2 + H_{vi} J^3 (J+1)^3 \qquad (6.15.4)$$

with

$$B_{vi} = B_e - \alpha_e (v + \tfrac{1}{2}) + \gamma_e (v + \tfrac{1}{2})^2 \qquad (6.15.5)$$

$$D_{vi} = D_e + \beta_e (v + \tfrac{1}{2}) \qquad (6.15.6)$$

$$H_{vi} = H_e \qquad (6.15.7)$$

The values of the constants occurring in these equations are to be found in Table 24.

TABLE 24

ROTATION CONSTANTS OF H_2 IN CM^{-1} ACCORDING TO HERZBERG (LOC. CIT.)

B_e	60.809	D_e	0.04648
α_e	2.993	β_e	—0.00134
γ_e	0.025	H_e	0.0000518

With the help of these values and the equations, it is a simple matter to calculate the data given in Table 25.

Using Equation (6.11.1) we can deduce the standard value of the rotational-nuclear-spin entropy of H_2 from Table 25:

$$s^0_{rs} = R\left(\ln 7.7240 + \frac{7.0724}{7.7240}\right) = 5.882$$

For the standard value of the translational entropy of H_2, we find from Equation (5.5.13) the value $s^0_{tr} = 28.080$ units. The total standard entropy of equilibrium hydrogen is thus $5.882 + 28.080 = 33.962$ cal/deg.mole. It followed from Equation (6.15.2) that equilibrium hydrogen at 298.15 °K was already virtually identical with normal hydrogen, so that the value calculated also applies to normal hydrogen. According to the derivation, this entropy includes the nuclear spin entropy. Consequently, it can not be used in combination with the calorimetric entropies of other substances. To be able to do this, it must first be reduced by the nuclear spin entropy $R \ln 4 = 2.755$ units. The standard value which must be included in entropy tables is thus $s^0_{H_2} = 31.207$ cal/deg.mole [1]).

Once again, it is very important to compare this statistical entropy with the entropy determined by calorimetric means. Originally, the measurements of specific heat only extended to a few degrees below the solidifying point of H_2 (14 °K). They led to a calorimetric entropy which was about $\frac{3}{4}R \ln 3$ cal/deg.mole smaller than the statistical value given above. It will be seen from the following that this was due to extrapolating to 0 °K from too high a temperature.

In normal hydrogen (ortho : para = 3 : 1) at very low temperatures, all the para-molecules are to be found in the rotational state $J = 0$, while the

---

[1]) The correct way of calculating this value was first indicated by Giauque and his collaborators. See: W. F. Giauque and H. L. Johnston, J. Amer. Chem. Soc. **50**, 3221 (1928); W. F. Giauque, J. Amer. Chem. Soc. **52**, 4816 (1930); C. O. Davis and H. L. Johnston, J. Amer. Chem. Soc. **56**, 1045 (1934).

TABLE 25

ENERGY LEVELS ε OF THE ROTATION OF H_2 AND THE CORRESPONDING VALUES OF $ge^{-\varepsilon hc/kT}$ AND $ge^{-\varepsilon hc/kT} \times \varepsilon hc/kT$ AT $T = 298.15$ °K

J	g	ε (cm^{-1})	$ge^{-\varepsilon hc/kT}$ ($T = 298.15$ °K)	$ge^{-\varepsilon hc/kT} . \varepsilon hc/kT$ ($T = 298.15$ °K)
0	1	0.00	1.0000	0.0000
1	9	118.45	5.0814	2.9047
2	5	354.27	0.9047	1.5466
3	21	705.32	0.6983	2.3767
4	9	1168.47	0.0320	0.1805
5	33	1739.73	0.0075	0.0626
6	13	2414.42	0.0001	0.0013
			7.7240	7.0724

ortho-molecules continue to rotate with 1 quantum. Since the rotation quanta in hydrogen are large and the crystal forces small, the crystallization is not able to prevent rotation. Therefore the ortho-molecules rotate with 1 quantum even in the crystal. This rotation level corresponds to $2J + 1 = 3$ eigenfunctions, i.e. each ortho-molecule can be in three different rotational states in the crystal. In the pictorial model each of these states corresponds to a particular orientation of the axis of rotation.

Due to interaction of the molecules, however, the state in which all the axes of rotation are similarly oriented is energetically the most favourable. In other words, a splitting of the levels occurs in the crystal. The separation of these levels is so small that the three rotational eigenfunctions are still distributed at random among the ortho-molecules at temperatures just below the freezing point.

Thus, if one calculates $\int c_p \mathrm{d} \ln T$ by applying a Debye extrapolation in the range between 12 °K and 0° K, one is extrapolating from a temperature at which the $\frac{3}{4}$ mole of ortho-hydrogen present in 1 mole of normal hydrogen, still possesses a rotational entropy of $\frac{3}{4}R \ln 3$ cal/deg. The extrapolation thus leads to a fictitious state at 0 °K, in which not only the nuclear spin entropy, but also this rotational entropy, is still present in the crystal. This explains the difference in value between the statistical and the calorimetric entropy.

It could easily be predicted that the remaining rotational entropy would be released by the solid hydrogen, if the measurements were extended to a sufficiently low temperature. This release would have to appear as an extra contribution to the specific heat, superimposed upon the Debye curve.

The predicted effect was found in 1953 in the laboratory of F. Simon [1]). Fig. 69 gives the results of measurements of the specific heat of 73 % ortho and 27 % para-hydrogen. The peak in the specific heat at 1.6 °K corresponds

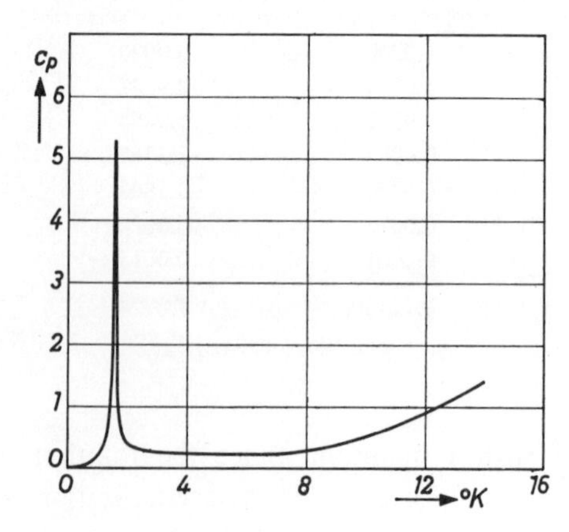

Fig. 69. Specific heat of 73 % ortho- and 27 % para-hydrogen in cal.deg^{-1}.mole^{-1} at temperatures below the melting point (14 °K). According to F. Simon, Yearbook of the Physical Society, London (1956).

to the expected entropy effect of $\frac{3}{4}R \ln 3$ cal/deg. mole of normal hydrogen.

Agreement between the statistical entropy (without nuclear spin entropy) and the calorimetric entropy is seen from the above to be obtained for H_2 only when the measurements of the specific heat extend downwards to circa 1 °K. In contrast to this, we have seen that for HCl a lower limit of measurement of 20 °K is quite sufficient (see Section 6.13).

The foregoing related to calorimetric measurements on normal hydrogen. If it were possible to carry out these measurements on equilibrium hydrogen, quite different results could be expected. The results obtained for normal hydrogen were not satisfactory when the lowest measuring temperature was 12 °K. But at that temperature equilibrium hydrogen is already almost completely in the para-state, i.e. in a state in which not only the whole rotational entropy, but also the whole nuclear spin entropy has already been released. In this case one would again find a calorimetric entropy which could not be used in combination with the calorimetric entropies of other substances, but this time the entropy would not be an amount $\frac{3}{4}R \ln 3$ too small, but an amount $R \ln 4$ too large.

---

[1]) F. Simon, Yearbook of the Physical Society, London (1956); R. W. Hill and B. W. A. Ricketson, Phil. Mag. **45**, 277 (1954).

6.16. THE ENTROPY OF OTHER HOMONUCLEAR GASES

In solid deuterium of normal composition (2 parts ortho-deuterium to 1 part para-deuterium) phenomena occur which are comparable to those occurring in solid H_2. It will be clear, in connection with what was discussed in Section 6.8, that this time it is the para-molecules and not the ortho-molecules which rotate with one quantum just under the freezing point in the crystal. If the specific heat is not measured to a sufficiently low temperature (e.g. to 10 °K), one would expect a difference of $\frac{1}{3}R \ln 3$ between the statistical entropy (without nuclear spin entropy) and the calorimetric entropy. This expectation is confirmed by experiment [1]).

For the heavier homonuclear molecules, such as N_2, Cl_2, Br_2, I_2, the rotation quanta are so much smaller and the crystal forces so much larger than in H_2 and D_2, that no rotation occurs in the solid state. Hence for these gases, complete agreement is found between the statistical and the calorimetric entropy. There is now no objection to using the Equation (6.8.7) to calculate the statistical entropy (even its exact value), since the rotational characteristic temperature lies far below the condensation temperature.

At first sight, there would appear to be some difficulty with a gas like Cl_2 [2]), which consists of a mixture of Cl_2^{35}, Cl_2^{37} and $Cl^{35}Cl^{37}$. Apart from the differences in entropy due to the differences in the molecular constants, the entropy of the heteronuclear gas $Cl^{35}Cl^{37}$ is greater than that of Cl_2^{35} or Cl_2^{37} by an amount equal to $R \ln 2$ per mole. Each $Cl^{35}Cl^{37}$ molecule has, in fact, twice as many rotational states as a homonuclear chlorine molecule. It is easy to see, however, that the conventions made previously imply that we must neglect this extra quantity of entropy. Let us consider, for example, the reaction between hydrogen and chlorine:

$$H_2 + Cl_2^{35} \rightarrow 2\,HCl^{35}$$
$$H_2 + Cl_2^{37} \rightarrow 2\,HCl^{37}$$
$$H_2 + Cl^{35}Cl^{37} \rightarrow HCl^{35} + HCl^{37}$$

As compared with the first two reactions, the left-hand side of the third reaction contains the above-mentioned extra entropy $R \ln 2$. On the right-hand side, however, an equal extra quantity of entropy occurs due to the mixing of HCl^{35} and HCl^{37}. It was agreed in Section 6.13 that the isotopic

---

1) H. L. Johnston and E. A. Long, J. Chem. Phys. **2**, 389 (1934); K. Clusius and E. Bartholomé, Z. physik. Chem. B. **30**, 258 (1935).

2) W. F. Giauque and R. Overstreet, J. Amer. Chem. Soc. **54**, 1731 (1932).

mixing entropy would be neglected and we are therefore obliged to neglect also the extra entropy of $Cl^{35}Cl^{37}$ since the entropy *change* which occurs during the course of a reaction is a well-defined one. Finally, thus, we ignore the existence of isotopes completely and treat a gas like Cl_2 as though it consisted exclusively of symmetrical molecules, each with an average mass and an average moment of inertia. Reasoning analogous to that used here for chlorine is also applicable to bromine which contains the isotopes Br^{79} and Br^{81} [1]).

Table 26 gives the exact values of the entropy of a number of homonuclear and heteronuclear diatomic gases in the perfect gas state at 298.15 °K and 1 atm.

TABLE 26

EXACT VALUES OF THE STATISTICAL ENTROPIES IN CAL/DEG.MOLE OF SOME DIATOMIC GASES IN THE PERFECT GAS STATE AT 298.15 °K AND 1 ATM (EXCLUDING THE NUCLEAR SPIN ENTROPY AND THE ISOTOPIC MIXING ENTROPY)

Gas	Translational entropy	Electronic-vibrational-rotational entropy	Total entropy
I_2	42.49	19.77	62.26
Br_2	41.11	17.53	58.64
Cl_2	38.69	14.59	53.28
O_2	36.32	12.68	49.00
N_2	35.92	9.85	45.77
D_2	30.14	4.46	34.60
H_2	28.08	3.13	31.21
NO	36.13	14.20	50.33
CO	35.92	11.38	47.30
OH	34.43	9.45	43.88
HI	40.45	8.89	49.34
HBr	39.08	8.38	47.46
HCl	36.71	7.93	44.64

6.17. CHEMICAL EQUILIBRIA

In many cases chemical equilibria between gases can be calculated accurately from the statistically determined state sums of the substances taking part in the reaction. Let us take as an example the simple reaction

$$H_2 \rightleftarrows 2\,H \tag{6.17.1}$$

[1]) W. G. Brown, J. Amer. Chem. Soc. **54**, 2394 (1932).

According to Section 3.4 the position of the equilibrium is given by

$$\Delta G_T^0 = -RT \ln \frac{p_H^2}{p_{H_2}} \qquad (6.17.2)$$

where

$$\Delta G_T^0 = 2\,\mu_T^0(H) - \mu_T^0(H_2) \qquad (6.17.3)$$

ΔG_T^0 is the change which occurs in the free enthalpy when 1 mole of perfect gaseous H_2 at 1 atm and T °K is transformed into 2 moles of perfect gaseous H at 1 atm and T °K. The symbols $\mu_T^0(H)$ and $\mu_T^0(H_2)$ indicate the molar free enthalpies of atomic and molecular hydrogen at a pressure of 1 atm and a temperature of T °K. The symbols p_H and p_{H_2} indicate the partial pressures of H and H_2 in the equilibrium mixture at T °K.

The position of the equilibrium at a certain temperature will thus be known if the values of $\mu^0(H)$ and $\mu^0(H_2)$ at that temperature are known. With the help of the spectroscopic data already discussed, these quantities can be calculated from Equation (5.10.2). In this connection it must immediately be pointed out that in the foregoing, when calculating a state sum, the zero-level for the energy was always chosen as that of the state of the molecule existing at absolute zero. It is obvious that this is no longer permissible when molecules are converted into one another, i.e. when a chemical reaction takes place. This already appears in the fact that the heat of reaction (in contrast to the reaction entropy) by no means approaches zero as the temperature approaches the absolute zero (cf. Section 5.13, small print).

If we take into account the zero-point energy of the molecule and indicate this by the symbol ε_0, Equation (5.11.2) becomes, for the "internal" part of μ:

$$\mu_{evr} = -RT \ln \sum g e^{-(\varepsilon_0 + \varepsilon)/kT} \qquad (6.17.4)$$

in which the various states are again collected into groups g of equal energy. Taking $\exp(-\varepsilon_0/kT)$ outside the summation symbol, (6.17.4) becomes:

$$\mu_{evr} = \frac{\varepsilon_0 R}{k} - RT \ln \sum g e^{-\varepsilon/kT}\,. \qquad (6.17.5)$$

In this equation R/k is Avogadro's number and consequently $\varepsilon_0 R/k$ is simply the zero-point energy u_0 of the perfect gas per mole:

$$\mu_{evr} = u_0 - RT \ln Z_{evr} \qquad (6.17.6)$$

In this equation Z_{evr} has the same significance as it had in all the preceding sections. It can thus be calculated in the familiar manner, after which $\mu_{evr} - u_0$ is known. The translational part of μ

$$\mu_{tr} = u_{tr} - Ts_{tr} + RT \tag{6.17.7}$$

is found by means of Equations (5.5.1) and (5.5.12). For each of the substances taking part in the reaction we then know

$$\mu_{tr} + \mu_{evr} - u_0 = \mu - u_0 \tag{6.17.8}$$

For (6.17.3) we write:

$$\Delta G_T^0 = 2(\mu_T^0 - u_0)_{\mathrm{H}} - (\mu_T^0 - u_0)_{\mathrm{H_2}} + 2u_0(\mathrm{H}) - u_0(\mathrm{H_2}) \tag{6.17.9}$$

The way to calculate $\mu_T^0 - u_0$ for both H and H_2 is already known from earlier discussions. The position of the equilibrium is therefore known from (6.17.2), once we know the value of

$$\Delta U_0 = 2u_0(\mathrm{H}) - u_0(\mathrm{H_2})$$

This is simply the dissociation energy of H_2 at absolute zero temperature.

The dissociation energies of molecules are deduced from thermochemical or spectroscopic data. For the most important diatomic molecules the spectroscopic values of the dissociation energies will be found in Table 18.

The dissociation equilibrium as a function of the temperature has been calculated for many diatomic gases in the way described. The agreement with the experimental values leaves nothing to be desired.

Also equilibria of the type

$$2\,\mathrm{HCl} \rightleftarrows \mathrm{H_2} + \mathrm{Cl_2}$$

can be calculated according to the method described. The required reaction energy ΔU_0 at absolute zero temperature (the change in the zero-point energy during the course of the reaction) can be calculated, in this case too, from the spectroscopic data in Table 18:

$$\Delta U_0 = 2D_0(\mathrm{HCl}) - D_0(\mathrm{H_2}) - D_0(\mathrm{Cl_2})$$

The accurate calculation of chemical equilibria between gases from exclusively spectroscopic data, is one of the most interesting results of statistical thermodynamics.

APPENDIX

1. Method of undetermined multipliers

In Section 2.12 we wanted to find the maximum of $\ln g$ that does not involve all possible variations $\delta n_0, \delta n_1, \delta n_2, \ldots$ of the distribution numbers $n_0, n_1, n_2, \ldots$, but only those variations that are subject to the constraining relations

$$\delta N = \sum \delta n_i \quad = 0 \tag{1}$$

$$\delta U = \sum \varepsilon_i \delta n_i = 0 \tag{2}$$

At this conditional maximum, $\delta \ln g$ is equal to zero and thus, according to Equation (2.12.4),

$$\sum \ln n_i \,.\, \delta n_i = 0 \tag{3}$$

Multiplying (1) and (2) by the constant parameters α and β, as yet undetermined, and subsequently adding both to (3) gives

$$\sum (\ln n_i + \alpha + \beta\varepsilon_i)\delta n_i = 0 \tag{4}$$

where α is a pure number and β has the dimensions of the reciprocal of an energy.

If each variation δn_i in (4) might be arbitrarily chosen, then the expression in parentheses would be equal to zero for each value of i. This is seen immediately because all variations but one, say δn_k, could then be made equal to zero, reducing (4) to

$$(\ln n_k + \alpha + \beta\varepsilon_k)\delta n_k = 0$$

Since δn_k is not equal to zero, the bracketed expression vanishes for any value of k.

The above argument loses its validity because of the constraining relations (1) and (2). They limit the arbitrariness of the variations δn_i. One may say that δn has lost two degrees of freedom. If all but two of the variations are chosen, the remaining two are fixed, i.e. any two δn's may be regarded as a function of the other variations. Here the unspecified character of the multipliers α and β comes to our rescue. If we regard δn_0 and δn_1 as the dependent variations, then the values of α and β can be so adjusted that

$$\ln n_0 + \alpha + \beta\varepsilon_0 = 0$$

$$\ln n_1 + \alpha + \beta\varepsilon_1 = 0$$

We thus eliminate the first two terms in the sum of Equation (4), which now only contains the independent variations δn_2, δn_3, δn_4, ... Instead of δn_0 and δn_1 we might have chosen any other two δn's as dependent variations. Hence, the only solution of (4) is that each expression in parentheses is zero. Thus, in general

$$\ln n_i + \alpha + \beta\varepsilon_i = 0 \tag{5}$$

For a more general and more exact discussion of Lagrange's method of undetermined multipliers the reader is referred to books on the calculus of variations. The authors, titles and publishers of four of these books are given below [1–4].

2. Calculation of isotopic gas equilibria

It was shown in Chapter 6 that chemical equilibria between perfect diatomic gases can be calculated accurately with the help of spectroscopic data relating to the rotational and vibrational states of the reacting molecules. No experimental data are needed for calculating equilibria between isotopic molecules. As an example we consider the equilibrium

$$\tfrac{1}{2}Cl_2^{35} + \tfrac{1}{2}Cl_2^{37} \rightleftarrows Cl^{35}Cl^{37}. \tag{6}$$

Since, at constant temperature and pressure, this reaction is not accompanied by a change in enthalpy or energy, the value of the equilibrium constant is found by a statistical argument. If one gram atom of Cl^{35} is added to one gram atom of Cl^{37}, chlorine molecules will form without any preference for $Cl^{35}Cl^{35}$, $Cl^{35}Cl^{37}$ or $Cl^{37}Cl^{37}$. Consequently the ratio of the numbers of these molecules will be 1 : 2 : 1. This follows immediately from our "experiments" with white and red billiard balls described in Chapter 2 (cf. Fig. 12, third line from top).

At a total pressure of 1 atm the partial pressure of $Cl^{35}Cl^{37}$ in the equilibrium mixture is 0.5 atm and thus

$$K_p = \frac{0.5}{\sqrt{0.25} \times \sqrt{0.25}} = 2 \tag{7}$$

1) L. E. Elsgolc, *Calculus of Variations*, Pergamon Press, Oxford, 1961.
2) P. Funk, *Variationsrechnung und ihre Anwendung in Physik und Technik*, Springer, Berlin, 1962.
3) L. A. Pars, *Introduction to the Calculus of Variations*, Heinemann, London, 1962.
4) I. M. Gelfand and S. V. Fomin, *Calculus of Variations*, Prentice-Hall, Englewood Cliffs, U.S.A., 1963.

From Equations (3.4.23) and (3.4.26) we have

$$\Delta G^0 = \Delta H^0 - T\Delta S^0 = -RT \ln K_p$$

or, since $\Delta H^0 = 0$:

$$\Delta S^0 = R \ln K_p = R \ln 2 \qquad (8)$$

It will be recalled (see Section 3.4) that the superscript (^{0}) in Equation (8) expresses the fact that the change in entropy, $R \ln 2$, corresponds to the formation of one mole of $Cl^{35}Cl^{37}$ from the separated gases $Cl^{35}Cl^{35}$ and $Cl^{37}Cl^{37}$ under standard conditions of temperature and pressure. In other words, Equation (8) shows that the entropy of the heteronuclear gas $Cl^{35}Cl^{37}$ is greater than the average value for the homonuclear gases by an amount $R \ln 2$ per mole.

If we start with a *mixture* of the homonuclear gases, then the difference in entropy between the reactants and the final product in Equation (6) is zero, the entropy of mixing being

$$\Delta S = -R(\tfrac{1}{2} \ln \tfrac{1}{2} + \tfrac{1}{2} \ln \tfrac{1}{2}) = R \ln 2$$

according to Section 1.17.

The entropy of the equilibrium mixture of the three gases ($\tfrac{1}{4}$ mole $Cl^{35}Cl^{35}$ plus $\tfrac{1}{4}$ mole $Cl^{37}Cl^{37}$ plus $\tfrac{1}{2}$ mole $Cl^{35}Cl^{37}$) must necessarily be greater than $R \ln 2$ since the increase of entropy is the only "driving force" of the reaction and since the entropy of the equilibrium mixture must remain unchanged in an infinitesimal process occurring at constant temperature and pressure. In order to calculate the equilibrium entropy with reference to the separated homonuclear gases, we imagine the gases to be present in a vessel of volume v such that, at the prevailing temperature, it can contain exactly one mole of Cl_2 at a pressure of 1 atm. The vessel is supposed to be divided into three compartments, one of volume $\tfrac{1}{2}v$ containing $\tfrac{1}{2}$ mole $Cl^{35}Cl^{37}$ and two of volume $\tfrac{1}{4}v$ containing $\tfrac{1}{4}$ mole $Cl^{35}Cl^{35}$ and $\tfrac{1}{4}$ mole $Cl^{37}Cl^{37}$, respectively. The entropy content of the first compartment is $\tfrac{1}{2}R \ln 2$, that of the other two is zero. By removing the partitions between the compartments the equilibrium mixture forms. This gives rise to an entropy of mixing

$$\Delta S_m = -R(\tfrac{1}{4} \ln \tfrac{1}{4} + \tfrac{1}{4} \ln \tfrac{1}{4} + \tfrac{1}{2} \ln \tfrac{1}{2}) = \tfrac{3}{2}R \ln 2$$

increasing the total entropy from $\tfrac{1}{2}R \ln 2$ to

$$\Delta S_{max} = 2R \ln 2 \qquad (9)$$

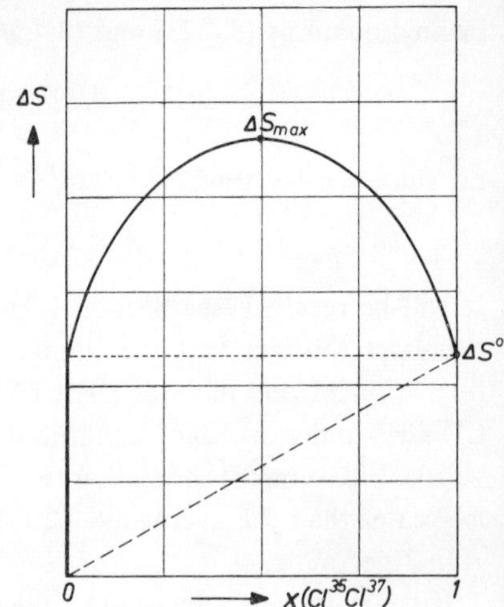

Fig. 70. Increase in entropy as a function of the extent of reaction (6) proceeding at constant temperature and pressure. The broken sloping line gives the total entropy of the separated gases and the full line that of the mixed gases, both with reference to the separated reactants as zero of entropy. $\Delta S^0 = R \ln 2$ is the standard entropy of reaction (6), $\Delta S_{\max} = R \ln 4$ is the entropy corresponding to the equilibrium between the three isotopic gases, and $x(Cl^{35}Cl^{37})$ is the mole fraction of the heteronuclear gas.

In Fig. 70 the entropy of the separated gases is given by the broken sloping line and that of the mixed gases by the full line, both as a function of the mole fraction of $Cl^{35}Cl^{37}$.

3. "Negative absolute temperatures", masers and lasers

Negative temperatures

In Sections 2.10 to 2.12 it has been shown that the equilibrium distribution of a number of quasi-independent localized particles among a set of energy levels is given by the exponential function (2.10.5), where $\beta = 1/kT$. When the number of levels is infinite, the particles will spread out over more and more levels as the temperature rises. This is shown by Fig. 16 for the simple case of equidistant levels. Equation (2.10.4) by itself would also permit an inverted exponential distribution in which the occupation number increases with the height of the levels. However, such an inverted exponential distribution is excluded by the accessory conditions (2.10.2) and (2.10.3).

An inverted exponential distribution does become admissible if each particle has only a limited number of levels at its disposal. According to the relation $\beta = 1/kT$, a system showing exponential population inversion can be formally characterized by a "negative absolute temperature". It must be remarked, however, that a distribution of this type can be realized only among a set of levels corresponding to part of the degrees of freedom of a system and, moreover, only in those cases where the establishment of equilibrium with the other degrees of freedom takes a relatively long time. Purcell and Pound [1]) were the first investigators to show population inversion in experiments on the magnetic resonance of the ^{7}Li nuclei in pure crystals of lithium fluoride, where the exchange of energy between the nuclear spins and the lattice vibrations is a slow process. As the quantum number of the nuclear spin of ^{7}Li is $\frac{3}{2}$ we get four lithium spin states of markedly different energy when a crystal of LiF is placed in a strong magnetic field. Using special techniques, it is possible to reverse the direction of the field so rapidly that the spins are unable to follow the change and remain in their original orientations. As a consequence, the spins that formerly had least energy now have most, or, in other words, the normal exponential population of the four spin levels has been transformed into an inverted exponential population. At room temperature this distribution decays to the equilibrium distribution with a relaxation time of several minutes.

The term "negative absolute temperature" is often used in the literature on masers and lasers (see below) to indicate inverted populations. However, it is not a felicitous expression to describe these non-equilibrium states. Moreover, the term could give the impression that the absolute zero of temperature is attainable, whereas in reality we do not pass through absolute zero when going from positive temperatures to "negative temperatures". The dividing line is at an equal distribution among the levels, i.e. at an infinite temperature.

Masers and lasers

The period between 1950 and 1960 witnessed the birth of the maser, one of the most interesting developments of modern science and technology. It is a device for amplifying electromagnetic radiation. Its principle is based on the inversion of the population of a pair of energy levels and on stimulated emission. By stimulated emission we mean emission which is caused not by the spontaneous transition of particles from the upper level down to the lower level, but by their enforced descent when they are struck by electromagnetic radiation of the transition frequency, i.e. the frequency corresponding to the energy difference ($\varepsilon_2 - \varepsilon_1$) between levels 2 and 1 (see Fig. 71). The foregoing explains the meaning of the word maser, which is an acronym derived

[1]) E. M. Purcell and R. V. Pound, Phys. Rev. **81**, 279 (1951).

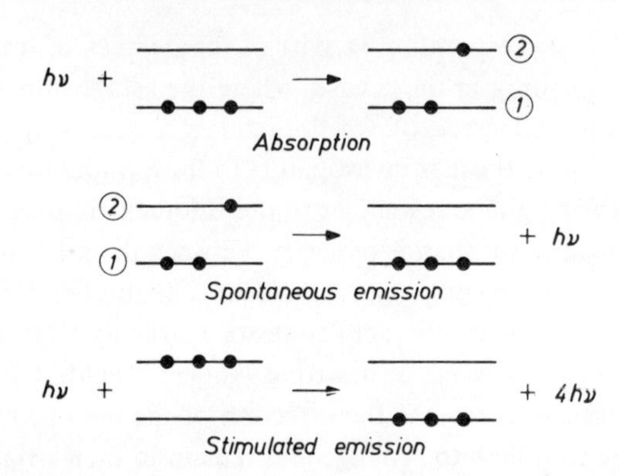

Fig. 71. The figure symbolizes the processes of absorption, spontaneous emission and stimulated emission for the simple case of a system consisting of three particles which can only be in two different energy states ε_1 and ε_2. By injecting a photon of the appropriate energy ($h\nu = \varepsilon_2 - \varepsilon_1$) into the system, one of the atoms can be raised from the ground state to the upper state (top). The excited atom may then emit a photon spontaneously and revert to the lower energy level (middle). When, by an artifice, all three atoms are brought into the excited state (bottom) one of them can be stimulated to emit a photon when it is struck by an injected photon of the same energy. The two photons may then knock out the two remaining atoms from the upper level releasing two more photons. The bottom part of the figure not only demonstrates the principle of stimulated emission but also that of amplification of micro-waves, or light, since one photon enters the system and four photons leave it.

from "microwave amplification by stimulated emission of radiation". The first masers, constructed in 1954 and 1955 operated at centimeter wavelengths [1,2]. Of much greater importance are those developed a few years later, which operate at optical wavelengths [3,4]. These optical masers, including those operating in the infrared region, are generally called lasers, where the letter l stands for light.

Population inversion is a prerequisite for amplifying radiation by stimulated emission. This is a consequence of the fact that particles in the lower of the two energy levels absorb radiation of the transition frequency, $(\varepsilon_2 - \varepsilon_1)/h$, with a probability equal to that for stimulated emission by particles in the upper level. Use is made of an artifice for increasing the population of the upper level and at the same time decreasing that of the lower level. In many optical and non-optical masers this continuous emptying of the lower level and filling of the upper level is brought about by irradiating the particles responsible for the maser action with photons of energies greater

[1] J. P. Gordon, H. J. Zeiger and C. H. Townes, Phys. Rev. **99**, 1264 (1955).
[2] N. G. Basov and A. M. Prokhorov, Discussions Faraday Soc., No. 19, 96 (1955)
[3] A. L. Schawlow and C. H. Townes, Phys. Rev. **112**, 1940 (1958).
[4] T. H. Maiman, Nature **187**, 493 (1960).

than $(\varepsilon_2 - \varepsilon_1)$. In the simplest case, this results in raising the particles from level 1 to a third level that is higher than level 2 (Fig. 72). Maser or laser action at a frequency corresponding to the energy difference between levels 2 and 1 is then possible if the excited particles loose their energy by first descending to level 2 and staying there for a relatively long time before loosing the rest of their excitation energy spontaneously.

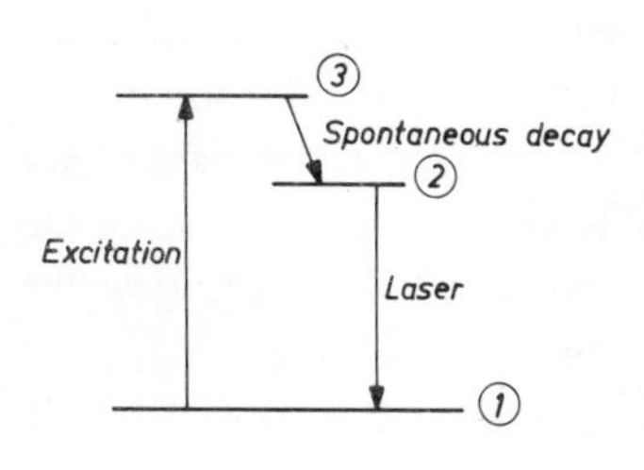

Fig. 72. In the simplest case of amplification by stimulated emission, particles in the ground level 1 are raised to a high level 3 by irradiation with photons of the appropriate energy. They fall spontaneously to a metastable level 2 where, under normal conditions, their average time of stay is relatively long. If enough radiation energy is "pumped in" the population of level 2 is made to exceed that of level 1 and particles in level 2 are stimulated to emit photons of energy $(\varepsilon_2 - \varepsilon_1)$ when hit by photons of the same energy. This stimulated emission brings the particles back to the ground level. In many lasers "level" 3 is in fact an energy band. In other kinds of optical masers the laser action takes place between two levels well above the ground level.

As an example we consider the ruby laser. Ruby consists of aluminium oxide in which a small fraction of the Al atoms have been replaced by chromium atoms. By irradiating the ruby with light of high frequencies the chromium atoms are brought into a band of high energies corresponding to level 3 in Fig. 72. Instead of descending directly to the ground level, the atoms first land at an intermediate level 2 and, under normal conditions, stay there for a few milliseconds before descending at random to level 1. If radiation power is supplied at a sufficiently high rate, an inversion of population is produced between levels 2 and 1. In that case, the first few photons that are released by transitions $2 \rightarrow 1$ stimulate other chromium atoms in level 2 to make the same transition and to give up additional photons. In other words, during the operation of the laser most particles in level 2 get no opportunity to descend spontaneously; long before their average time of stay has expired they emit a photon of energy $(\varepsilon_2 - \varepsilon_1)$ after being hit by a photon of exactly the same energy.

A laser is generally not used as a device for amplification of incident light but as a *generator of radiation*. The great interest of the laser as a light source is the fact that it can produce a very powerful, directional beam of extremely coherent light. This is achieved in practice by adding parallel mirrors to the ends of its working element. In the ruby laser this element is a cylinder of ruby whose end faces are at right angles to its axis. These faces are accurately plane and parallel and covered with a layer of silver. After population inversion by intense irradiation of the crystal (see above) photons of energy $(\varepsilon_2 - \varepsilon_1)$ are emitted in all directions. Most of them pass out of the crystal but those that happen to be emitted parallel to the axis of the cylinder are reflected by the mirrors and, on their way back and forth between the ends of the crystal, stimulate other atoms in level 2 to emit photons in the same direction. If this amplification by stimulated emission is great enough to make up for the various losses an electromagnetic wave will build up in the working element, which may be regarded as a special kind of resonator. If one of the silver films is so thin as to be slightly transparent, part of the light is transmitted through it. The emergent beam is highly directional since the waves, before leaving the laser, have made several passages between the mirrors without deviating very far from the axis of the crystal.

Most important is the fact that the laser is a powerful source of highly coherent light in contrast with the conventional light sources which spread their output over a wide range of frequencies and directions. The laser light is of nearly one frequency, and all the points in the beam bear a definite and fixed phase relationship to one another, i.e. the light emitted by the laser has both time coherence and space coherence. Since the radiation of the laser is very intense and its linewidth very small, the power emitted per unit bandwidth corresponds to that of black-body radiation at enormously high temperatures (10^{20} to 10^{30} °K).

Many other lasers, using solids, liquids or gases as working elements, have been developed since the construction of the first ruby laser. Initially they could only be operated with short pulses, but many of them can now be operated continuously. In contrast to the amplification occurring in the ruby laser, amplification in the newly developed lasers is not based on transitions between a higher level and the ground level, but on those between two levels well above the ground level. As a consequence, depopulation of the lower level is easier to achieve and less energy is required to get population inversion. As an example, we mention that the operation of the infrared CO_2 gas laser is based on transitions between two vibrational states of the CO_2 molecules. This laser which contains nitrogen and water vapour in addition to the carbon dioxide, also exemplifies the subtlety of recent laser design. The nitrogen is added for accelerating the raising of the CO_2 molecules to the upper laser level and the water vapour for accelerating their descent from the lower laser level. The mechanism of the CO_2 laser can be regarded as a chain consisting of four processes: (a) excitation of vibrations of N_2 molecules by *electron impact* in an electrical gas discharge, (b) transfer of vibrational energy from N_2 molecules to CO_2 molecules, raising the latter to the upper laser level, (c) stimulated emission of infrared radiation as the excited CO_2 molecules drop to a lower vibrational level (identical with the lower laser level), (d) descent of the CO_2 molecules to the ground level by transfer of their vibrational energy to H_2O molecules which very quickly convert the vibrational energy into translational energy by collisions with other H_2O molecules.

The laser has two basic fields of application. In the first field its emitted radiation is used as heat energy. Materials with a high melting point, e.g. tungsten, can easily be cut with a focused laser beam. Surgical applications of the laser, e.g. the treatment of malignant tumors and retinal defects, are based on the fact that it is possible not only to limit the generation of heat to a localized, accurately-focused spot, but, in addition, to generate it in so short a time that little heat is transferred to the surrounding tissues. In the second field of applications use can be made of the exceptional qualities of the emitted radiation, especially its time and space coherence. Because of its high frequency the coherent light is, in principle, able to transmit an enormous volume of information. Infrared lasers might be used for all kinds of infrared research, particularly in the less-explored far infrared region.

Though ten years ago the laser was not yet in existence, the number of articles and books on the subject already runs into thousands. We mention here only a few [1–12], where the reader can find extensive surveys of the literature. Röss' book [6] published in 1966 gives more than three thousand references.

---

1) G. Troup, *Masers and Lasers*, Methuen, London, 1963.
2) O. S. Heavens, *Optical Masers*, Methuen, London, 1964.
3) R. Saltonstall, *Laser Technology*, Hobbs, Dorman & Co., New York, 1965.
4) B. A. Lengyel, *Introduction to Laser Physics*, Wiley, New York, 1966.
5) Monte Ross, *Laser Receivers*, Wiley, New York, 1966.
6) D. Röss, *Laser Lichtverstärker und -Oszillatoren*, Akad. Verlagsges., Frankfurt a.M., 1966.
7) P. E. McGuff, *Surgical Applications of Laser*, Thomas, Springfield, U.S.A., 1966.
8) C. G. B. Garrett, *Gas Lasers*, McGraw-Hill, New York, 1967.
9) H. A. Elion, *Laser Systems and Applications*, Pergamon Press, Oxford, 1967.
10) A. L. Schawlow, *Optical Masers*, Scient. American, June 1961.
11) W. J. Witteman, *Sealed-off High Power CO_2 Lasers*, Philips tech. Rev. **28**, 287 (1967).
12) T. A. Osial, *Industrial Laser Applications*, Instruments & Control Systems **40**, 101, 1967.

4. Thermodynamics of superconductors

The electrical resistivity of metals generally decreases at decreasing temperature. Below a few degrees Kelvin the resistivity of many metals, e.g. copper, silver, platinum, iron, nickel and cobalt, reaches a value which is almost independent of temperature. This *residual resistivity* is not a characteristic property of the metal in question, but is caused by collisions of the conduction electrons with impurity atoms, vacancies, dislocations and other lattice imperfections.

Many other metals including lead, tin, mercury, niobium, tantalum and zirconium, show a quite different behaviour. Far from reaching gradually a constant residual value, their electrical resistivity vanishes suddenly at the so-called *transition temperature* T_c which is characteristic of the particular metal concerned. This remarkable phenomenon is called *superconductivity* and was observed first by Kamerlingh Onnes in Leiden in 1911, three years after the first successful liquefaction of helium by the same investigator [1]. Fig. 73 shows schematically the difference in behaviour of a superconducting and a non-superconducting metal.

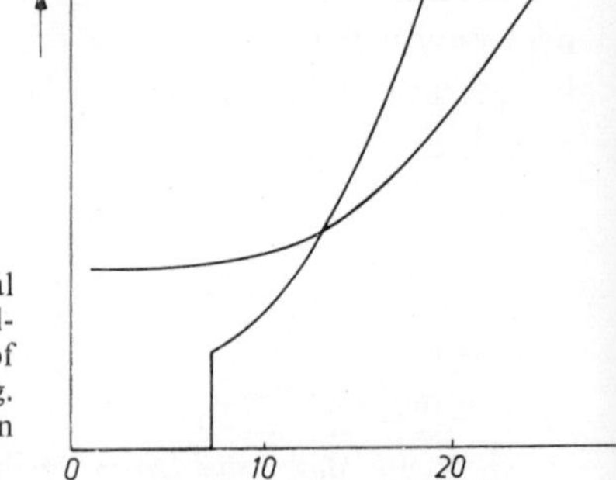

Fig. 73. At decreasing temperature the electrical resistivity of many metals (e.g. Pt) falls to an almost constant value which depends on the purity of the metal. The resistivity of a superconductor (e.g. Pb) disappears abruptly at the so-called transition temperature which is 7.2 °K for lead.

At present it is known that in addition to many metallic elements several hundreds of compounds and alloys become superconducting at low temperatures. The highest transition temperature so far found (20.05 °K) is that of the intermetallic compound Nb_3Al in which part of the aluminium has been replaced by germanium [2].

[1] Historical references are given by C. J. Gorter, Revs. mod. Phys. **36**, 3 (1964).
[2] B. T. Matthias, Science **156**, 645 (1967).

By exposing a superconducting ring of lead to a changing magnetic field at liquid helium temperature (4.2 °K) it is possible to induce a persisting current, i.e. a current that does not show any detectable decay in a period of several years. From experiments of this kind it was possible to place an upper limit of 10^{-23} ohm·cm on the resistivity of a superconductor. This value may be compared with that of very pure copper which lies between 10^{-9} and 10^{-10} ohm·cm at 4.2 °K. It is therefore not surprising that copper is often used as an insulating material for superconductors. At present there is little doubt that the resistance of a superconductor to direct current is not only very small, but actually zero.

The normal, resistive, state of a superconductor can be restored not only by increasing its temperature, but also by exposing it to a magnetic field. Here we will consider only the most simple case, where the superconductor is a pure, well-annealed metal showing a complete Meissner effect (see below). If a superconductor of this type ("type I") has the form of a long cylindrical rod and is placed in a uniform magnetic field parallel to its length, then the transition will take place at a definite field strength, the so-called *critical field* $\mathbf{H}_c$ which depends on the temperature and is characteristic of the metal in question [1]). The relation between $\mathbf{H}_c$ and T is given approximately by the parabolic equation

$$\mathbf{H}_c/\mathbf{H}_{c0} = 1 - (T/T_c)^2 \tag{10}$$

where $\mathbf{H}_{c0}$ is the critical field at $T = 0$ and T_c the critical temperature at $\mathbf{H} = 0$ (Fig. 74).

The Meissner effect

The applicability of reversible thermodynamics to the phenomenon of superconductivity is based on the Meissner effect, according to which the magnetic flux in the interior of a type I superconductor is always zero, regardless of whether the sample is cooled in the presence or the absence of an applied magnetic field [2]). This implies that the magnetic transition between the normal and the superconducting state is reversible. It is a phase transition in the usual thermodynamic sense of the phrase, so that Fig. 74 may be regarded as the phase diagram of a superconductor in the $\mathbf{H},T$ plane. When

[1]) The symbol $\mathbf{H}$ for the field strength is printed in bold type, not to emphasize its vector character which is of little importance in our discussions, but to distinguish it from the symbol H used to denote the enthalpy of a system.

[2]) W. Meissner and R. Ochsenfeld, Naturwiss. **21**, 787 (1933).

the parabolic phase boundary in this diagram is crossed by lowering the field strength, or the temperature, or both, the superconducting phase becomes stable and the magnetic flux is expelled. Crossing the phase boundary in the reverse direction stabilizes the normal phase and allows the external field to penetrate.

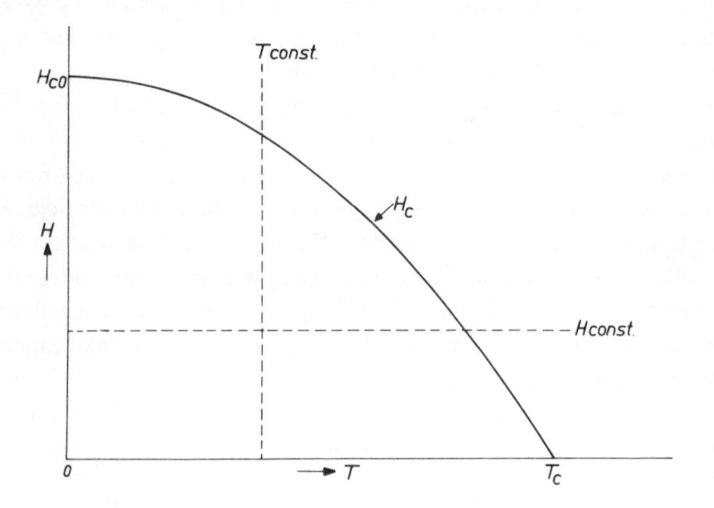

Fig. 74. Phase diagram of a superconductor in the **H**,T plane. The phase boundary gives the critical field $\mathbf{H}_c$ as a function of temperature. The superconducting phase is stable below this threshold curve, the normal phase is stable above it.

The transition temperature and the critical field are affected to a small extent by pressure. Taking the influence of pressure into acccount it will be clear that the diagram under discussion (Fig. 74) is a cross-section at constant pressure of a three-dimensional ($\mathbf{H}$,T,p) diagram in which the transition line is extended into a transition surface. A section of this surface at a few hundred or even a few thousand atmospheres does not differ greatly from that at one atmosphere.

A magnetic field is prevented from entering a superconducting sample by shielding currents which are induced in a very thin surface layer. Consequently the field strength, the induction and the magnetization in the interior of a superconductor are all three equal to zero. In the thermodynamic treatment of the phase change under discussion it is more convenient to ignore the surface currents and to make use of an alternative description which leads to identical results. In this description the magnetic induction B at the interior of a superconductor is not excluded by surface currents, but cancelled by a magnetization per unit volume

$$I_s = -\frac{\mathbf{H}}{4\pi} \tag{11}$$

In this equation the subscript s stands for superconducting and **H** is the strength of the uniform magnetic field into which our long cylindrical superconductor is placed with its axis parallel to the field. Substituting Equation (11) in the well-known relation

$$B = \mathbf{H} + 4\pi I \tag{12}$$

results indeed in $B = 0$. This means that a superconductor behaves as a perfect diamagnetic substance with susceptibility $-1/4\pi$ (in gaussian units).

When at constant pressure and temperature the field strength increases, the type I superconductor will revert to its normal state as soon as the critical field is exceeded. In this state the susceptibility of the metal is so small that in its interior B and **H** are equal:

$$B_n = \mathbf{H} \tag{13}$$

$$I_n = 0 \tag{14}$$

where the subscript n denotes the normal, non-superconducting state. In accordance with Equations (11) to (14) Fig. 75 gives the values of $-4\pi I$ and B as functions of the applied field.

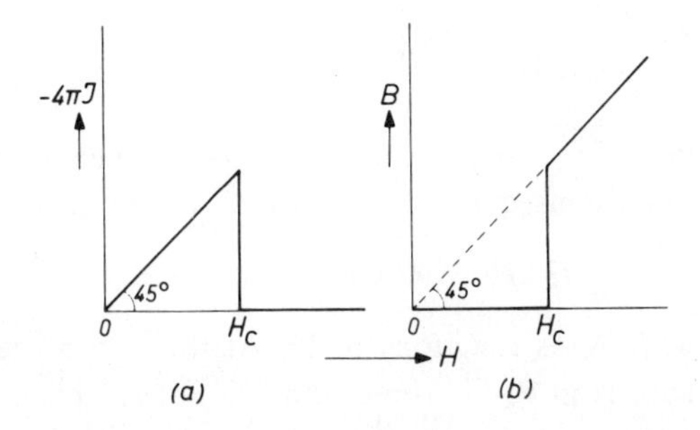

Fig. 75. Apparent magnetization per unit volume, I, and induction, B, inside a long cylindrical superconductor of type I, as functions of the strength, **H**, of a uniform field parallel to the axis of the cylinder

Application of the thermodynamic laws [1])

At constant pressure and temperature the state of equilibrium is reached as soon as the free enthalpy or Gibbs free energy, G, is at its minimum. In the case under consideration this function takes the form

$$G = U - TS + pV - \mathbf{H}M \tag{15}$$

where V is the volume of the sample and $M = IV$ (cf. Section 3.3). From Equation (15) it follows that

$$\mathrm{d}G = \mathrm{d}U - T\mathrm{d}S - S\mathrm{d}T + p\mathrm{d}V + V\mathrm{d}p - \mathbf{H}\mathrm{d}M - M\mathrm{d}\mathbf{H}. \tag{16}$$

Combination of the first and second laws gives

$$\mathrm{d}U = T\mathrm{d}S - p\mathrm{d}V + \mathbf{H}\mathrm{d}M \tag{17}$$

where $T\mathrm{d}S$ is the quantity of heat taken up reversibly by the metal from the surroundings, and $-p\mathrm{d}V$ and $\mathbf{H}\mathrm{d}M$ are the volume work and the magnetic work performed on the metal by the surroundings. Combining (16) and (17) the differential equation for G becomes

$$\mathrm{d}G = -S\mathrm{d}T + V\mathrm{d}p - M\mathrm{d}\mathbf{H} \tag{18}$$

According to (11) the magnetization in the superconducting state is given by

$$M_s = I_s V_s = -\frac{V_s \mathbf{H}}{4\pi} \tag{19}$$

At constant values of p and T we obtain from (18) and (19):

$$\mathrm{d}G_s = -M_s \mathrm{d}\mathbf{H} = \frac{V_s \mathbf{H} \mathrm{d}\mathbf{H}}{4\pi}$$

or, integrating:

$$G_s(\mathbf{H}) = G_s(0) + \frac{V_s \mathbf{H}^2}{8\pi} \tag{20}$$

For the normal state, G is virtually independent of $\mathbf{H}$ since M_n is generally vanishingly small. As a consequence, at constant values of p and T:

$$G_n(\mathbf{H}) = G_n(0) = G_n \tag{21}$$

Fig. 76 shows G_s and G_n as functions of $\mathbf{H}$. At the critical field, $\mathbf{H}_c$, the superconducting phase is in equilibrium with the normal phase:

$$G_s(\mathbf{H}_c) = G_n(\mathbf{H}_c) = G_n(0) \tag{22}$$

---

[1]) Cf.: C. J. Gorter and H. B. G. Casimir, Physica **1**, 306 (1934).

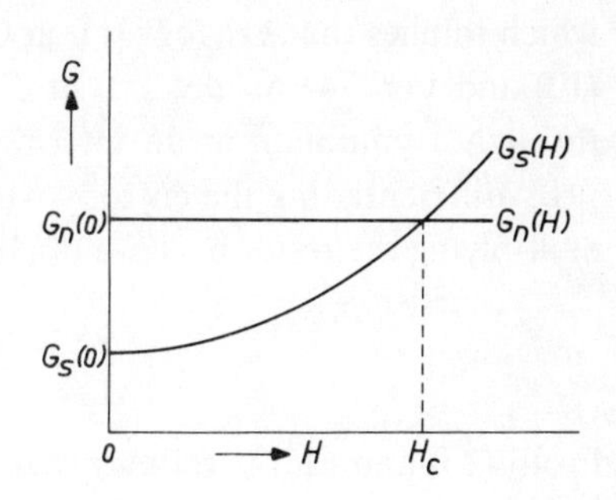

Fig. 76. Schematic representation of the Gibbs free energy of a metal in the superconducting state, G_s, and that in the normal state, G_n, as a function of the magnetic field strength, **H**, at constant values of p and T. At the temperature to which the curves relate the material is superconducting in zero magnetic field, so that $G_s(0)$ is lower than $G_n(0)$.

From (20) and (22) we obtain the relation

$$G_s(0) - G_n(0) = -\frac{V_s \mathbf{H}_c^2}{8\pi}. \tag{23}$$

Since, according to (18),

$$\left(\frac{\partial G}{\partial T}\right)_{p,\mathbf{H}} = -S,$$

differentiation of (23) yields

$$S_s(0) - S_n(0) = \frac{V_s \mathbf{H}_c}{4\pi} \frac{\partial \mathbf{H}_c}{\partial T} \tag{24}$$

From (18) it follows also that

$$\left(\frac{\partial S}{\partial \mathbf{H}}\right)_{p,T} = \left(\frac{\partial M}{\partial T}\right)_{p,\mathbf{H}}$$

Consequently S_s and S_n are virtually field-independent, the former because $\partial M_s/\partial T$ is zero apart from a minute effect caused by thermal expansion, and the latter because $M_n \cong 0$. Equation (24) may therefore be written in the simple form

$$\Delta S = S_s - S_n = \frac{V_s \mathbf{H}_c}{4\pi} \frac{\partial \mathbf{H}_c}{\partial T}. \tag{25}$$

Since $\partial \mathbf{H}_c/\partial T$ is always negative (cf. Fig. 74), it follows that the entropy of the superconducting phase is lower than that of the normal phase. In accordance with this conclusion, the microscopic theory of superconductivity shows that the conduction electrons in the superconducting phase are in a state of higher order than in the normal phase [1]). The entropy difference, ΔS, vanishes at the critical temperature, T_c, where $\mathbf{H}_c = 0$ and $\partial \mathbf{H}_c/\partial T$ is finite. According to the third law (cf. Section 2.17) it must also vanish at 0 °K

[1]) J. Bardeen, L. N. Cooper and J. R. Schrieffer, Phys. Rev. **108**, 1175 (1957).

which implies that $\partial \mathbf{H}_c/\partial T = 0$ at 0 °K. This is in agreement with Equation (10) and Fig. 74. As $\Delta S = 0$ at $T = 0$ and at $T = T_c$, its value must pass through a minimum at an intermediate temperature.

By differentiating the entropy with respect to T, at constant pressure, and multiplying the result by T we obtain the heat capacity at constant pressure:

$$T\left(\frac{\partial S}{\partial T}\right)_p = \left(\frac{\mathrm{d}Q}{\mathrm{d}T}\right)_p \equiv C_p \qquad (26)$$

From (25) and (26) it follows that

$$C_{p,s} - C_{p,n} = \frac{V_s T}{4\pi}\left[\left(\frac{\partial \mathbf{H}_c}{\partial T}\right)^2 + \mathbf{H}_c\left(\frac{\partial^2 \mathbf{H}_c}{\partial T^2}\right)\right] \qquad (27)$$

In the absence of a magnetic field (i.e. for $T = T_c$ and $\mathbf{H}_c = 0$) we obtain

$$C_{p,s} - C_{p,n} = \frac{V_s T_c}{4\pi}\left(\frac{\partial \mathbf{H}_c}{\partial T}\right)^2_{T = T_c} \qquad (28)$$

which is known as Rutgers' formula. This discontinuity in heat capacity at $\mathbf{H} = 0$ can be measured calorimetrically but, according to (28), it can also be calculated from the measured phase boundary curve, i.e. from the slope of the curve at $T = T_c$ and the value of T_c (cf. Fig. 74). For many metals the agreement is very satisfactory.

In the presence of a magnetic field the phase change takes place at a lower temperature. As the difference in entropy between the two phases passes through an extremum at decreasing temperature, ΔC must change sign at a field strength somewhere between zero and $\mathbf{H}_{c0}$.

The above thermodynamic treatment applies only to *type I superconductors* which show a complete Meissner effect, i.e. perfect diamagnetism. The magnetization curve in Fig. 75(*a*) relates to this type of superconductor, having the form of a long cylinder placed in a uniform magnetic field with its axis parallel to **H**.

Type II superconductors have a magnetization curve as sketched in Fig. 77. All superconductive alloys with high values of the electrical resistivity in the normal state belong to this group. Many of them offer promise of useful technical applications. At increasing field strength the magnetic flux starts to penetrate these superconductors at a field $\mathbf{H}_{c1}$ which is lower than the thermodynamic critical field $\mathbf{H}_c$ defined by Equation (23) and determinable by calorimetric measurements. The flux invasion takes place in the form of millions of flux "filaments", each containing one flux quantum $hc/2e$. Above $\mathbf{H}_{c1}$ the specimen remains electrically superconducting. Only at a much higher field, in many cases 10^5 gauss or more, the flux penetration is complete and all superconductivity has disappeared. This field is called the *upper critical field* and denoted by $\mathbf{H}_{c2}$. Between $\mathbf{H}_{c1}$ and $\mathbf{H}_{c2}$ the specimen is said to be in the *mixed state*.

Not all type II superconductors can be used in the construction of superconducting magnet coils. What are needed are *hard superconductors* containing many lattice imperfections (dislocations, precipitates, crystal boundaries, etc.) acting as pinning points for the flux filaments. As a consequence of this pinning, hard superconductors are capable

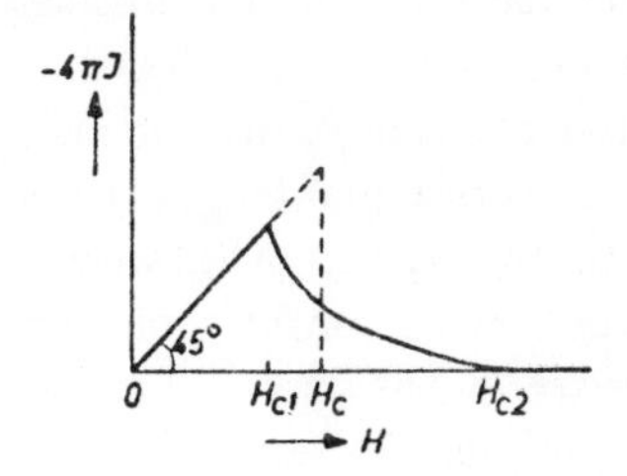

Fig. 77. Apparent magnetization per unit volume inside a long cylindrical superconductor of type II as a function of the strength of a uniform field parallel to the axis of the cylinder. Above $\mathbf{H}_{c2}$ the specimen is a normal conductor except for possible surface effects. Between $\mathbf{H}_{c1}$ and $\mathbf{H}_{c2}$ it is in a mixed state.

of carrying very heavy currents without exhibiting measurable electrical resistivity. In the future superconducting magnets producing fields of the order of magnitude of 10^5 gauss in a large volume will probably play an important part in MHD power generation, i.e. in the direct magnetohydrodynamic conversion of heat into electrical energy without the intermediary of rotating machinery.

For more details on the theory and potential technical applications of superconductivity the reader is referred to a number of books and articles[1-8]).

5. Entropy and information

The concept of information

The concept of information plays an essential part in modern science. It is of great importance not only in communications, that is the transmission of messages by means of acoustical, mechanical, electrical or optical signals, but also in the fields of computers, automation and heredity.

Information removes uncertainty, or rather, *may* remove uncertainty. For the communication engineer only the latter is of importance. He is

[1]) D. Shoenberg, *Superconductivity*, Cambridge University Press, 1960.
[2]) E. A. Lynton, *Superconductivity*, Methuen, London, 1962.
[3]) J. R. Schrieffer, *Theory of Superconductivity*, Benjamin, New York and Amsterdam, 1964.
[4]) P. G. de Gennes, *Superconductivity of Metals and Alloys*, Benjamin, New York and Amsterdam, 1966.
[5]) V. L. Newhouse, *Applied Superconductivity*, Wiley, New York, 1964.
[6]) J. Volger, *Superconductivity*, Philips tech. Rev., **29**, 1 (1968).
[7]) A. L. Luiten, *Superconducting magnets*, Philips tech. Rev., **29**, 309 (1968).
[8]) M. H. Cohen (Ed.), *Superconductivity in Science and Technology*, University of Chicago, 1968.

interested only in how much information can be contained in a message, and how much difficulty is involved in transmitting it. Its significance to the receiver or its intrinsic value do not concern him. For him all messages of equal length (i.e. symbol content) are equivalent.

For the person who wishes to transmit a message the choice of the symbols is of considerable importance. If the message to be transmitted consists of a series of 1000 symbols, chosen from an alphabet of 32 symbols (26 letters + some punctuation symbols) then the message will be one out of a total of 32^{1000} possible messages. If the message were to contain only one symbol, there would be 32 possibilities; if the message were 2 symbols long, then each of the 32 symbols might be followed by any of 32 symbols, so that there would be 32^2 possibilities, etc. In normal practice, as regards conversation, telephony and telegraphy, one is restricted to those series of symbols that constitute an intelligible message in one of the known languages or in an accepted code. For a message of 1000 symbols one then still has the choice of a multitude of possibilities, but considerably less than 32^{1000}.

Let us, for the time being, consider the symbols as independent of each other and occurring with equal probability; then the general expression for a number q of possible messages of length N symbols, selected from m different symbols is $q = m^N$. Let another message have a length of M symbols, then the number of possible selections is $q' = m^M$. If the two messages are transmitted in direct succession, and are considered as a single message, then a choice has been made out of $Q = m^N \times m^M$ possible messages. On the other hand it is only logical to claim that the total service rendered by the telegraph office is the sum of the two separate services. These services consist of the transmission of information. We must, therefore, find a definition of the concept "quantity of information" I as a function of the number of possible messages, in such a form that:

$$I(m^N) + I(m^M) = I(m^{N+M}).$$

To express the additive character of the function, the quantity of information in a message of a given length is defined as the logarithm of the number of possible messages of that length. For the case considered here we can put:

First quantity of information:	$\log q \;\; = N \log m$
Second quantity of information:	$\log q' \; = M \log m$
Total quantity of information:	$\log Q = (N + M) \log m.$

Any separate symbol from the series of m equally probable and mutually independent symbols thus represents a quantity of information

$$i = \log m = -\log p, \tag{29}$$

where $p = 1/m$ is the *a priori* probability of the symbols. The choice of the base of the logarithms is quite arbitrary since this merely determines the magnitude of the unit. If a choice of two equally probable possibilities is transmitted, then the quantity of information supplied is, according to equation (29), $i = \log 2$. This is the simplest situation conceivable, and it is therefore reasonable to choose this quantity of information as the unit: $\log 2$ assumes the value unity if 2 is chosen as the base of the logarithms. This has in fact been generally adopted in communication theory. This unit of information is called a "bit" (short for "binary digit"). Equation (29) appropriately allots a higher numerical value to information according as the freedom of choice is greater, i.e. according as the result is more uncertain (less probable).

Information content per symbol

A single symbol from the language containing $32 = 2^5$ equally probable and mutually independent symbols, supplies, according to equation (29), 5 bits of information. The symbols of a language are, in fact, not equally probable; moreover they cannot always occur independently. In order to estimate the actual information content per symbol, we shall first of all get rid of the restriction that the symbols are equally probable. Later we shall also drop the restriction that they are independent.

Let us begin by considering an imaginary language in which the symbols occur with unequal probability and are independent of each other. For

TABLE 27

PROBABILITIES OF OCCURRENCE OF THE VARIOUS LETTERS IN THE ENGLISH LANGUAGE
(Taken from Fletcher Pratt, *Secret and Urgent*, Blue Ribbon Books, N.Y., 1942)

Letter	Probability	Letter	Probability
e	0.131	m	0.025
t	0.105	u	0.025
a	0.086	g	0.020
o	0.080	y	0.020
n	0.071	p	0.020
r	0.068	w	0.015
i	0.063	b	0.014
s	0.061	v	0.0092
h	0.053	k	0.0042
d	0.038	x	0.0017
l	0.034	j	0.0013
f	0.029	q	0.0012
c	0.028	z	0.00077

any given language the probabilities are immediately apparent from the frequencies with which the symbols occur in a number of random texts in that language. The frequency of letters in the English language is given in Table 27.

It is found that letter frequencies are only slightly influenced by the subject and the author of the text.

The quantity of information given by a single letter we shall define by analogy with equation (29) as $-\log_2 p$, p representing the relative frequency with which this letter occurs. The question now arises how much information will be given on the average by a symbol from an English text. There is a chance p_e that it will be an e, in which case the quantity of information is $-\log_2 p_e$. There is further a chance p_t that it will be a t, in which case the quantity of information is $-\log_2 p_t$, etc. The average amount of information per symbol is therefore

$$H = -\sum_{\nu} p_\nu \log_2 p_\nu \quad \text{bits.} \tag{30}$$

If we evaluate in this way the average quantity of information per letter of the English language from the probabilities given in Table 27, we obtain 4.16 bits/letter. If the 26 letters had had equal probabilities, then H would have been $\log_2 26 = 4.70$ bits/letter. The quantity of information per letter has therefore been reduced [1]), even for an independent choice. Moreover, the probability of a letter is also somewhat dependent on the letters earlier written down. We shall illustrate this with two examples: (1) The probability of the letter u is not particularly great, but the probability that a u follows a q approaches unity. (2) If part of a message contains the following symbols: "Arrive Saturday mor...." then it is highly probable that the next letters will be "ning", although "ose", "e", "eover" or "occo" are not inconceivable. Many other examples of correlation could be given. They all show that the number of different meaningful messages which are possible in a message of a given length is reduced by the presence of correlation.

An artifice to compute roughly the quantity of information per symbol is the following: let p_ν in equation (30) represent the probability of the occurrence of super-symbols each of length A symbols. As we make A larger, the intercorrelations between the super-symbols will become smaller. The super-symbols, if sufficiently long, can therefore be chosen nearly independently of each other. In a very long message of N super-symbols there will be on the average $p_1 N$ specimens of the first super-

---

[1]) When determining the probability per symbol we should actually also take the spacings and punctuations into account. This has not been done here, in order to make use of existing tables.

symbol, $p_2 N$ specimens of the second super-symbol, etc. For the information per super-symbol we then obtain, according to equation (30):

$$-\sum_{\nu} p_{\nu}(A) \log p_{\nu}(A).$$

Since each super-symbol comprises A symbols, the average information per symbol is represented by:

$$H = -\lim_{A \to \infty} \frac{1}{A} \sum_{\nu} p_{\nu}(A) \log p_{\nu}(A) \quad \text{bits.} \tag{31}$$

An increasingly closer approximation to H can be reached by choosing successively 1, 2, 3, . . . for A, i.e. by determining the frequencies with which single symbols, symbol pairs, symbol triples, etc. occur in the language. Another estimate of the quantity of information per symbol can be arrived at by establishing experimentally how many times a person has to guess before he has found the next letter in a given text, and then the next letter, etc. There have not yet been enough experiments of this kind made to establish accurately the actual information content per symbol of a language, but it has been found that for the modern European languages the information content is around 1.5 bits per symbol.

Relation between information and entropy

In statistical thermodynamics the most general definition of the entropy of a system is given by the equation

$$S = -k \sum_{\nu} p_{\nu} \ln p_{\nu}, \tag{32}$$

where the p_{ν}'s are the probabilities of the different microstates of the system and k is Boltzmann's constant. If all microstates have the same probability p, then their number is given by $m = 1/p$, and equation (32) becomes

$$S = -\frac{k}{p}(p \ln p) = -k \ln p = k \ln m. \tag{33}$$

It is obvious that the information equations (29) and (30) are similar to the statistical-thermodynamic equations (32) and (33). Following Shannon [1]) the quantity H defined by equation (30) is therefore often designated "entropy". One then speaks of the entropy of a language as

[1]) C. E. Shannon, "A mathematical theory of communication", *Bell System tech. J.* **27,** 379 and 623 (1948).

its information content per symbol. This practice is not to be recommended, since the relationship between H and S is only of a formal mathematical nature, due to their common statistical background. The thermodynamic concept of entropy, however, also has physical implications. This finds its expression in the fact that equation (32) contains Boltzmann's constant, giving S the dimension cal. per degree. The quantity H, on the other hand, is a dimensionless non-physical quantity.

In principle it is always possible to determine the entropy of a system in two different ways, the statistical one making use of equation (32), and the calorimetric one making use of the well-known classical formula

$$S = \int \frac{dQ_{rev}}{T}, \tag{34}$$

where T is the absolute temperature and dQ_{rev} refers to quantities of heat reversibly exchanged with the surroundings of the system. Experience shows that both formulae lead to the same result (see preceding chapters). The often-used expressions "statistical entropy" and "calorimetric entropy" may cause confusion since they do not refer to different kinds of entropies, but to different ways of determining and calculating the same physical quantity.

It follows that the concept of entropy cannot be dissociated from two other concepts, those of temperature and quantity of heat. If it made sense to talk of the entropy of a language, then it would also make sense to ask how many calories have to be supplied to that language to raise its temperature from 20 to 100 °C. This absurd result follows from the misuse of the term entropy in information theory.

That information and entropy are quite different concepts can easily be seen by visualizing a book containing a lot of information, and a "dummy" of it containing exactly the same amounts of paper and printer's ink. On the pages of the dummy the ink is present as irregular spots covering the same total area as the letters in the real book. The dummy contains no information at all, but it will undoubtedly have the same heat capacity as the real book and thus, according to equation (34), the same entropy at a given temperature.

Transmission of information and negentropy

Of great importance in communication engineering is the efficiency of the *transmission* of information; in the form of electric signals over a cable for example. The thermal noise (cf. page 62) interferes with the signals so that they arrive at the receiving end of the cable in a more or less mutilated condition. The information received is consequently less than it would be if no noise were present. This phenomenon is often wrongly called degradation of information; in reality only the signals degrade. All the information is still present in the source (man or machine).

Since the intensity of thermal noise in a communication channel depends on the temperature, more energy is needed to transmit signals at higher temperatures. It is therefore obvious that the laws of thermodynamics may be applied to the degradation of the signals [1]). By ignoring the difference between information and signal, some investigators were induced to the following conclusion. The tendency of the entropy to increase is equivalent to the tendency of the information to decrease; apart from the sign there is no difference between entropy and information or, in other words, information = negentropy. As indicated above, this conclusion contains at least three errors: (a) the tendency of the entropy to increase applies only to an isolated system, (b) there is no degradation of information, but of the signals conveying it from the source to the receiver, (c) information and entropy cannot be compared; the first is dimensionless and the second has the dimension cal. per degree.

By suitably chosen definitions it is of course possible to give information and entropy the same dimensions. As we saw above, Shannon arrived at this result by introducing entropy as a dimensionless quantity giving the amount of information directly, but having no direct connection with physical entropy. Brillouin and others, who are more interested in theoretical physics than in communication engineering, do just the reverse by giving information the dimension cal. per degree. Brillouin [2]) says: "When however we are mainly interested in physical problems, we want to use similar units for information and entropy". By this artifice he arrives at a number of further statements. For instance, when considering a gas that can be in many different micro-states ("complexions") he writes: "We may sometimes possess some special information on the system under consideration; for instance, we may know how it was constituted at a given time, and hence know the original distribution of densities and velocities. The knowledge of such additional information allows us to state more accurately the structure of the system, to decrease the number of elementary complexions, and to decrease probability and entropy. *Any additional piece of information increases the negentropy of the system*". The italics are Brillouin's. Taking into consideration that information and negentropy are introduced as similar quantities by Brillouin, the whole quotation amounts to the tautology: When we know

[1]) J. D. Fast and F. L. Stumpers, *Philips tech. Rev.* **18,** 201 (1956/57).

[2]) L. Brillouin, *Scientific uncertainty and information*, Academic Press, New York (1964).

that a system has not yet reached its state of maximum entropy, then it is still possible for its entropy to increase.

That entropy and information, apart from their common statistical background, are two entirely different concepts follows also from what is known about genetic information. It is now generally agreed that this information is coded in the molecules of DNA (deoxyribonucleic acid). These polymer molecules are of enormous length, their monomer units being the so-called nucleotides. Usually two polynucleotide chains are twisted about each other in a regular helix. There are only four sorts of nucleotides, and the uniqueness of any one nucleic acid lies in the order in which they are arranged.

To be able to transfer information the configuration of the DNA molecule in a living cell has to be permanent. The information of heredity is given by this permanent serial order; the entropy, on the other hand, by the many different states of vibration and internal rotation compatible with the permanence and integrity of the molecule.

INDEX

CAMROSE LUTHERAN COLLEGE LIBRARY
QC 318 .F25
22,671

Date Due

BJJH

PRINTED IN U.S.A. CAT. NO. 24 161 BRODART